Lecture Notes of the Institute for Computer Sciences, Social Informatics and Telecommunications Engineering 344

More information about this series at http://www.springer.com/series/8197

Ding Wang · Weizhi Meng ·
Jinguang Han (Eds.)

Security and Privacy in New Computing Environments

Third EAI International Conference, SPNCE 2020
Lyngby, Denmark, August 6–7, 2020
Proceedings

 Springer

Editors
Ding Wang
Nankai University
Tianjin, China

Weizhi Meng 🄳
Technical University of Denmark
Lyngby, Denmark

Jinguang Han 🄳
Queen's University Belfast
Belfast, UK

ISSN 1867-8211   ISSN 1867-822X (electronic)
Lecture Notes of the Institute for Computer Sciences, Social Informatics
and Telecommunications Engineering
ISBN 978-3-030-66921-8   ISBN 978-3-030-66922-5 (eBook)
https://doi.org/10.1007/978-3-030-66922-5

This Springer imprint is published by the registered company Springer Nature Switzerland AG
The registered company address is: Gewerbestrasse 11, 6330 Cham, Switzerland

Preface

This book contains the papers that were selected for presentation and publication at the third edition of the 2020 European Alliance for Innovation (EAI) International Conference on Security and Privacy in New Computing Environments (SPNCE). This conference has brought together researchers, developers and practitioners around the world who are leveraging and developing security and privacy in new computing environments. Due to COVID-19, SPNCE was held fully online.

SPNCE received 63 submissions this year, and each paper was reviewed by at least three reviewers. A total of 31 papers were accepted with camera-ready copy, and 4 papers were accepted with no camera-ready copy. The conference sessions were: Session 1–Network Security; Session 2–System Security; Session 3–Machine Learning; Session 4–Authentication and Access Control; Session 5–Cloud Security; Session 6–Cryptography; Session 7–Applied Cryptography. Aside from the high-quality technical paper presentation, the technical program also featured two keynote speeches and three technical workshops. The two keynote speakers were Prof. S.M. Yiu from the University of Hong Kong and Prof. Shui Yu from the University of Technology Sydney, Australia. The three workshops organized were on Security and Privacy Issues for Intelligent Edge Environments (SPIIEE), Security and Reliability Challenges in the Mobile Internet of Things (SRCMIoT) and Security and Privacy in Computing and Communications with Smart Systems (SPCCSS). The SPIIEE workshop aims to address the new dimension of threats in Intelligent Edge Environments. The SRCMIoT workshop aims to bring together state-of-the-art contributions on security- and reliability-related research in the Mobile Internet of Things. The SPCCSS workshop aims to show state-of-the-art contributions on security and privacy in Computing and Communications with Smart Systems.

For the success of SPNCE 2020, we would like to first thank the authors of all submissions and all the PC members and external reviewers for their great efforts in selecting the papers. For the conference organization, we would like to thank the steering chairs, Dr. Imrich Chlamtac, Dr. Jin Li and Dr. Yang Xiang, and the Technical Program Committee Chairs, Prof. Ding Wang, Prof. Xiao Wang and Prof. Jinguang Han, who completed the peer-review process of the technical papers and made a high-quality technical program. Finally, we thank everyone else for their contribution to the program of SPNCE 2020.

November 2020   Weizhi Meng
                                                                                                        Sokratis Katsikas

Conference Organization

Steering Committee

Imrich Chlamtac	University of Trento, Italy
Jin Li	Guangzhou University, China
Yang Xiang	Deakin University, Australia

Organizing Committee

General Chairs

Weizhi Meng	Technical University of Denmark, Denmark
Sokratis Katsikas	Norwegian University of Science and Technology, Norway

TPC Chair and Co-chairs

Ding Wang	Nankai University, China
Xiao Wang	Northwestern University, USA
Jinguang Han	Queen's University Belfast, UK

Sponsorship and Exhibit Chairs

Debiao He	Wuhan University, China
Kuo-Hui Yeh	National Dong Hwa University, Taiwan

Local Chair

Wei-Yang Chiu	Technical University of Denmark, Denmark

Workshops Chairs

Xiaochun Cheng	Middlesex University London, UK
Hongyu Yang	Civil Aviation University of China, China

Publicity and Social Media Chairs

Kim-Kwang Raymond Choo	The University of Texas at San Antonio, USA
Zheli Liu	Nankai University, China
Shujun Li	University of Kent, UK

Publications Chairs

Marios Anagnostopoulos	Norwegian University of Science & Technology, Norway
Chengyu Wang	Beijing University of Posts and Telecommunications, China

Web Chairs

Zengpeng Li	Lancaster University, UK
Wenjuan Li	The Hong Kong Polytechnic University, China

Technical Program Committee

Weizhi Meng	Technical University of Denmark, Denmark
Sokratis Katsikas	Norwegian University of Science and Technology, Norway
Debiao He	Wuhan University, China
Kuo-Hui Yeh	National Dong Hwa University, Taiwan
Xiaochun Cheng	Middlesex University, UK
Hongyu Yang	Civil Aviation University of China, China
Kim-Kwang Raymond Choo	The University of Texas at San Antonio, USA
Zheli Liu	Nankai University, China
Shujun Li	University of Kent, UK
Zengpeng Li	Qingdao University, China
Jeremiah Blocki	Purdue University, USA
Haipeng Cai	Washington State University, USA
Kai Chen	Institute of Information Engineering, Chinese Academy of Sciences, China
Long Cheng	Clemson University, USA
Rongmao Chen	National University of Defense Technology, China
Xiaofeng Chen	Xidian University, China
Yi Deng	State Key Laboratory of Information Security, China
Ashok Kumar Das	Indian Institute of Technology Kharagpur, India
Fuchun Guo	University of Wollongong, Australia
Fei Gao	Beijing University of Posts and Telecommunications, China
Yong Guan	Iowa State University, USA
Chun Guo	Shandong University, China
Jinguang Han	Queen's University Belfast, UK
Marko Holbl	University of Maribor, Slovenia
Qiong Huang	South China Agricultural University, China
Xinyi Huang	Fujian Normal University, China
Zhicong Huang	Alibaba, China

Chunfu Jia	Nankai University, China
Linzhi Jiang	University of Surrey, UK
Anca D. Jurcut	University College Dublin, Ireland
Mukul R. Kulkarni	University of Massachusetts Amherst, USA
Muhammad Khurram Khan	King Saud University, Saudi Arabia
Xiapu Luo	The Hong Kong Polytechnic University, Hong Kong
Jin Li	Guangzhou University, China
Jingqiang Lin	Institute of Information Engineering, Chinese Academy of Sciences
Juanru Li	Shanghai Jiao Tong University, China
Kaitai Liang	University of Surrey, UK
Qi Li	Tsinghua University, China
Hongwei Li	University of Electronic Science and Technology of China, China
Jingwei Li	University of Electronic Science and Technology of China, China
Junzuo Lai	Jinan University, China
Xiao Lan	Sichuan University, China
Joseph Liu	Monash University, Australia
Kangjie Lu	University of Minnesota, USA
Ximeng Liu	Fuzhou University, China
Qingchuan Zhao	Ohio State University, USA
Zhe Liu	Nanjing University of Aeronautics and Astronautics, China
Zengpeng Li	Qingdao University, China
Zhen Ling	Southeast University, China
Mark Manulis	University of Surrey, UK
Chunguang Ma	Shandong University of Science and Technology, China
David A. Mohaisen	University of Central Florida, USA
Jianting Ning	Singapore Management University, Singapore
Longjiang Qu	National University of Defense Technology, China
Dimitrios Papadopoulos	HKUST, China
Bo Qin	Renmin University of China, China
Michael Scott	MIRACL Labs, Ireland
Chao Shen	Xi'an Jiaotong University, China
Chunhua Su	The University of Aizu, Japan
Jian Shen	Nanjing University of Information Science & Technology, China
Kun Sun	George Mason University, USA
Shifeng Sun	Monash University, Australia
Qingni Shen	Peking University, China
Salman Salamatian	MIT, USA
Ni Trieu	Oregon State University, USA
Yuan Tian	University of Virginia, USA
Junfeng Tian	Hebei University, China

Contents

Authentication and Access Control

Cloud Security

Cryptography

Applied Cryptography

Network Security

A Characterisation of Smart Grid DoS Attacks

Dilara Acarali[1(✉)], Muttukrishnan Rajarajan[1], Doron Chema[2],
and Mark Ginzburg[2]

[1] School of Mathematics, Computer Science and Engineering,
City University of London, London, UK
{dilara.acarali.2,r.muttukrishnan}@city.ac.uk
[2] Technical Team, L7 Defense, BeerSheva, Israel
{doron,marik}@l7defense.com
https://www.city.ac.uk/about/schools/mathematics-computer-science-
engineering, https://www.l7defense.com/

Abstract. Traditional power grids are evolving to keep pace with the
demands of the modern age. Smart grids contain integrated IT systems
for better management and efficiency, but in doing so, also inherit a
plethora of cyber-security threats and vulnerabilities. Denial-of-Service
(DoS) is one such threat. At the same time, the smart grid has particular
characteristics (e.g. minimal delay tolerance), which can influence the
nature of threats and so require special consideration. In this paper,
we identify a set of possible smart grid-specific DoS scenarios based on
current research, and analyse them in the context of the grid components
they target. Based on this, we propose a novel target-based classification
scheme and further characterise each scenario by qualitatively exploring
it in the context of the underlying grid infrastructure. This culminates
in a smart grid-centric analysis of the threat to reveal the nature of DoS
in this environment.

Keywords: Smart grid · Cyber-security · DoS · DDoS

1 Introduction

The digital age has caused an increased dependency on electricity, and consequently, given rise to an increased demand on power systems. As a result,
traditional power grids have had to evolve to deal with this inflated demand. A
smart grid is a traditional power grid integrated with information communication systems. The former is referred to as the physical or operational technology
(OT), and the latter is called the cyber or information technology (IT). In practice, this means that IT networks gather data from field systems and deliver
them to a central command centre via local controllers. That data can then be
used to regulate physical grid components and to make management decisions.

© ICST Institute for Computer Sciences, Social Informatics and Telecommunications Engineering 2021
Published by Springer Nature Switzerland AG 2021. All Rights Reserved
D. Wang et al. (Eds.): SPNCE 2020, LNICST 344, pp. 3–21, 2021.
https://doi.org/10.1007/978-3-030-66922-5_1

Whilst this can greatly improve efficiency, the IT network also makes the smart grid more vulnerable to malicious activity by expanding the previously limited attack surface. Grid components are now remotely accessible, and grid processes are now dependent on data flows through communication channels that can be disrupted. Furthermore, the cyber and the physical systems are highly interconnected and interdependent, meaning that faults or attacks at one point can cause a chain of effect across the wider smart grid.

This work is focused on Denial-of-Service (DoS), a well-known cyber-attack targeting availability, designed to hinder normal system processes. It is popular because, despite being relatively simple, a successful DoS attack can cause a large degree of disruption. DoS attack methodology may consist of a) flooding, where a channel/device is overwhelmed with data, b) the exploitation of vulnerabilities or quirks in systems and protocols, or c) both. A DoS attack launched by multiple dispersed individuals (e.g. in a botnet) is known as Distributed DoS (DDoS). Whilst disruption resulting from physical tampering can also be explored, we consider DoS predominantly as a cyber threat.

DoS attacks in conventional networks are well-studied, but the smart grid has particularities that influence both methodology and results. In this paper, we first identify and then characterise smart grid DoS scenarios to build up a picture of how this threat manifests in this new environment. Identification is achieved via a detailed survey of existing research, which then forms the basis of a new classification scheme organised by potential targets. This differs from conventional approaches and is designed to link targeted grid components with likely DoS attack methods, providing a reference for researchers and defenders. To our knowledge, a survey and classification of smart grid-specific DoS scenarios is novel to this work.

The research presented in this paper is part of the European Union's *Energy Shield* project [1], commissioned in recognition of the energy industry's transition from traditional systems to smart grids. The aim of this project is to develop a defence toolkit for EPES (Electrical Power and Energy System) operators [1] to protect critical infrastructure from cyber-attacks, including DDoS. This work was conducted to provide a foundation of understanding of the smart grid-centric DoS threat on which the project can further build.

The rest of the paper is structured as follows: Sect. 2 provides a background explanation of smart grid architecture, including domains, flows, and key components. Then, the smart grid DoS research survey is presented in Sect. 3. Based on the information in 2 and 3, Sect. 4 presents our classification scheme and characterisations of the identified DoS scenarios. This is followed by the discussion of our findings in Sect. 5, with an analysis of related works given in Sect. 6. We then conclude in Sect. 7.

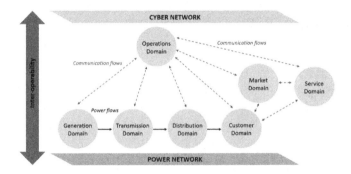

Fig. 1. Smart grid domains.

2 Smart Grids Background

2.1 Smart Grid Domains

The smart grid follows a domain model architecture, which means it is split into multiple domains, each handling a different function. A domain will generate/process either power flows (in the OT), communication flows (in the IT), or both. Generally, data on power flows is collected and shared as communication flows. Figure 1 provides an illustration of the core domains as they relate to each other and the flows between them. Note that this paper focuses on communication flows as the main target of DoS attacks, but the characteristics of these are influenced by the nature of the underlying power flows.

The core domains of the traditional power grid, responsible for the production and delivery of electricity as a utility, are described first. These are the ones that contain power flows. The Customer domain deals with the delivery of power to customer premises. It also handles the collection of usage data. Power is received from the Distribution domain, which is responsible for dissemination. This domain also transforms the power received from the Transmission domain, which is where bulk energy is carried between geographically distributed locations (e.g. a power plant and a city). The Generation domain houses power plants where electricity is 'produced'. If distributed power generation exists (i.e. customers generate their own electricity), this is also handled within the Customer and Distribution domains.

The rest are the communication-focused domains. Operations is the main control hub and the core of the communication network, responsible for the collection of monitoring data and the dissemination of control commands. The Service domain houses the providers who deliver electricity as a utility, whilst the buying/selling of electricity is handled within the Markets domain. Both Service and Markets make use of the IT to provision services and to bill customers. However, it should be noted that there is a separation between the IT that controls the grid, and the corporate TCP/IP networks of energy companies and service providers.

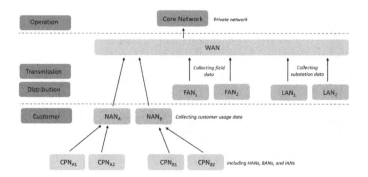

Fig. 2. Smart grid IT network hierarchy with CPNs (Customer Premise Networks).

2.2 Smart Grid Structure

The traditional power grid (i.e. the OT) is hierarchical in the way it transmits and distributes electricity to users. One or more power plants generate electricity. This is then carried in bulk through heavy-duty transmission lines to many geographically dispersed substations. Here, transformers convert (i.e. step-down) and transfer energy to distribution lines, which then branch out to deliver power to a large number of individual customer premises. Hence, there are a larger number of component systems at the bottom then at the top, which means that a single problem higher up in the power hierarchy affects many systems lower down.

The IT network is also hierarchical (Fig. 2). For a given region and/or provider, a single core management system, such as SCADA (Supervisor Control and Data Acquisition), contains various master controls to monitor, analyse, and regulate grid operations. This core network receives inputs from distributed monitoring devices (like RTUs and PLCs, described in the next sub-section) sitting in FANs (Field Area Networks), NANs (Neighbourhood Area Network) and substation LANs (Local Area Networks) operating throughout the Generation, Transmission, and Distribution domains. NANs amalgamate smaller customer networks, including HANs (Home Area Networks), BANs (Business Area Networks), and IANs (Industry Area Networks). Each contain devices that connect to upload usage data. As with the OT, the IT hierarchy means that higher level issues impact a large number of lower level systems. Furthermore, there is a variation in DoS attack surface (smaller at the top, bigger at the bottom).

In addition to this, connectivity and dependency exist between the IT and OT, which means that malicious activity in one will have some influence on the other. In other words, assuming that DoS attacks are targeted at IT devices/flows, when some function of the IT network is denied, there will be a corresponding impact, related to that function, on the OT. Furthermore, disruption to a particular grid process can have subsequent effects on other processes, and may escalate into general grid instability. This is a unique characteristic of

Table 1. Smart grid sub-systems and components, their domains, and their functions.

Systems	Domains	Function
Smart Meter	- Customer	Located in customer premises, collects usage data for operations (via the AMI) and provides pricing information to users
Remote Telemetry Unit (RTU)	- Distribution - Transmission - Generation	Located in substations, collects data on grid processes from PLCs for automated monitoring and control processes
Phasor Measurement Unit (PMU)	- Distribution - Transmission - Generation	Located in substations, collects data on electrical phasor patterns for synchronisation of grid supply and demand
Programmable Logic Controller (PLC)	- Distribution - Transmission - Operations	Field device, collects telemetry data on grid processes, communicated to RTUs and LFC to generate control signals
Master Telemetry Unit (MTU)	- Operations	Part of SCADA or WAMS, processes data received and aggregated from RTUs
Phasor Data Concentrator (PDC)	- Distribution - Transmission - Generation - Operations	Part of SCADA or WAMS, processes data received and aggregated from PMUS
Load Frequency Control (LFC)	- Generation	Located alongside generators, minimises fluctuations in energy input and output for frequency balance
Supervisory Control & Data Acquisition (SCADA)	- Operations	Control system, responsible for manipulating grid topology, monitoring processes, and maintaining functionality
Wide Area Management System (WAMS)	- Operations	Control system, uses information gained from PMU/PDC data to monitor and react to grid instabilities/issues

critical infrastructures, as DoS within conventional IT networks does not typically have the potential for this level of widespread disruption.

2.3   Smart Grid Components

Within the IT network, there are a number of sub-systems responsible for different grid processes. The key sub-systems are:

– **Advanced Metering Infrastructure (AMI):** A two-way communication network between smart meters sitting within customer premises and utility servers in the core network. The AMI enables the collection of usage data (analysed for load forecasting and pricing models), as well as the delivery of relevant customer services.

- **Phasor Control:** Consists of PMUs (Phasor Measurement Units) which measure electrical signals and monitor phasor patterns, and PDCs (Phasor Data Concentrators) which aggregate and process this data for monitoring, fault response, and command generation. The purpose of this is to ensure that the frequency of the electricity in the grid is synchronised and the grid remains stable.
- **Telemetry Control:** Consists of RTUs (Remote Telemetry Units, distributed across the domains), which collect telemetry data from grid components, and MTUs (Master Telemetry Units, connected to core systems), which receive that data and process it for management and topology manipulation. This supports efficient power generation and transfer. RTUs connect multiple PLCs (Programmable Logic Controllers), which connect to field devices.
- **Load Frequency Control:** Consists of PLCs (that connect to field devices) and RTUs which collect data on the performance of various processes. This is communicated to LFCs (Load Frequency Controls), which use the data to manage generators to maintain a stable frequency in the grid.
- **Core Control Systems:** Includes SCADA (Supervisor Control and Data Acquisition) systems and WAMS (Wide Area Management Systems), which act as the central management point for the grid, where all data is aggregated, processed, and used for human-lead decision-making and for automatically generated controls.

Table 1 provides a summary of typical sub-systems and components, which domains they sit in, and what they do. Note that this list is not exhaustive. For the scope of this paper, we have focused on the most common sub-systems which rely on communication flows. Other systems include those responsible for pricing models, for distributed power generation management, and external data systems used for load prediction (e.g. weather forecasts).

3   Smart Grid DoS Survey

The purpose of this survey is to answer the following research question: *According to the literature, what are the DoS attack possibilities in smart grids, given the smart grid's unique characteristics?* The answer to this question must consider attack method, attack target, and attack impact. The survey methodology used is as outlined in [4], and the review protocol was to search IEEEXplore, Science Direct, and Google Scholar with a set of DoS and smart grid search term pairs. Specifically, we aimed to identify recent works that defined, modelled, simulated, or discussed DoS scenarios. Note that SCADA or WAMS are not considered, as these systems are common across critical infrastructures and are considered a separate field of study. The scenarios identified in the surveyed works are summarised here.

Wang et al. (2017) [23] explored the adversarial interaction between smart grid defenders and attackers, anchored on an AMI DDoS attack. The AMI is modelled as a tree with many smart meters connecting to aggregators in layers.

Traffic from the meters travels up these layers to a base station, which then relays it to the AMI core. In this study, the attacks were targeted at smart meters and aggregators, assuming the attack source to be a botnet. The authors found that, depending on where the target sits, certain communication paths become saturated, with nodes attached to these paths consequently being knocked offline. Meanwhile, the AMI's tree structure eventually causes "downstream" nodes to lose core connectivity [23]. Honeypots embedded within the AMI core were used to derive optimal attack and defence strategies.

As with [23], Guo and Ten (2015) [8] also studied a botnet-driven AMI DDoS scenario. They created a two-staged model combining botnet formation and attack launch. Three actor categories were considered: attackers, victims, and agents (i.e. smart meters converted into bots). The authors posited that it is reasonable to assume that a population of smart meters will have similar vulnerabilities, and hence, will be susceptible to the spread of an automated malware targeting firmware or communication functions [8]. They simulated a UDP flood against a 2-layer AMI topology, with each bot generating packets at a rate of 2Mbps [8] and reported that growth in the bot population directly correlated with an increased number of dropped packets and longer end-to-end delays [8].

Similarly, Sgouras et al. (2017) [20] investigated the AMI impact of a botnet-launched DDoS attack. They modelled the AMI as multiple residential smart meters connected to a central control server, and posited that the Internet is a likely channel for communication between control servers and aggregators. Based on this, they suggested that a botmaster could sniff traffic to determine the server IP and then use a botnet (external to the smart grid network) to launch a TCP SYN DDoS attack at great scale. Using this proof-of-concept, the authors were able to demonstrate how Internet-connectivity exposes the grid to the outside world and can make it susceptible to attacks from remote adversaries.

Asri and Pranggono (2015) [3] also investigated botnet-based DDoS attacks against the AMI. With similar assumptions to [20], they modelled the AMI as a collection of households containing smart meters connecting to a central utility server via the Internet. An external botnet can then connect to the utility server to flood it in a UDP storm attack, targeting many random ports. The server will try to initiate applications on random ports that do not exist and respond with ICMP 'Destination Unreachable' packets [3]. With simulations, the authors showed that the entire grid could be compromised with a large-enough DDoS attack, though the effect on the power supply network was not immediate. Only after the server had been knocked offline was an impact observed.

Sgouras et al. (2014) [19] considered four different AMI DoS setups: 1) DoS on a smart meter, 2) DDoS on a smart meter, 3) DoS against an AMI utility server, and 4) DDoS on an AMI utility server. Comparing the DoS and DDoS scenarios, they found that the targeted smart meter suffered from significantly increased queue lengths under the latter. In fact, they observed that queue lengths reached maximum levels much faster. The DoS attack against the server caused a drop in the number of TCP packets delivered to smart meters, leading to some service degradation. In comparison, the DDoS attack on the server reportedly diminished connections with almost 90% of the smart meters [19].

Hoffman and Bumiller (2019) [9] proposed a special AMI DoS attack called Denial-of-Sleep. This is where a battery-powered device is prevented from entering sleep mode (i.e. a low-power state intended to conserve energy), thereby significantly reducing its lifespan. Smart meters enter sleep mode when they are not forwarding measurement data or receiving traffic from other nodes [9]. Two sleep protocols are identified: S-/T-mode (where the device transmits and waits some time for a response before sleeping) and C-mode (where the device sleeps immediately after completing its transmission). The authors used an abstract version of a TLS (Transport Layer Security) handshake to model Denial-of-Sleep attacks in the context of the C-mode sleep protocol and reported that a small number of attacks of relatively short length can significantly deplete batteries [9].

Chatfield et al. (2018) [5] studied jamming, which they categorised as a form of DoS used to disrupt the wireless networks within smart grids. They defined two possible scenarios. The first is where a jamming attack produces lots of radio signals on the same frequency as legitimate communications, causing delays and increased latency for control messages. The second is where the degradation caused by a jamming attack interferes with standard protocol processes and leads to constant retransmissions, further congesting the network and, in the case of routing protocols, causing network instability [5]. In their AMI model, they used the received signal strength of nodes to differentiate between normal and attack scenarios, and related attack effectiveness to the distance between attacker and victim nodes.

Pedramnia and Rahmani (2018) [17] explored AMI-based DoS against cellular LTE (Long-Term Evolution) networks. They identified signalling attacks, where bearer assignment is exploited to prevent legitimate use. A malicious bearer request is sent, and once an assignment is made, that bearer is left unused. It expires, triggering another assignment process, and the pattern is repeated. LTE-specific jamming attacks may be possible too. The authors also discussed SMS link saturation, where device user panels (where customers receive updates) are flooded. This drowns out legitimate messages, causes buffer overflows, or leads to delays [17]. Lastly, DoS against NAT (Network Address Translation) systems are highlighted, specifically for NAT64. These include NAT overflow (where malicious mapping requests block legitimate use), NAT wiping (where TCP-RST messages are used to delete mappings), and NAT breaking (where spoofed IPs make NAT requests and ignore server responses, forcing those IPs onto blacklists) [17].

Yi et al. (2016) [27] defined a new AMI DoS technique called the puppet attack. This seeks to exploit the use of DSR (Dynamic Source Routing) and the way that mesh networks are formed. One or more nodes are selected as 'puppets', and receive attack commands. DSR uses route requests (RREQ) and route replies (RREP) to build address lists amongst nodes. The attack makes a puppet node erase addresses from the list, causing path errors [27]. This then triggers another round of path discovery and list building, thereby stopping the

Table 2. DoS types and targets identified by surveyed works.

Ref	Premise	Target	Impact
[23]	Defender vs. attacker interactions	AMI	Channel saturation; downstream nodes lose core connectivity
[8]	Botnet-launched UDP flood DDoS	AMI	Botnet growth increases dropped packets and end-to-end delays
[20]	Botnet-launched TCP-SYN DDoS	AMI	Load fluctuations and system availability diminished
[3]	UDP storm DDoS against residential AMI	AMI	Possible complete compromise; delayed impact on power network
[19]	DoS/DDoS attack impact in the AMI	AMI	Dropped TCP packets, service degradation, and diminished connectivity
[9]	Denial-of-Sleep attacks against battery-powered AMI nodes	AMI	Increased battery depletion in affected nodes
[5]	Jamming attack detection in wireless networks	AMI	Delays on legit frequencies; protocol process interruption
[17]	Signalling, SMS link saturation, and NAT attacks against cellular LTE networks	AMI	Repeated bearer assignments; reduces devices' interface functionality. NAT-based disruptions
[27]	Puppet attack against mesh networks	AMI	Corrupted address lists, causing path discovery cycles
[25]	Resilient routing model for PMUs-WAMS traffic	PMUs	Delays/packet dropping within PMU data channels
[21]	Nash Equilibrium-based DoS-resilient routing	PMUs	Reduced relay nodes functionality; delays/reduced connectivity
[28]	Interface and PLC-targeting attacks exploiting query replies	PLCs	Delays increase with attack length, management SW becomes non-functional
[14]	DoS switching strategies against LFCs	LFCs	Attack impact is maximised via start time selection and attack length

network from settling into a routing structure. This contrasts with standard DoS which relies on crafted packets or exhaustion of resources. Meanwhile, puppet attacks undermine the structure and functionality of the mesh itself.

Wei and Kundar (2012) [25] explored DoS attacks targeting communication channels between distributed PMUs and the WAMS. Specifically, the authors suggested that PMU data rates may vary with network congestion. They modelled a hierarchical network covering the cyber (IT) and physical (OT) networks. PMUs in the cyber network collect data on a particular generator node in the physical network. Local controllers obtain data from PMUs to create control signals to be applied to the generator nodes. Generators are grouped into clusters, with a single PMU and local controller in charge of each cluster. Hence, power flow depends on control signals, and control signals depend on PMU data. The DoS attack then targets the PMUs' communication channels with the aim of causing delays or packet drops. The authors used this attack model to propose a flocking-based scheme to route traffic around DoS-affected regions.

Srikantha and Kundur (2015) [21] also studied attacks on PMUs to enhance resiliency against DoS attacks. They modelled the smart grid as a pair of hierarchical, inter-connected directed graphs, populated with relay nodes (RNs) responsible for transmitting data. Some function as PMUs (collecting data) and some as cyber-actuators (sending control signals). At the top of the hierarchy is the root node that transmits control data downstream, whilst PMUs send measurement data upstream. The authors then experimented with DoS attacks that target one or more RNs in the tree. This causes delays and disruption for the downstream/upstream movement of traffic as RNs are rendered non-functional. As with [25], they proposed a routing system to allow the topology to morph around the attacked nodes.

Yilmaz et al. (2018) [28] explored the possibility of DoS attacks against PLCs, suggesting that a PLC can be targeted both from within and outside of its own IP network, as long as its IP address is known. Furthermore, the authors highlighted that PLCs reply to any queries from any source, further increasing their potential for exploitation [28]. A testbed was built, consisting of PLC devices and some PCs running a) TIA (Totally Integrated Automation) portal management software, b) DoS tools, and c) attack detection systems [28]. Attacks were then simulated against both the PLCs and the TIA portal. The results showed that the PLCs' ping response delay continued to increase the longer the attack was sustained. Meanwhile, the TIA portal became non-functional. The authors noted that the network was quickly disrupted even with a small number of attackers [28].

Finally, Liu et al. (2013) [14] investigated DoS attacks designed to disrupt the delivery of telemetry data from RTUs to LFC systems. This would prevent the LFC from generating accurate command signals for physical grid components, potentially causing further issues. The authors modelled DoS as a switched system and suggested that attacks can have maximum impact if attackers select the optimal switching strategy. They identified this to be a sequential attack over multiple intervals [14]. DoS attacks were simulated with different starting times, revealing that impact was more significant for those launched before power systems have fully converged. This period was therefore highlighted as one of increased vulnerability [14].

The survey is summarised in Table 2 and the findings are analysed in the next section.

4   Smart Grid Denial-of-Service Characterisation

Based on the architecture of smart grids (discussed in Sect. 2) and the literature survey (presented in Sect. 3), we propose a set of smart grid DoS scenarios, classified in a target-based structure. This contrasts with existing classification schemes which tend to focus on attack methodology. The main categories are:

- *A:* **Network-Targeting**
  - *A.1:* Saturation Scenarios (aiming to use up channel resources)
  - *A.2:* Exploit Scenarios (aiming to manipulate standard processes)
- *B:* **Device-Targeting**
  - *B.1:* Exhaustion Scenarios (aiming to use up device resources)
  - *B.2:* Compromise Scenarios (aiming to manipulate a device)

In A scenarios, the aim is to disrupt the network itself, either by blocking communications through full consumption of channel capacity ($A.1$), or the blocking of standard operations by preventing normal protocols from functioning as intended ($A.2$). Meanwhile, the aim of B scenarios is to disrupt the operation of particular network nodes so that they can be manipulated and cannot function as normal. This may be achieved by overwhelming the capacity of a device ($B.1$) or by exploiting some vulnerability in it ($B.2$). Note that $B.1$ scenarios are similar in concept to $A.1$. In the following, each scenario is described in terms of the smart grid sub-systems that may be targeted, allowing us to consider subsequent impact possibilities given the smart grid's multifaceted and interconnected infrastructure. Key points for effective defence and mitigation are also highlighted.

4.1   Network Saturation Scenarios ($A.1$)

The AMI may use a number of communication technologies, including WiFi, WSNs, cellular networks, or the Internet. Hence, it can be targeted in several different ways. Saturation may be attempted using typical flooding attacks (e.g. ICMP flood, UDP flood, HTTP flood) on any layer of the TCP/IP protocol stack, as seen in conventional networks. This type of scenario was explored by [8] and [3]. Jamming attacks may also be used to similar effect against wireless channels, as cited by [5]. Similarly, where SMS communication is used to push information to device interfaces, SMS link saturation may be employed [17]. Therefore, $A.1$ scenarios in the AMI can be characterised by a reduced upstream flow of usage data and downstream flow of service data. This could consequently lead to load estimation errors, bad pricing models, and reduced service quality for customers. Furthermore, as modelled in [23], the hierarchical structure of the AMI can cause a larger number of nodes lower down in the chain to lose connectivity to the core, further exasperating the issue.

Meanwhile, the Load Frequency Control sub-system may use SCADA protocols like ICCP (Inter Control Center Protocol) [12] or the IEC 61850 protocol stack [15] running on top of TCP/IP infrastructure [7,12] to ensure RTU-to-core communication. Flooding-style attacks can therefore be deployed here too, as examined by [14]. $A.1$ scenarios in the LFC system would be characterised by the untimely or reduced sharing of telemetry data by RTUs. This could result in incorrect control signal generation and consequently, the incorrect operation of physical grid devices, which may escalate into grid instability. Furthermore, it should be noted that ICCP and IEC 61850 do not provide robust and secure authentication mechanisms [7,12], leaving these communications vulnerable to malicious influence. The mechanisms for how flooding may be achieved on channels using these protocols (and for DoS in general) is an area for further study.

The survey also threw up the threat of *A.1* scenarios against the Phasor Control sub-system, which may use IEC 61850 protocols too [15]. This attack possibility was explored by [25], where the authors examined the relationship between PMU data and control signals. The results of their experiments suggest that the prevention of timely measurement readings due to flooded communication channels can result in incorrect control signals, which in turn may lead to fluctuations in frequency and ultimately, an unstable grid. This assessment on the impact of disrupted PMU flows was also supported by the results of [21]. A similar angle may be considered for the Telemetry Control and Core Control sub-systems as well.

Saturation scenarios on TCP/IP connections may be dealt with using anomaly detection to identify increases in traffic volume. This is also applicable against low-rate DoS attacks [26]. Similarly, IDS (Intrusion Detection Systems) should be used to identify suspicious activity, including jamming attacks [5]. For external attack sources, traffic from suspected IPs can then be blocked. Both anomaly detection and IDS (Intrusion Detection Systems) should be deployed at each layer of the cyber hierarchy [8]. Honeypots are suggested by [23] and can absorb attack impact. Saturation attacks can also be prevented by minimising the number of device and server interfaces with remote access. For the disruption caused by ongoing DoS attacks, Li et al. [13] suggested the use of predictive algorithms and historical data to estimate correct values and maintain grid stability. Meanwhile, SCADA and other control layer protocols need to introduce stronger authentication and improved security.

4.2   Network Exploit Scenarios (*A.2*)

Wireless ad-hoc network architectures are designed to be self-forming so that nearby nodes can organise themselves to define routes for data transmission. As stated previously, mesh networks may be deployed in the AMI (and possibly within the wider distribution and transmission domains for field sensors). However, the protocols used may be exploited to prevent these networks from forming and/or stabilising. This possibility was explored by [27], who defined the puppet attack against the DSR protocol. Other ad-hoc routing protocols like RPL (Routing Protocol for Low-Power) or OLSR (Optimized Link State Routing) may also be vulnerable to similar attacks. The result of an *A.2* scenario in the AMI may therefore be characterised as a customer domain network which has failed to converge and so no smart meter data can be delivered. Jamming attacks can similarly disrupt the normal operation of wireless network protocols [5].

As suggested in [17], LTE-based cellular networks may be used as an AMI architecture. In this type of setup, bearers are created to link devices to the data network. Bearer assignment may be exploited in a method similar in concept to TCP SYN; channels are opened to the target and then not used. As a result legitimate access is denied [17]. Once again, this would lead to the denial of AMI services.

Another possible *A.2* scenario is the exploitation of AMI NAT systems, as identified by [17]. Despite the introduction of IPv6, IPv4 is still widely used. This means that NAT is required for both translating between private and public IPv4 addresses, and for IPv4-to-IPv6 mappings [11,17]. For example, if field sensors using IPv6 attempt to send data to core networks still operating on IPv4, the corruption of NAT64 mappings would deny such transmissions. In cases such as this and in LTE networks, *A.2* scenarios will be characterised by a lack of connectivity between endpoints. We did not identify *A.2* scenarios for the other sub-systems during our survey and suggest this as an area for further study.

To avoid network disruption, routing protocols should be secure against tampering attempts, and robust enough to re-converge efficiently. Such an approach was proposed by [21] whose topology configuration scheme is designed to maintain routing between PMUs and RNs. Similarly, [27] suggested that corrupted nodes be identified by their abnormal communications and isolated, given that WSN nodes depend on their neighbours for their network connectivity. For field networks, physical security is needed to prevent illegitimate devices from joining ad-hoc networks. Anomaly detection applied to network exploitation needs to monitor protocol activity rather than general communication traffic. For SMS links, [17] cited the use of machine learning with bearer-related data to achieve this. They also suggest that migrating more widely to IPv6 could reduce the need for NAT mappings, thus reducing the risk of this type of DoS [17].

4.3   Device Exhaustion Scenarios (*B.1*)

Certain devices may be specifically targeted. An example in the Phasor Control sub-system is the relay nodes between PMUs and control systems, as highlighted by [21]. Given the large number of data collection points, relay nodes are used to ensure end-to-end delivery, sometimes also acting as aggregators. A targeted DoS attack on these devices can therefore disrupt the whole sub-system's communication flows, again leading to inaccurate measurements being collected. Furthermore, it is plausible to assume that relay node services in other sub-systems, such as the AMI, Load Frequency Controls, and Telemetry Controls, can be denied in this way too.

PLCs (and their interfaces) may also be actively targeted, as highlighted by [28]. Like relay nodes, PLCs have a central role in data collection and aggregation, but are more closely integrated with core control systems. Therefore, deliberate disruption of PLCs will significantly impact management decisions. The same scenario may be applicable to RTUs too. However, devices such as these, which are found deeper within the smart grid architecture, should theoretically be more difficult to access and would require insider access or skilled adversaries. Therefore, *B.1* scenarios in such systems will be characterised by the likely presence of skilled attackers or malicious insiders, and the hindering of data delivery to the core.

Finally, direct attacks within the AMI may be directed at smart meters, overwhelming their processing capacity as demonstrated by [23] and [19]. Unlike attacks on the AMI channels, forcing smart meters offline has the benefit of preventing both their communications with appliances and with head nodes directing traffic to the core. The location of smart meters within customer premises also exposes them as easier targets. Furthermore, [23] and [19] both demonstrated that AMI aggregator nodes and utility servers can be directly targeted. Attacks may be in the form of TCP SYN flood, HTTP flood, or other similar device-focused TCP/IP methods. Therefore, these *B.1* scenarios will be characterised as disabled devices within the AMI tree, with diminished connectivity for all the nodes served by them.

B.1 is the device-focused equivalent of *A.1*, so similar defensive measures may be applied. The aim is to knock particular devices offline, so suspicious traffic directed at those devices can be identified using anomaly detection. Anomaly profiles should be derived from the normal activity of those devices, with consideration of typical performance values, and critical nodes may be prioritised to reduce the overheads introduced by this. Predictive algorithms may again be applied to control systems to reduce the impact of lost telemetry data [13]. To deal with possible insider threats, [28] recommends strict monitoring and regulation of user privileges and activities.

4.4   Device Compromise Scenarios (*B.2*)

A possible compromise scenario is where smart grid devices are recruited into botnets. Whilst this is less probable in the deeper areas of the grid (given that those devices may be running proprietary software and may be difficult to physically access), it should not be completely disregarded as a possibility where malicious insiders are considered. Bot compromise is most likely to occur within edge systems (like the AMI), where some devices sit within customer premises and may be exposed to public networks. As highlighted in [8], a population of smart meters (with similar manufacturers and models) could see the spread of bot code via an automated malware. IoT-based botnets (made up of smart appliances) are also a factor. Whilst a botnet compromise is itself not necessarily a DoS scenario, it provides a platform from which to launch such attacks. Additionally, bots provide backdoor access and can serve as vectors for smaller and more targeted DoS campaigns. Botnet-based DDoS, where smart meters are compromised, was studied by [8], and similar botnet scenarios were considered by [20, 23] and [3].

Another *B.2* scenario, primarily targeting the field sensors, is the forced depletion of physical resources on the target device. This is especially problematic in low power networks such as those in WSNs. For example, the Denial-of-Sleep attack defined by [9] is designed to drain the energy of battery-powered sensors and control devices to render them non-functional. Whilst the impact of such an attack may be slower to manifest, it could have a longer-term impact on service quality as depleted nodes would need to be physically changed. The loss of nodes in the AMI would reduce its overall accuracy and functionality, but Denial-of-Sleep could plausibly be used against any type of battery-powered field device.

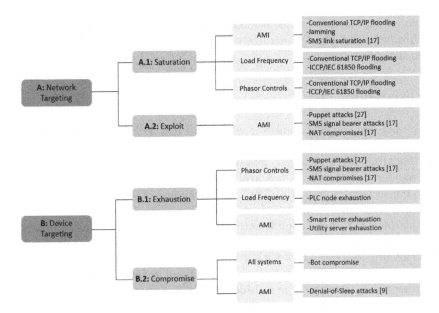

Fig. 3. Smart grid DoS attack classification tree.

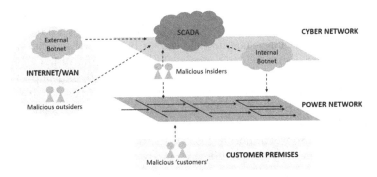

Fig. 4. Smart grid DoS attack sources.

B.1 scenarios may therefore be characterised by the presence of grid devices over which the operators do not have full control.

To stop nodes from being recruited into a botnet, devices must be regularly updated and patched with the latest software to minimise vulnerabilities. This can be challenging for difficult-to-access field sensors with limited capacity, which highlights the need for security-by-design in such devices. Login credentials must also be properly configured, as default passwords can be used to gain access [2]. Battery depletion can be mitigated by designing more energy-efficient devices, and connecting nodes to the main power where possible. To deal with Denial-of-Sleep, [9] suggested the need for an additional security layer for key exchanges to prevent meters being forced into FAC (Frequent Access Cycle).

Figure 3 summarises and depicts the smart grid DoS scenario classifications described here, and Fig. 4 illustrates where DoS attacks sources may sit in relation to the smart grid.

5 Discussion

In this work, we sought to uncover where and how DoS attacks may manifest in the smart grid. Given that this is still a relatively new technology, there have thankfully been very few real-world attacks thus far, for which (to our knowledge) detailed data is not available. Therefore, we turned to the available cyber-security literature and our understanding of smart grid architectures (combined with what we know of typical DoS attacks) for answers. We believe that this proved to be a sound methodology, as through the survey, we were able to enumerate and identify the sub-systems which could be targeted and how. This provides defenders with a means to prioritise their methods. It also provides future researchers with a reference point for what has been done so that other sub-systems may be considered as well. For example, areas for future DoS research may include attacks against generators and distributed power generation, and those originating from within the corporate networks of service provides or the Markets domain.

Through the survey, we were able to identify that the research community considers the AMI to be the most likely target of DoS attacks, with botnets being the most likely source. This suggests the need for better AMI security measures, and relates to other open research topics such as the security of IoT devices and WSNs. Meanwhile, the definitions of Denial-of-Sleep by [9] and puppet attacks by [27] demonstrate how the particularities of the smart grid can create space for new attack scenarios. Both scenarios were set in the AMI, once again highlighting the need for resilient security channels, as well as security-by-design in IoT technologies. It is also worth noting that these attacks may be feasible wherever mesh networks or battery-powered devices are used across the grid domains.

The potential vulnerability of other systems, despite their locations deeper within the grid network, is also apparent in the works of [25] and [21] who observed the impact of DoS on PMUs, and [28] and [14] who did the same for PLCs and LFCs, respectively. These components are likely harder to impact remotely and without specialist knowledge, but may still be targeted by skilled attackers (e.g. those working on behalf of nation states). The location of such components, higher up within the grid hierarchy, also means that the impact of any successful attack will be felt throughout the grid. Therefore, these systems must be secured appropriately.

Furthermore, we believe that the proposed classification provides a novel perspective for smart grid DoS. There are many classifications of DoS attacks in the existing literature, but most focus on the methodology. This is appropriate for the high-level characterisation of generic DoS attacks, but we argue that in the context of smart grids, it is beneficial to characterise attacks by target and impact, as this helps to align them with the IT and OT layers. Simply, this view enumerates the grid's vulnerable points. Furthermore, it makes a distinction between attacks against the channel and against the device. This is not always clear in exploit-based DoS scenarios, but is significant in smart grids because the domain of attack and the domain of impact can be different. By determining where an attack is intended to cause damage, we can work towards expanding typical DoS defence methods.

Due to scope restrictions, the survey itself was not exhaustive and can also be expanded - the smart grid is a complex web of sub-systems spanning the IT, OT, and domains. Therefore, another possibility for future work is to build upon this survey to uncover more grid systems which may be vulnerable to DoS attacks, both known and novel. We assume that as smart grids become more well-established, the research into their security will also grow. Finally, as mentioned in Sect. 4.1, protocols developed for critical infrastructure may be further studied to uncover new DoS attack methods and possible exploits.

6   Related Work

The novelty of this work comes from our singular focus on DoS scenarios for smart grids, with special consideration for smart grid sub-systems, supported by a survey of the existing literature on the subject. Whilst there are many works considering the different types of DoS attack and the cyber-security challenges faced by smart grids, few works focus on both at the same time. In addition to this, most existing classification schemes deal with methodology whilst this work aims to highlight the relationship between attack targets and attack impact.

Huseinovic et al. [10] developed a taxonomy of possible DoS attacks based on available literature, followed by a discussion of defensive strategies. Similar to this work, they explored different taxonomy perspectives, including one that considers which grid applications are targeted. However, they put more emphasis on analysing the security measures against each attack type and did not provide in-depth characterisations as we have. Otuoze et al. [16] also considered an alternative classification perspective by looking at threat sources, identifying both technical (i.e. infrastructural, operational, data) and non-technical (environmental, policy) sources, but did not provide details on DoS scenarios.

Wang and Lu [24] contributed a thorough survey of cyber-security challenges in the smart grid, with a section dedicated to DoS threats. However, they did not examine different sub-systems to characterise and classify attacks as we have. El Mrabet et al. [6] conducted a similar survey but also did not offer characterisations. In their extensive cyber-security survey, Tan et al. [22] did consider different smart grid components, but organised them differently as sub-systems (AMI, SCADA/WAMS) and data-generating devices (smart meters, PMUs, etc). Meanwhile, they did not look in detail at DoS scenarios. Our work sits alongside this to provide an alternative perspective.

Lastly, Ramanauskaite and Cenys [18] created a detailed taxonomy of DoS attack types and the defences against them, including considerations of attack source, exploited vulnerabilities, method, target type, and rate. However, smart grids were not in the scope of their work and so they did not consider the grid-specific attack targets. As with [22], we believe our work sits alongside this to help focus in on smart grid-specific DoS threats.

7   Conclusions

Smart grids are designed to make the generation and provision of power services more efficient and sustainable through the integration of IT technologies. However, this exposes the OT to cyber threats as seen in conventional networks and the Internet. DoS and DDoS attacks are among the most prevalent of these threats. This work provides a survey and a summary of the grid sub-systems that may be targeted, and characterises several possible DoS scenarios, alongside a target-based classification scheme to support this new perspective. Overall, we hope to have highlighted areas of smart grid vulnerability and set a foundation for better DoS defence and mitigation.

Acknowledgment. This work is funded by and a part of Energy Shield, a project under the European Union's H2020 Research and Innovation Programme.

References

1. Energy Shield (2019). https://energy-shield.eu/. Accessed 18 Dec 2019
2. Antonakakis, M., et al.: Understanding the Mirai Botnet. In: 26th USENIX Security Symposium (USENIX Security 2017), pp. 1093–1110 (2017)
3. Asri, S., Pranggono, B.: Impact of distributed denial-of-service attack on advanced metering infrastructure. Wireless Pers. Commun. **83**(3), 2211–2223 (2015). https://doi.org/10.1007/s11277-015-2510-3
4. Brereton, P., Kitchenham, B.A., Budgen, D., Turner, M., Khalil, M.: Lessons from applying the systematic literature review process within the software engineering domain. J. Syst. Softw. **80**(4), 571–583 (2007)
5. Chatfield, B., Haddad, R.J., Chen, L.: Low-computational complexity intrusion detection system for jamming attacks in smart grids. In: 2018 International Conference on Computing, Networking and Communications (ICNC), pp. 367–371. IEEE (2018)
6. El Mrabet, Z., Kaabouch, N., El Ghazi, H., El Ghazi, H.: Cyber-security in smart grid: survey and challenges. Comput. Electr. Eng. **67**, 469–482 (2018)
7. Elgargouri, A., Virrankoski, R., Elmusrati, M.: IEC 61850 Based Smart Grid Security, July 2015. https://doi.org/10.1109/ICIT.2015.7125460
8. Guo, Y., Ten, C.W., Hu, S., Weaver, W.W.: Modeling distributed denial of service attack in advanced metering infrastructure. In: 2015 IEEE Power & Energy Society Innovative Smart Grid Technologies Conference (ISGT), pp. 1–5. IEEE (2015)
9. Hoffmann, S., Bumiller, G.: Identification and simulation of a denial-of-sleep attack on open metering system. In: 2019 IEEE PES Innovative Smart Grid Technologies Europe (ISGT-Europe), pp. 1–5. IEEE (2019)
10. Huseinovic, A., Mrdovic, S., Bicakci, K., Uludag, S.: A taxonomy of the emerging denial-of-service attacks in the smart grid and countermeasures. In: 2018 26th Telecommunications Forum (TELFOR), pp. 1–4. IEEE (2018)

11. IETF: Stateful NAT64: Network Address and Protocol Translation from IPv6 Clients to IPv4 Servers (2011). https://tools.ietf.org/html/rfc6146. Accessed Mar 2020

12. Knapp, E.: Industrial network protocols, Chap. 4. In: Knapp, E. (ed.) Industrial Network Security, pp. 55–87. Syngress, Boston (2011). https://doi.org/10.1016/B978-1-59749-645-2.00004-5

13. Li, Y., Zhang, P., Ma, L.: Denial of service attack and defense method on load frequency control system. J. Franklin Inst. **356**(15), 8625–8645 (2019)

14. Liu, S., Liu, X.P., El Saddik, A.: Denial-of-Service (DoS) attacks on load frequency control in smart grids. In: 2013 IEEE PES Innovative Smart Grid Technologies Conference (ISGT), pp. 1–6. IEEE (2013)

15. Mattioli, R., Moulinos, K.: Communication Network Interdependencies in Smart Grids (2015). https://www.enisa.europa.eu/publications/communication-network-interdependencies-in-smart-grids. Accessed Mar 2020

16. Otuoze, A.O., Mustafa, M.W., Larik, R.M.: Smart grids security challenges: classification by sources of threats. J. Electr. Syst. Inf. Technol. **5**(3), 468–483 (2018)

17. Pedramnia, K., Rahmani, M.: Survey of DoS attacks on LTE infrastructure used in AMI system and countermeasures. In: 2018 Smart Grid Conference (SGC), pp. 1–6. IEEE (2018)

18. Ramanauskaite, S., Cenys, A.: Taxonomy of DoS attacks and their countermeasures. Open Comput. Sci. **1**(3), 355–366 (2011)

19. Sgouras, K.I., Birda, A.D., Labridis, D.P.: Cyber attack impact on critical smart grid infrastructures. In: ISGT 2014, pp. 1–5. IEEE (2014)

20. Sgouras, K.I., Kyriakidis, A.N., Labridis, D.P.: Short-term risk assessment of botnet attacks on advanced metering infrastructure. IET Cyber-Phys. Syst. Theory Appl. **2**(3), 143–151 (2017)

21. Srikantha, P., Kundur, D.: Denial of service attacks and mitigation for stability in cyber-enabled power grid. In: 2015 IEEE Power & Energy Society Innovative Smart Grid Technologies Conference (ISGT), pp. 1–5. IEEE (2015)

22. Tan, S., De, D., Song, W.Z., Yang, J., Das, S.K.: Survey of security advances in smart grid: a data driven approach. IEEE Commun. Surv. Tutor. **19**(1), 397–422 (2017)

23. Wang, K., Du, M., Maharjan, S., Sun, Y.: Strategic honeypot game model for distributed denial of service attacks in the smart grid. IEEE Trans. Smart Grid **8**(5), 2474–2482 (2017)

24. Wang, W., Lu, Z.: Cyber security in the smart grid: survey and challenges. Comput. Netw. **57**(5), 1344–1371 (2013)

25. Wei, J., Kundur, D.: A flocking-based model for DoS-resilient communication routing in smart grid. In: 2012 IEEE Global Communications Conference (GLOBECOM), pp. 3519–3524. IEEE (2012)

26. Xiang, Y., Li, K., Zhou, W.: Low-rate DDoS attacks detection and traceback by using new information metrics. IEEE Trans. Inf. Forensics Secur. **6**(2), 426–437 (2011)

27. Yi, P., Zhu, T., Zhang, Q., Wu, Y., Pan, L.: Puppet attack: a denial of service attack in advanced metering infrastructure network. J. Netw. Comput. Appl. **59**, 325–332 (2016)

28. Yilmaz, E.N., Ciylan, B., Gönen, S., Sindiren, E., Karacayılmaz, G.: Cyber security in industrial control systems: analysis of DoS attacks against PLCs and the insider effect. In: 2018 6th International Istanbul Smart Grids and Cities Congress and Fair (ICSG), pp. 81–85. IEEE (2018)

Security and Privacy in 5G Applications: Challenges and Solutions

Qin Qiu[1] ⓘ, Sijia Xu[1] ⓘ, and Shengquan Yu[2(✉)] ⓘ

[1] China Mobile Communications Group Co., Ltd., Beijing 100032, China
{qiuqin,xusijia}@chinamobile.com
[2] Advanced Innovation Center for Future Education, Beijing Normal University, Beijing 102206, China
yusq@bnu.edu.cn

Abstract. 5G is a new generation mobile network that enables innovation and supports progressive change across all vertical industries and across our society. 5G usage scenarios face new security risks due to the technology used and the characteristics of the specific application scenario. The security risks have become a key factor affecting the development of 5G convergence services. First we summarize the technical characteristics and typical usage scenarios of 5G. Then, we analyze the security and privacy risks faced by 5G applications, and give the system reference architecture and overall security and privacy solutions for 5G applications. Based on the three major applications scenarios of eMBB, uRLLC, and mMTC, we also provide specific suggestions for coping with security and privacy risks.

Keywords: 5G · Security · Privacy · eMBB · uRLLC · mMTC · MEC · Network slicing

1 Introduction

The fifth-generation mobile networks (5G) is a new generation mobile network that enables innovation and supports progressive change across all vertical industries and across our society [1]. 5G mobile communication technology is based on a new architecture [2]. The 3rd Generation Partnership Project (3GPP) has provided complete system specifications for 5G network architecture, see Fig. 1. Components of the core network can be instantiated multiple times to support virtualization technologies and network slicing. The architecture is driven by the motivation to remove the data overlay that has been traditionally used in previous generations of mobile networks [3].

The introduction of new key technologies such as Network Function Virtualization (NFV), Software Defined Network (SDN), network slicing, Multi-access Edge Computing (MEC) [5], mm Wave Communication [6] and massive MIMO [7] will greatly improve the network's support for various applications. The International Telecommunication Union (ITU) identifies three new usage scenarios of 5G (depicted in Fig. 2),

D. Wang et al. (Eds.): SPNCE 2020, LNICST 344, pp. 22–40, 2021.
https://doi.org/10.1007/978-3-030-66922-5_2

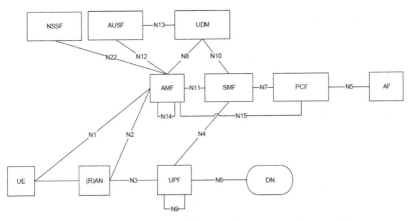

Fig. 1. 3GPP 5G system architecture for non-roaming cases [4]

which are enhanced mobile broadband (eMBB), ultra reliable and low latency communications (uRLLC), and massive machine type communications (mMTC), and proposes eight key performance indicators (KPI) [7]. Regarding these KPIs, 5G has high performances, reaching 10 times the peak rate of 4G, shortening the transmission latency to milliseconds and handling a million concurrent connections per square kilometer [8, 9]. The above-mentioned new features enable 5G to extend the traditional human-to-human communication to intelligent interconnections of man-to-things, and things-to-things. The rich and diverse 5G applications and their broad development prospects will change the traditional mode of social production, people's lifestyle and social governance, and start a new era of ubiquitous and intelligent internet. The European Union even predict

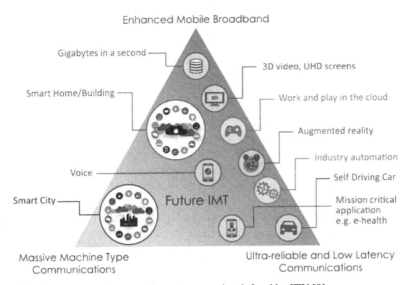

Fig. 2. 5G main usage scenarios defined by ITU [8]

that 5G will become the backbone of vital societal and economic functions – such as energy, transport, banking, and health, as well as industrial control systems [10]. According to HIS Markit [11], 5G will generate a global economic output worth \$13. 2 trillion and create 22.3 million jobs by 2035.

5G applications face new security risks due to the technology used and the characteristics of the specific application scenarios, and that has become a key factor affecting the development of 5G convergence services. This paper makes contributions in the following aspects:

1. Summarizes the technical characteristics and typical application scenarios of 5G, including the features of large bandwidth, low latency, and high-volume connection introduced by new technologies such as edge computing and network slicing, and introduce smart manufacturing, smart traffic, smart grid and smart campus enabled by these technologies.
2. Analyze the security and privacy risks faced by 5G applications, including privacy leakage in the eMBB scenario, DDoS attacks in the uRLLC scenario, and remote control in the mMTC scenario;
3. Provide the system reference architecture and overall security and privacy solutions for 5G applications, including the device layer, network layer, platform layer, and service layer, and provide security and privacy goals and corresponding solutions layer by layer;
4. Provide specific suggestions for security and privacy risks for typical application scenarios, including deployment of edge computing node in the eMBB scenario, preventing application data from tampering/falsification/replay attacks in the uRLLC scenario, and lightweight equipment authentication in the mMTC scenario, etc.

The abbreviations in Table 1 are applied in this paper.

2   Usage Scenarios and Applications of 5G

The ITU [9] divides the main 5G use cases into three categories:

- *eMBB* focuses on applications with extremely high bandwidth requirements. The main applications include 4K/8K ultra high definition mobile video and immersive AR (augmented reality) and VR(virtual reality) services. It meets people's needs for a digital life.
- *uRLLC* focuses on services that are extremely sensitive to latency, such as autonomous driving/assisted driving, remote control, and industrial Internet. It meets people's needs for the digital industry.
- *mMTC* covers scenarios with requirements for high connection density, such as smart transportation, smart grids, and smart cities. It meets people's needs for a digital society.

Based on the above three types of usage scenarios, 5G enables a variety of intelligent applications, including smart manufacturing, smart traffic, smart grids, smart campus,

Table 1. Abbreviations

Abbreviations	Explanation
5G	5th Generation Mobile Network
AF	Application Function
AI	Artificial Intelligence
AMF	Access and Mobility Management Function
API	Application Programming Interface
AUSF	Authentication Server Function
CPE	Customer Premise Equipment
DDoS	Distributed Denial of Service
eMBB	enhanced Mobile Broadband
EMS	Element Management System
IMSI	International Mobile Subscriber Identity
IoT	Internet of Things
IPS	Intrusion Prevention System
LTE	Long Term Evolution
MANO	Management and Orchestration
MEC	Multi-access Edge Computing
mMTC	massive Machine Type Communications
NEF	Network Exposure Function
NFV	Network Function Virtualization
NSSAI	Network Slice Selection Assistance Information
NSSF	Network Slice Selection Function
PCF	Policy Control Function
PDCP	Packet Data Convergence Protocol
RAN	Radio Access Network
RBAC	Role Based Access Control
SBA	Service Based Architecture
SDN	Software Defined Network
SMF	Session Management Function
SUCI	Subscription Concealed Identifier
UDM	Unified Data Management
UE	User Equipment
UPF	User Plane Function

(*continued*)

Table 1. (*continued*)

Abbreviations	Explanation
uRLLC	Ultra-Reliable and Low Latency Communications
VR/AR	Virtual Reality/Augmented Reality
WAF	Web Application Firewall
WLAN	Wireless Local Area Network

etc. We will analyze how the new technologies enable these applications in detail (see Fig. 3). In Fig. 3, the blue points are the typical 5G applications and the grey points are some specific use cases of these applications.

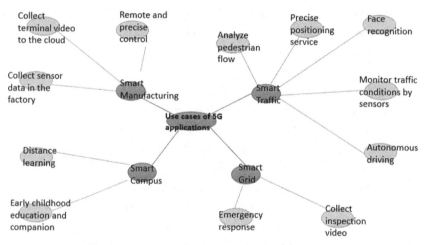

Fig. 3. Classifications of use cases of 5G applications

2.1 5G Enabled Smart Manufacturing

Smart manufacturing, today, is the ability to continuously maintain and improve performance, with intensive use of information, in response to the changing environments [12]. The applications of 5G technology in the field of intelligent manufacturing can be divided into three categories:

- utilizing 5G low-latency features, network slice, edge computing and other new technologies to ensure network quality for remote and precise control, such as engineering machinery remote control, AGV control, robot control, and on-site production line equipment control, etc.
- using 5G high-bandwidth features and edge computing technology, collecting terminal-side video to the cloud for deep analysis, such as defect detection, OCR

decoding, AR assistance, VR complex assembly, production safety behavior analysis, 5G PLC, etc.
- using 5G mass-connection, high-bandwidth characteristics and edge computing technology, collecting sensor data in the factory and transmitting it to the cloud for deep analysis, such as 5G large-scale data collection.

2.2 5G Enabled Smart Traffic

Smart traffic covers vehicles, road infrastructure, traffic management facilities, transportation planning, digital transportation platforms, and various transportation-based applications [13]. The applications of 5G technology in the transportation industry [14] can be divided into 5 categories:

- based on the user's access to the 5G base station, analyze the pedestrian flow within the coverage of the base station, such as: smart train station traffic transfer linkage, smart subway passenger flow analysis.
- based on the 5G base station's precise positioning function, to provide precise positioning services for vehicles and people, such as high-precision positioning, high-precision indoor navigation.
- based on 5G high-bandwidth transmission capabilities, using high-definition video capture and transfer back to the application platform to perform face recognition, such as passenger behavior safety analysis, and passengers exit without perception of smart train station.
- based on the 5G massive connection characteristics, connect various types of traffic sensors and other IoT devices, to analyze the health status of traffic infrastructure, and timely alert traffic conditions by analyzing various types of data received, such as infrastructure monitoring and inspection, smart subway inspections and maintenance, warning and management of smart roads.
- based on the high bandwidth, low latency, and massive connection characteristics of 5G, new technologies such as network slicing and edge computing are used to meet the high requirements of unmanned and remotely controlled driving, such as autonomous driving (autonomous driving of public transportation, freight logistics, special vehicles, autonomous driving in networked parks, 5G road test field and connected remote low-speed driving), smart ports (remotely controlled bridges, intelligent tally, unmanned transportation system), smart airport (wireless dispatch platform, driverless shuttle connection).

2.3 5G Enabled Smart Grids

Smart grid uses two-way flows of electricity and information to create a widely distributed automated energy delivery network [16]. The applications of 5G technology in the smart grid industry [15] can be divided into 2 categories:

- The application that is based on 5G low-latency features, slicing, edge computing and other new technologies, ensure emergency response of the power grid, such as distribution network differential protection, distribution network PMU, and precise load control.

- The application that is based on 5G mass-connection, high-bandwidth characteristics, and network slicing, edge computing technology, collect inspection video and transmit to the cloud for deep analysis, such as distribution automation of FTU, DTU and TTU, advanced metering, intelligent inspection, power grid emergency communications.

2.4 5G Enabled Smart Campus

Smart campus refers to a smart campus based on the Internet of Things, which integrates work, study and life. This integrated environment takes various application service systems as the carrier, and fully integrates teaching, scientific research, management and campus life. The usage scenarios of 5G in smart campuses can be summarized into two categories:

- using 5G high-bandwidth features, network slicing, and edge computing technologies for distance learning and AR content dissemination, while also ensuring campus security, such as remote teaching and research, holographic projection public courses, cloud AR interaction teaching, smart examination room, safe campus;
- use 5G slicing technology to carry out applications such as early childhood education and companion robots and 5G infant growth assessment.

3 Security and Privacy Issues in 5G Applications

3.1 eMBB Scenario

Currently, 4K/8K high-definition video and mobile roaming immersive services based on virtual reality and augmented reality have become the main application forms of eMBB, which mainly includes the following security risks:

Failure of Monitoring Means. eMBB applications produce huge volumes of traffic which would make it extremely difficult for security devices such as firewalls and intrusion detection systems deployed in existing networks to ensure adequate security protection when it comes to traffic detection, radio coverage, and data storage [17].

User Privacy Leakage. eMBB services (such as VR/AR) contain a large amount of user privacy information, such as personal information or identification, device identification, and address information, etc., and the openness of 5G networks has increased the probability of leakage of private information, for example, the situation that different application slices based on the same infrastructure network. On the other hand, due to the development of data mining technology, the way of extracting private information has become more powerful and efficient. It can associate the device identification with the user identification (such as the user's application identification), and thereby mines the user's network behavior. Therefore, eMBB business has great risk of user privacy leakage.

3.2 uRLLC Scenario

Low latency and high reliability are the basic requirements for uRLLC services. For example, if the internet of vehicles is subject to security threats in communications, it may cause danger of life. Therefore, uRLLC services require high-level security protection measures and should not add additional communication delays. The main security risks are as follows [27]:

DDoS Attacks. Attackers can use DoS/DDoS attacks to cause network congestion or wireless interference to cause communication interruptions. As a result, uRLLC services cannot run properly, service data transmission delays cannot be guaranteed, and even communication being cut off.

Data Security Risks. Attackers use vulnerabilities in devices and protocols along network data transmission paths (5G air interfaces, core networks, and the Internet) to tamper with/forge/replay application data [14], causing the drop of data transmission reliability and harm to normal application operations. For example, in the situation of Internet of Vehicles, vehicle may be out of control.

3.3 mMTC Scenario

The 5G mMTC scenario supports IoT applications with massive devices being connected. Due to the low cost, mass deployment, and limited resources (such as processing, storage, energy, etc.) of the Internet of Things [18], the following security risks are common to IoT devices:

Counterfeit Terminals. The IoT terminal has limited resources and weak processing and computing capabilities. Therefore, it is likely that access authentication will not be performed or a simple method will be adopted for access authentication (see the schemes in [19, 20]), which brings opportunities for counterfeit terminals. Illegal terminal equipment can use the loopholes in the authentication mechanism to impersonate legitimate terminals to access the IoT application platform and report fake data, causing confusion for the operation of the IoT application.

Data Tamperings. The amount of data perceived by IoT terminals is small, but it is of great significance. Attackers can tamper with application data by exploiting weaknesses along data transmission paths (air interfaces, backhaul links, and the Internet). Therefore, it is necessary to prevent attackers from tampering with the data exchanged between the terminal and the network, and to ensure the authenticity and integrity of application data.

Data Eavesdropping. The data collected by IoT terminals deployed in special environments (such as home environments and medical environments) is very sensitive and involves user privacy information. By exploiting weaknesses along data transmission paths (air interfaces, backhaul links, and the Internet), attackers eavesdrop on application data, leading to user privacy breaches. Therefore, it is necessary to prevent attackers from eavesdropping on the data exchanged between the terminal and the network, and ensure the privacy of application data.

Remote Controls. Attackers can remotely access and control IoT terminals through software and hardware interfaces by taking advantages of the simplicity of IoT terminals and weak security protection capabilities, then use the captured terminals to launch attacks that interfere with the normal operation of IoT services [21–25].

Based on the above analyses, typical security and privacy risks of use cases in smart manufacturing, smart traffic, smart grids and smart campus are listed as below, see Table 2.

Table 2. The security and privacy risks of typical applications

Typical applications	Specific use cases	Risks examples
Smart manufacturing	AR assistance, VR complex assembly	Failure of monitoring means
	Collecting sensor data of IoT device	Data tampering & Data eavesdropping
Smart traffic	Connected vehicles	DDoS attacks& Data security risks
	Passenger behavior safety analysis	Failure of monitoring means
Smart grids	Distribution network differential protection and precise load control	DDoS attacks
	Customized network slice to satisfy the low time latency requirement	Counterfeit terminals & Management of network slices
Smart campus	Distance learning and AR content dissemination	Failure of monitoring means& User privacy leakage

4 Security and Privacy Solutions in a Systematic View

4.1 Reference Architecture of 5G Application Systems

As shown in Fig. 4, 5G applications can be modeled into of four layers – the terminal layer, network layer, platform layer, and service layer [26, 28] – from the bottom up.

- **The terminal layer** involves mobile phone terminals, and VR/AR terminals for individual users (to C), as well as industrial control terminals, CPEs, and various sensors for vertical industries (to B).
- **The network layer** is an end-to-end 5G network, including the radio base stations (RBS), MEC, the bearer network, the 5G Core network, and 5G network slices from base stations to the core network.

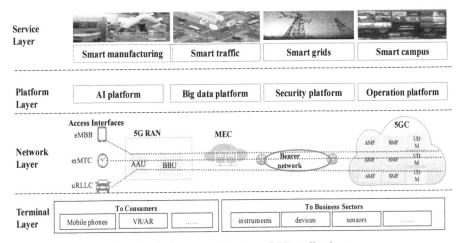

Fig. 4. Reference architecture of 5G applications

- **The platform layer** contains public IT platform systems such as the AI platform, big data platform, operation platform, and security platform. It is recognized that innovative technologies, such as artificial intelligence and big data, are good means to improve security and privacy capabilities [47, 48].
- **The service layer** is composed of 5G enabled smart application systems such as smart manufacturing, smart traffic, smart grids and smart campus.

Each layer has corresponding security goals and solutions, as shown in Table 3.

4.2 Solutions on Terminal Layer

A large number of 5G terminals have low power consumption, as well as limited computing and storage resources, which makes deployment of complex security policies and control over the software difficult. Consequently, these limitations make the terminals become easy and likely targets to be hacked [23].

Firstly, DDoS attacks initiated by terminals need to be prevented and resisted. Such DDoS attacks may be initiated by hacked terminals, or be unintentionally caused when a large number of terminals trigger control-plane signaling registration at the same time due to software defects or network faults. It is recommended that a set of security defense mechanisms to be built at the network level for attack detection and self-protection (such as active flow control) to ensure that any DDoS attacks can be detected the first time to prevent major global negative effects on network services. Besides, some proactive and preventive measures are recommended in terms of terminal exception handling and signaling registration.

Secondly, for the prevention of risks brought by terminal hacking, it is recommended that certain security capabilities such as SSH security login, TLS transmission encryption, and built-in security chip being built in terminals in terms of access authentication [24, 25] on the management and O&M plane as well as encryption protection on the signaling/data plane.

Table 3. Security and privacy solutions for 5G applications

Layer	Targets	Security and privacy Solutions
Terminal layer	Prevent and defend against DDOS attacks initiated by exploited terminals	• Attack detection and self-protection mechanisms (such as active flow control) • Proactive preventive measures, e.g. terminal exceptions handling and signaling registration
	Prevent various damage caused by exploited terminals to industry production and applications	• Access authentication [49, 50] on the operation and maintenance side • Encryption protection on the signaling/data plane (such as SSH secure login, TLS transmission encryption, and built-in security chip, etc.)
Network layer	Base station air interface security	• Defense Eavesdropping and tampering of user data from air interface • Defense DDOS attack from air interface • Pseudo base station detection [52]
	MEC security	• Physical environment security control • Enterprise and operator network isolation
	5GC security	• Manage operation and maintenance plane security • Network north-south border security • East-west security within the network • Cloud-based security of the core network
	Bearer network security	• Network redundant design • Account authority management and access authentication • Increase security measures on control protocols • User plane security encryption
	5G slicing security [43]	• Isolation between slices • Secure access and use of slices • Privacy protection

(*continued*)

Table 3. (*continued*)

Layer	Targets	Security and privacy Solutions
Platform layer	Communication interface security	• Routine maintenance of various account passwords • Encryption of communication interfaces such as TLS, etc.
	Platform data security	• Data availability, integrity and privacy
Service layer	Software security of the application	• Vulnerability scanning of the software • Software operation logging • Highly available disaster recovery of software systems (such as dual-machine backup)
	Security operation and maintenance management of the application system	• Security constraints and controls for application system, e.g. multi-factor authentication for important sensitive operations • Physical security control (personal access control) of O & M operations office/machine room, etc.

4.3 Solutions on Network Layer

From the perspective of network components, the noteworthy aspects of network layer security include security in the RAN base station air interfaces [51], MECs, 5G Core, bearer networks, and 5G slices.

• Base Station Air Interface Security

Air interfaces between 5G UEs and base stations mainly have to deal with three types of security threats:

– Countermeasures of User data eavesdropping and tampering over air interfaces. For the prevention of this type of security threat, SUCI encryption and encryption for air-interface PDCP data packets can be enabled.
– Countermeasures of DDoS attacks over air interfaces. For the prevention of this type of attacks, a DDoS detection and defense system can be deployed so that base stations can implement flow control in the case of mega DDoS attacks.
– Countermeasures of Malicious attacks and interference from pseudo base stations, such as spam short messages or valuable and sensitive information eavesdropping by such rogue base stations. For this type of attack, a unified rogue base station detection system can be deployed around the network so that rogue base stations on the network can be detected and located the first time.

• MEC Security

To avoid physical attacks and cross-network penetration and infection of network, 5G networks need to focus not only on the physical security control of MEC, but also on the isolation between enterprise networks and operators' 5G networks. Security facilities, such as firewalls and IPS, are recommended for network boundary protection [29–35].

• 5G Core Security

The security of the 5G Core has the top priority of the security of the entire 5G network. Security protection measures for the 5G Core need to be considered from the following aspects.

– The security of Operation & Management plane. For MANO, EMS and other systems on the O&M plane, an access security control system is recommended to avoid unauthorized management and O&M access, and ensure secure and compliant O&M operations. In addition, to prevent security risks such as viruses and OS vulnerabilities introduced by O&M terminals, desktop cloud terminals can be used.
– The security of the southbound-northbound boundary of the network. It is recommended that security facilities, such as firewalls, sandboxes, WAF, IPS, and anti-DDoS, being deployed in a centralized manner at the exit boundary of the data center to prevent possible security threats from external networks.
– The security of eastbound-westbound security inside the network. It is recommended that certain specific security measures to be deployed, such as network micro-segmentation and whitelist ACL, and network traffic probe collection and analysis.
– Cloud and virtualization security. Security threats caused by vulnerabilities in the OS software itself need to be prevented. VM escape threats must also be a focus where an attacked VM penetrates to the upper layer and causes risks to 5G core NEs. It is recommended that host security scanning and hardening being routinely implemented, and monitoring software being deployed at the hypervisor level of certain servers to prevent VM escape attacks [36–39].

• Bearer Network Security

The security of the bearer network needs to be protected in the following aspects:

– In network planning and design, redundancy design needs to be adopted to avoid single points of failure. In addition, on the management plane of the bearer network, permission management and access authentication of accounts and passwords need to be implemented.
– On the protocol control plane of the bearer network, security measures such as MD5 authentication or SSL encryption can be configured to avoid possible routing protocol attacks such as BGP routing hijack attacks.

– On the user plane of the bearer network, IPsec security encryption can be deployed to ensure the integrity of network data packets, to prevent illegal traffic interception or network replay attacks.

• 5G Slice Security

The security of 5G network slicing [43] needs to be protected by the following measures:

– Isolation between slices. The failure of one slice must not affect other slices.
– Secure access and use of slices. Access to a corresponding 5G network slice requires dual authentications and authorizations by the slice user (such as a government agency or an industrial mining enterprise) and the operator, ensuring legal access and use of slice resources. Moreover, the privacy protection of Network Slice Selection Assistance Information (NSSAI) needs to be provided.

4.4 Solutions on Platform Layer

The platform layer covers various intelligent analysis and processing AI platforms, big data platforms, and IT middleground [47, 48]. The security of this layer includes the following aspects.

• The security of communications interfaces. Human-machine communication involves the control of account password login and operation permissions of different systems, while machine-to-machine communication involves API invoking, information collection and transmission, and the transfer of operation instructions between platform systems and other related upstream or downstream component systems or NEs. In general, communication interface security at the platform layer mainly focuses on the routine maintenance and management of various accounts and passwords (such as regular password changes and password complexity requirements), and the encryption of communications interfaces (such as TLS).
• The security of platform data. Big data is usually used in 5G applications. The security of data at the platform layer involves the security of various basic data collected and stored by the big data platform (especially data involving user privacy or sensitive information on public safety), including data availability, integrity, and privacy. Availability is guaranteed by technologies such as data redundancy. Integrity is guaranteed by technologies such as data verification. For privacy, as the data amount is usually huge, more effective access control and security audit are required.

4.5 Solutions on Service Layer

The security of the service layer consists of various application system software security, secure O&M of application systems.

• Application system software security mainly involves scans for vulnerabilities and the improvement of software security (including the application software itself,

OS databases, and other software systems), software operation logging, and software system high availability (HA) disaster recovery deployment (such as dual-host backup).

- Secure O&M of application systems focus more on the operation and use of application systems, and the security constraints and control of information on the operation management personnel, for example, application system login accounts and passwords, multi-factor authentication for important and sensitive operations, permission-based operation access control (available operations and function menus vary with different levels of accounts), and physical security control of personnel access of O&M operations offices and equipment rooms.

5 Countermeasures Against Security and Privacy Risks in 5G Applications

Based on the systematic security and privacy solutions proposed above, the following specific security measures are recommended for 5G application service developers and providers in different application scenarios [40–42]. The related layers in the reference architecture to deploy these countermeasures are also suggested (see Table 4).

5.1 eMBB Scenario

- Deploy application traffic monitoring at edge computing [35] nodes and support suspension of high-risk services in specific cases
- The secondary authentication and key management mechanism are used to perform secondary identity authentication and authorization between the terminal and the eMBB application service platform to ensure the authenticity of the terminal and platform identity and the legality of application. At the same time, negotiate and manage the service layer key between the two sides to encrypt and protect user data,thus preventing attackers from eavesdropping;
- In applications with high security requirements, the user plane of the 5G network can be protected by physical isolation or encryption to ensure the security of user data transmission between network functions;
- The network slicing or data dedicated line is used between the operator's 5G core network and the eMBB application service platform to establish a secure data transmission channel to ensure the security of user business data transmission.

5.2 uRLLC Scenario

- Establish a two-way identity authentication mechanism between the user terminal and the application server to prevent fake users from establishing connections.
- Deploy anti-DDoS capabilities to prevent network congestion, wireless interference, and communication link disruptions.
- Through the security capabilities deployed at edge computing, as well as data integrity protection, time stamp, serial number and other mechanisms, to prevent application data from being tampered/falsified/replayed and ensure the reliability of data transmission [32];

Table 4. Countermeasures against Security and privacy risks in 5G Applications

Risks	Countermeasures	Related layer
eMBB scenario		
Failure of effective monitoring means	• Application traffic monitoring at edge computing [35] nodes, suspension of high-risk services in specific cases	Network layer
User privacy leakage risk	• Perform secondary identity authentication and authorization between the terminal and the eMBB application service platform • Negotiate and manage the service layer key between the two sides to encrypt and protect user data • Physical isolation or encryption • Network slicing [43] or data dedicated line	Terminal layer/network layer/service layer
uRLLC scenario		
DDoS attack risk	• Two-way identity authentication mechanism between the user terminal and the application servers • Deploy anti-DDoS capabilities	Network layer and terminal layer
Data security risks	• Through the security capabilities deployed at edge computing [26], as well as data integrity protection, time stamp, serial number and other mechanisms [44]	Network layer
mMTC scenario		
Counterfeit terminal	• Using lightweight [45, 49, 50] security algorithms, simple and efficient security protocols to implement two-way authentication	Terminal layer
Data tampering and Eavesdropping	• Encrypt and protect the integrity of sensitive application data generated by IoT terminals [44]	Terminal layer
Remote control	• Deploy security monitoring methods [47, 48] to timely detect and prevent massive IoT devices from being controlled	Terminal layer

5.3 mMTC Scenario

- Using lightweight security algorithms, simple and efficient security protocols to implement two-way authentication between IoT terminals and the network to ensure that the access terminals are secure and reliable.
- Encrypt and protect the integrity of sensitive application data generated by IoT terminals to prevent attackers from eavesdropping, tampering, forging, and replaying business data on the transmission path
- Deploy security monitoring methods [47, 48] to timely detect and prevent massive IoT devices from being controlled, to prevent these devices from being used maliciously, such as launching DDoS attacks on air interfaces and service platforms, etc., causing network congestion and causing mMTC services to fail [46].

6   Conclusions

5G is deeply integrated with social life and vertical industries, and the security and privacy of the 5G ecosystem is largely influenced by application developers, service providers, as well as network operators and equipment suppliers. The achievement of security and privacy in 5G applications requires a comprehensive and systematic design, as well as the deployment of proper security measures according to the specific application scenarios and the needs of the industry.

In the future, in line with the continuous development of 5G applications, the security level will continue to improve. On the one hand, along with the evolution of 5G technology, the changes in application requirements, and the development of security offensive and defensive technologies, stakeholders should continue to enrich 5G security solutions, including the adoption of 5G network slicing, authentication capabilities, and other network capabilities to provide security for upper-layer applications; on the other hand, 5G network security will continue to develop in the direction of intelligence, providing a flexible and customizable security capability to facilitate vertical industries to choose security capabilities and management methods that match industry needs.

Acknowledgments. This paper is supported by the construction project of the Joint Laboratory for Mobile Learning, Ministry of Education-China Mobile Communications Corporation (no. ML2012934).

References

1. Bedo, J., Ayoubi, S., Filippou, M., et al.: 5G innovations for new business opportunities. In: Mobile World Congress, Barcelona, Spain. 5G Infrastructure Association, Mobile World Congress 2017, 5G IA Event (2017)
2. TS 22.261. Technical Specification Group Services and System Aspects; Service Requirements for the 5G system; Stage 1, 3GPP
3. 5G Network Architecture and Security, DCMS Phase 1 5G Testbeds & Trials Programme (2018)
4. TS 23.501, System Architecture for the 5G System, 3GPP

5. GS MEC-002. MEC Technical Requirements, ETSI
6. Niu, Y., Li, Y., Jin, D., et al.: A survey of millimeter wave communications (mmWave) for 5G: opportunities and challenges. Wirel. Netw. **21**(8), 2657–2676 (2015)
7. Gavrilovska, L., Rakovic, V., Atanasovski, V.: Visions towards 5G: technical requirements and potential enablers. Wirel. Pers. Commun. **87**(3), 731–757 (2015). https://doi.org/10.1007/s11277-015-2632-7
8. Setting the Scene for 5G: Opportunities & Challenges. https://www.itu.int/en/ITU-D/Documents/ITU_5G_REPORT-2018.pdf
9. IMT Vision – Framework and Overall Objectives of the Future Development of IMT for 2020 and Beyond, ITU-R M.2083-0
10. European Commission, Commission Recommendation of 26.3.2019 Cybersecurity of 5G networks
11. IHS Markit: How 5G technology contribute to the global economy The 5G Economy (201 9)
12. Jung, K., Kulvatunyou, B., Choi, S., et al.: An Overview of a Smart Manufacturing System Readiness Assessment (2017)
13. Bao, Z.: Discussing 5G network technologies in smart traffic construction. China ITS J. **226**(01), 81–82+102 (2019)
14. Basudan, S., Lin, X., Sankaranarayanan, K.: A privacy-preserving vehicular crowdsensing based road surface condition monitoring system using fog computing. IEEE Internet of Things J. (2017)
15. Saxena, N., Roy, A., Kim, H.: Efficient 5g small cell planning with eMBMS for optimal demand response in smart grids. IEEE Trans. Ind. Inform. **13**(3), 1471–1481 (2017)
16. Fang, X., Misra, S., Xue, G., et al.: Smart grid—the new and improved power grid: a survey. IEEE Commun. Surv. Tutor. **14**(4), 944–980 (2012)
17. CAICT, IMT 2020(5G) Promotion Group. 5G security Report (2020)
18. Fan, K., Gong, Y., Liang, C., et al.: Lightweight and ultralightweight RFID mutual authentication protocol with cache in the reader for IoT in 5G. Secur. Commun. Netw. **9**(16) (2016)
19. Wang, D., Wang, P.: Two birds with one stone: two-factor authentication with security beyond conventional bound. IEEE Trans. Dependable Secure Comput. **15**(4), 708–722 (2018)
20. He, D., Wang, D., Wu, S.: Cryptanalysis and improvement of a password-based remote user authentication scheme without smart cards. Inf. Technol. Control **42**(4), 170–177 (2013)
21. GTI. 5G Network Security Consideration white paper v1.0. GTI (2019)
22. ENISA Threat Landscape for 5G Networks, November (2019)
23. Ji, X., Huang, K., Jin, L., et al.: Review of 5G security technology. Mob. Commun. **43**(01), 40–45+51 (2019)
24. Wang, D., Ma, C.G., Zhang, Q.M., et al.: Secure password-based remote user authentication scheme against smart card security breach. J. Netw. **8**(1), 148–155 (2013)
25. Wang, D., Zhang, X., Zhang, Z., et al.: Understanding security failures of multi-factor authentication schemes for multi-server environments. Comput. Secur. **88**, 1–13 (2020)
26. Ahmad, I., Kumar, T., Liyanage, M., et al.: Overview of 5G security challenges and solutions. IEEE Commun. Stand. Mag. **2**(1), 36–43 (2018)
27. Schneider, P., Günther, H.: Towards 5G Security. IEEE Trustcom/bigdatase/ispa. IEEE Computer Society (2015)
28. Zhang, K., Ni, J., Yang, K., et al.: Security and privacy in smart city applications: challenges and solutions. IEEE Commun. Mag. **55**(1), 122–129 (2017)
29. ISO/IEC 23188 Information technology – Cloud Computing – Edge Computing Landscape
30. ITU-T X.5Gsec-netec "Security capabilities of network layer for 5G edge computing"
31. ITU-T X.5Gsec-ecs "Security Framework for 5G Edge Computing Services"
32. ETSI GS MEC 003 V1.1.1 Mobile Edge Computing (MEC); Framework and Reference Architecture (2016)

33. ETSI GS MEC-IEG 004 V1.1.1 Mobile-Edge Computing (MEC); Service Scenarios
34. ETSI GS MEC 002 V1.1.1 Mobile Edge Computing (MEC); Technical Requirements
35. Zhang, J., Zhao, Y., Chen, B., et al.: Research on edge computing data security and privacy protection. J. Commun. (2018)
36. ETSI GS NFV-SEC 001 V1.1.1 Network Functions Virtualisation (NFV); NFV Security; Problem Statement (2014)
37. ETSI GS NFV-SEC 003 V1.1.1: Network Functions Virtualisation (NFV); NFV Security; Security and Trust Guidance (2014)
38. ETSI GS NFV-SEC 012 V3.1.1 Network Functions Virtualisation (NFV) Release 3; Security; System architecture specification for execution of sensitive NFV components (2017)
39. ITU-T X.1038: Security requirements and reference architecture for software-defined networking
40. Zhang, B., Yuan, J., Qiu, Q., et al.: Research on 5G security technology and development. In: Proceedings of "5G+" China Mobile Science and Technology Association, pp 1–5 (2019)
41. Fan, N., Liu, G., Shen, J.: Analysis of mobile network security for operators in the initial stage of 5G commercialization. China Inf. Secur. **7**, 85–87 (2019)
42. China Mobile 5G Joint Innovation Center. White Paper on 5G Security for the Medical Industry (2019)
43. China Mobile 5G Joint Innovation Center. 5G Slicing Security White Paper for Vertical Industries (2018)
44. Ferrag, M.A., Maglaras, L., Argyriou, A., et al.: Security for 4G and 5G cellular networks: a survey of existing authentication and privacy-preserving schemes. J. Netw. Comput. Appl. S1084804517303521 (2017)
45. Das, A.K., Zeadally, S., Wazid, M.: Lightweight authentication protocols for wearable devices. Comput. Electr. Eng. S0045790617305347 (2017)
46. Gope, P., Lee, J., Quek, T.: Resilience of DoS attacks in designing anonymous user authentication protocol for wireless sensor networks. IEEE Sens. J. **17**(2), 498–503 (2017)
47. Wang, Y., Chu, W., Fields, S.: Detection of intelligent intruders in wireless sensor networks. Future Internet **8**(1), 2(2016)
48. Buczak, A., Guven, E.: A survey of data mining and machine learning methods for cyber security intrusion detection. IEEE Commun. Surv. Tutor. **18**(2), 1153–1176 (2016)
49. Duan, X., Wang, X.: Authentication handover and privacy protection in 5G hetnets using software-defined networking. Commun. Mag. IEEE **53**(4), 28–35 (2015)
50. Luo, H., Wen, G., Su, J., Huang, Z.: SLAP: succinct and lightweight authentication protocol for low-cost RFID system. Wirel. Netw. **24**(1), 69–78 (2016). https://doi.org/10.1007/s11276-016-1323-y
51. Ku, Y., Lin, D., Lee, C., et al.: 5G radio access network design with the fog paradigm: confluence of communications and computing. IEEE Commun. Mag. **55**(4), 46–52 (2017)
52. Shao, J., Zhu, D., Jin, H., et al.: A joint detection method for identifying pseudo base station based on abnormal access parameters. DEStech Transactions on Engineering and Technology Research (2017). https://doi.org/10.12783/dtetr/iceta2016/7000

Alarm Elements Based Adaptive Network Security Situation Prediction Model

Hongyu Yang[1]($\boxtimes$) (iD), Le Zhang[1] (iD), Xugao Zhang[1] (iD), Guangquan Xu[2] (iD), and Jiyong Zhang[3]

[1] School of Computer Science and Technology, Civil Aviation University of China, Tianjin 300300, China
yhyxlx@hotmail.com
[2] College of Intelligence and Computing, Tianjin University, Tianjin 300350, China
[3] Swiss Federal Institute of Technology in Lausanne, 1015 Lausanne, Switzerland

Abstract. To improve network security situation prediction accuracy, an adaptive network security situation prediction model based on alarm elements was proposed. Firstly, we used the entropy correlation method to generate the network security situation time series according to Alarm Frequency (AF), Alarm Criticality (AC) and Alarm Severity (AS). Then, the initial situation predicted value is calculated through sliding adaptive cubic exponential smoothing. Finally, based on the error state, we built the time-varying weighted Markov chain to predict the error value and modify the initial predicted value. The experimental results show that the network security situation prediction results of this model have a better fit with the real results than other models.

Keywords: Network situation · Alarm element · Entropy correlation · Cubic exponential smoothing · Time-varying weighted Markov · Predicated value

1 Introduction

With the rapid development of computer network technology, great changes have taken place in the network structure and interaction scenarios, making network security a key issue in the field of work and life. In order to avoid losses in all aspects, the evaluation of network security is particularly important. According to the historical information in the network structure, it is the hotspot of network security research to complete the prediction of network security situation and the evaluation of security status. The network security situation prediction forms a nonlinear time series through the factors affecting network security. Based on the historical data and the current network state, the network security status is predicted in a future period through a specific mathematical model, which facilitates network management personnel to detect threats in time and take corresponding protective measures. Currently used network security situation prediction methods include gray prediction method, time series based prediction and neural network based prediction [1].

© ICST Institute for Computer Sciences, Social Informatics and Telecommunications Engineering 2021
Published by Springer Nature Switzerland AG 2021. All Rights Reserved
D. Wang et al. (Eds.): SPNCE 2020, LNICST 344, pp. 41–53, 2021.
https://doi.org/10.1007/978-3-030-66922-5_3

Zhang et al. [2] proposed a network security situation prediction model based on grey neural network with multi-group chaotic particle optimization. The model improves the prediction of network conditions by establishing a nonlinear mapping relationship between the influence factors of the security situation and the situation and using the multi-group chaotic particles to optimize the key parameters of the gray neural network. Xiao et al. [3] proposed a network security situation prediction method based on MEA-BP. The method uses the Mind Evolution Algorithm (MEA) to optimize the weight and threshold of the BP neural network to improve the prediction accuracy and efficiency of the security situation, but the standardization of historical data is not perfect. Sun et al. [4] proposed a Markov prediction model based on complex networks. The model constructs the transformation relationship of network security status into a complex network, and uses the weighted Markov chain to predict the security situation, which can reflect the security status of the network to a certain extent, but the state transition probability matrix constructed by the multi-state network is too large. Zhou et al. [5] proposed a network multi-node security situation prediction model based on improved G-K algorithm. The model extracts the main factors affecting network security by grey entropy correlation method. Based on this, the Kalman filter equation is established to improve the accuracy of security situation prediction. Zhang et al. [6] reduced the training complexity of neural network situation prediction model by improving convolutional neural network, and improved the efficiency of network security situation prediction, but the quality of extracted features needs to be improved.

In view of the uneven quality of historical data in the above network security situation prediction methods and the lack of accuracy of the above methods for multi-peak changes in network security situation prediction, in order to improve the accuracy of network security situation prediction, this paper proposes an adaptive network security situation prediction model based on alarm elements.

2   Network Security Situation Prediction Model

The network security situation prediction model proposed in this paper is shown in Fig. 1.

The network security situation prediction process is designed as follows:

Step 1: Generating a non-linear time series of safety situation values by using an entropy correlation method based on the network alarm information;

Step 2: Using a sliding window to divide the network security situation value sequence segment, and each time a security situation value is updated, the sliding window slides backward by one unit;

Step 3: Establish a three-dimensional exponential smoothing prediction model based on the safety situation sequence in the sliding window, and adaptively adjust the static smoothing coefficient $\alpha \ \square$ to improve the prediction accuracy of the module;

Step 4: Calculate the error between the predicted and actual value of the safety situation in the sliding window, and divide the error into n error intervals, which are recorded as n error states. Using a time-varying weighted Markov chain to predict the error value and correct the situation predictor;

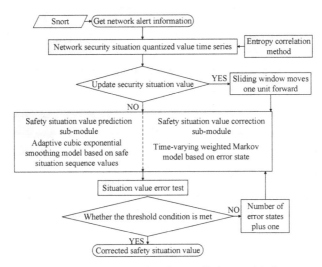

Fig. 1. Network security situation prediction model diagram

Step 5: Check the error. If the threshold condition is not met, return to step 4 and divide the error state into $n + 1$; if the threshold condition is met, obtain the next cycle safety situation value according to steps 1–4.

3   Quantification of Network Security Situation Assessment

First, the alarm information is acquired based on the Snort intrusion detection system. Then, the entropy correlation method is used to calculate the network security situation value in each quantization period. The specific method is designed as follows:

The network security situation quantified value is determined by Alarm Frequency (AF), Alarm Criticality (AC), and Alarm Severity (AS). Referring to the alarm quality quantification method [7], the observation vector obtained by this method based on the alarm quality can effectively improve the data source and the evaluation accuracy. Therefore, the alarm with the highest quality quantization value is selected as the basis for quantifying the network security situation in each cycle.

Let a total of T quantization periods, the network security situation quantization value of the i-th cycle is V_i ($i = 1, 2, \cdots, T$), and the alarm with the highest quality quantization value of the i-th cycle is H_i, then define $V_i = V_i (AF_{Hi}, AC_{Hi}, AS_{Hi})$ ($i = 1, 2, \cdots, T$), where

$$AF_{Hi} = \frac{H_i \text{ number of alarms in } T_i}{\text{Number of all alarms in } T_i} \tag{1}$$

AC_{Hi} is the critical degree of the alarm H_i, indicating the possibility that the occurrence of the alarm H_i causes a change in the network security state. When AC_{Hi} is higher, it means that the network security state is more likely to change. If the alarm AC_{Hi} is an alarm that has occurred in the period i, its priority is set to 1; If the alarm is generated in the period $i - N$ to the period $i - 1$, the priority is 2; If there is no alarm in the

AC_{Hi} from period i-N to period $i - 1$, the priority is set to 3. According to the intrusion detection alarm aggregation association algorithm [8], the algorithm aims at acquiring intrusion-detection alerts and relating them together to expose a more condensed view of the security issues raised by intrusion-detection systems, this paper takes $N = 2$.

AC_{Hi} is the severity of the alarm and indicates the negative impact of the alarm on the network. The greater the severity, the greater the impact of H_i on the network security status. In this paper, the severity of the alarm is divided into low, general, and high, and the corresponding priority values are 1, 2, and 3.

In order to quantify the network security situation of the period i, the comment support matrix P is set as shown in Table 1. Let $X_1 = AF_{Hi}$, $X_2 = AC_{Hi}$, $X_3 = AS_{Hi}$, then X_1, X_2, X_3 correspond to the three quantitative indicators of the alarm H_i with the highest alarm quality in period i, namely alarm frequency, alarm criticality and alarm severity; p_{ij} indicates The degree to which the i-th indicator supports the j-th comment $(i, j \in 1, 2, 3)$.

Table 1. Comment support matrix diagram

Index	Low	General	High
X_1	P_{11}	P_{12}	P_{13}
X_2	P_{21}	P_{22}	P_{23}
X_3	P_{31}	P_{32}	P_{33}

Among them, X_1's support for each comment is determined according to Table 2.

Table 2. X_1 comment interval table

Comment	Low	General	High
Frequency interval C_j	[0, 0.3)	[0.3, 0.7)	[0.7, 1]

Let $X_1 = x$, then calculate the support of the index for each comment by formula (2):

$$P_{ij} = \frac{1 - |x - ((a_j + b_j)/2)| + (b_j - a_j)/2}{\sum_{j=1}^{3} [1 - |x - ((a_j + b_j)/2)| + (b_j - a_j)/2]} \tag{2}$$

Where a_j and b_j respectively correspond to the lower end point and the upper end point of the interval $c_j, j = 1, 2, 3$.

The support for X_2 and X_3 for each comment is shown in Table 3.

Table 3. X_2, X_3 Comment support scale table

Priority	Low	General	High
1	0.5	0.333	0.167
2	0.25	0.5	0.25
3	0.167	0.333	0.5

Use Eq. (3) to calculate the absolute entropy of each indicator of the alarm:

$$H_i = -\sum_{j=1}^{n} P_{ij} \ln P_{ij} \tag{3}$$

When $P_{i1} = P_{i2} = \ldots = P_{in}$, $H_{max} = \ln n$, the relative entropy values of the alarm indicators are:

$$\mu_i = -\frac{1}{\ln n} \sum_{j=1}^{n} P_{ij} \ln P_{ij} \tag{4}$$

The larger the relative entropy value of an indicator is, the smaller the influence of the indicator on the quantized value of the alarm is, the weight of the corresponding indicator is represented by $(1 - \mu_i)$, namely:

$$\tau_i = \frac{1}{n - \sum_{i=1}^{n} \mu_i}(1 - \mu_i) \tag{5}$$

Where $\tau_i \in [0, 1]$ and $\tau_1 + \ldots + \tau_n = 1$. τ_i is the entropy weight coefficient of the index X_i. The vector of the comment weight is $W = (w_{low}, w_{normal}, w_{high}) = (1/5, 1/3, 7/15)$ [9]. Then the network security situation value operator [10] of period i is:

$$V_i = \mu \cdot \tau \cdot P \cdot W^T \tag{6}$$

Among them, μ is the correction factor, this paper takes $\mu = 10000$. The higher the security situation value, the greater the threat to the network and the less optimistic the network security situation.

4 Network Security Situation Prediction Sub-module

4.1 Sliding Window Mechanism

In view of the failure problem of the exponential smoothing prediction method under long time series, this paper defines the length of the historical data sequence based on the three-index exponential prediction by the sliding window mechanism.

Let the sliding window width be L (L is a positive integer), and the current network security posture values are arranged in chronological order as $V_1, V_2, \ldots, V_m$ (m is a positive integer), then the sliding window mechanism is designed as follows:

(1) If the number of safety situation values in the sliding window is k ($1 \leq k \leq m$), the sequence of safety situation values in the width of the sliding window is V_1', V_2', ..., V_k'. If $k + 1 \leq L$, the window does not move, predicting the $m + 1$th safety situation value and waiting for a new safety situation value to enter the window.

(2) If $k + 1 > L$, the sliding window moves forward by one unit when a new safety situation value is added to the sequence, and the safety situation value of the $m + 1$th period is predicted based on the sequence value in the new window.

4.2 Adaptive Cubic Exponential Smoothing Model

Let the network security situation value of m period currently have V_1, V_2, ..., V_m, and there are k security situation values in the sliding window width. If $m \leq L$, then $V_1' = V_1$ and $V_k' = V_m$; if $t > L$, then $V_1' = V_{m-L+1}$, and $V_k' = V_m$. This paper proposes a network security situation prediction model based on the cubic exponential smoothing method:

$$V_{t+T}^1 = a_t + b_t T + c_t T^2 \tag{7}$$

$$a_t = 3s_t^{(1)} - 3s_t^{(2)} + s_t^{(3)} \tag{8}$$

$$b_t = \frac{\alpha}{2(1-\alpha)^2}\left[(6-5\alpha)s_t^{(1)} - 2(5-4\alpha)s_t^{(2)} + (4-3\alpha)s_t^{(3)}\right] \tag{9}$$

$$c_t = \frac{\alpha^2}{2(1-\alpha)^2}\left(s_t^{(1)} - 2s_t^{(2)} + s_t^{(3)}\right) \tag{10}$$

$$s_t^{(1)} = \alpha X_t + (1-\alpha)s_{t-1}^{(1)} \tag{11}$$

$$s_t^{(2)} = \alpha s_t^{(1)} + (1-\alpha)s_{t-1}^{(2)} \tag{12}$$

$$s_t^{(3)} = \alpha s_t^{(2)} + (1-\alpha)s_{t-1}^{(3)} \tag{13}$$

In Eqs. (7)–(13), V_{t+T}' is the predicted value of the $t + T$ safety situation, T is the predicted lead period, and X_t is the actual value of the safety situation in the tth period. $S_t^{(1)}$, $S_t^{(2)}$, and $S_t^{(3)}$ are the first, second, and third smoothing indices of the t-th period; a_t, b_t, c_t are the prediction coefficients of the t-th period; $s_{t-1}^{(1)}$, $s_{t-1}^{(2)}$, and $s_{t-1}^{(3)}$ are the initial values of the first, second and third exponential smoothing of the tth period. In this paper, the initial value of the smoothing index is $s_0^{(1)} = s_0^{(2)} = s_0^{(3)} = (V_1' + V_2' + V_3')/3$; α is the static smoothing coefficient, and $\alpha \in [0, 1]$, its value indirectly affects the final prediction accuracy. Generally, when the actual value sequence shows a horizontal trend, $\alpha \in [0.05, 0.2]$; when the actual value sequence fluctuates, but the long-term fluctuation is small, $\alpha \in [0.3, 0.5]$; when the actual value sequence fluctuates greatly, it is obvious When rising or falling, $\alpha \in [0.6, 0.8]$. The larger the value of x, the greater the impact of the forward data on the predicted value. In order to adapt to the sliding window mechanism caused by the change of the actual value sequence, in this paper, we propose to minimize the sum of the absolute errors of predicted and actual values, and aim at this to obtain the optimal dynamic solution of α. The optimal dynamic solution process for α is designed as follows:

Step 1. It is assumed that the k network security situation actual values in the current sliding window constitute a vector $V' = (V_1', V_2', \cdots, V_k')$, and the static smoothing coefficient α initial value is 0.

Step 2. It is known that $s_0^{(1)} = s_0^{(2)} = s_0^{(3)} = (V_1' + V_2' + V_3')/3$, and $X_1 = V_1'$. From Eqs. (11)–(13), $s_t^{(1)}, s_t^{(2)}, s_t^{(3)}$ ($t = 0, 1, \cdots, k$) are obtained, and from the Eqs. (8)–(10), a_t, b_t, c_t are obtained ($t = 0, 1, \cdots, k$).

Step 3. Let $t = 0, 1, \cdots, k - 1$, let the lead prediction period $T = 1$, and obtain the predicted value sequence $V_1 = (V_1^1, V_2^1, \cdots, V_k^1)$ based on the current static smoothing coefficient from Eq. (7).

Step 4. The sum of the absolute values of the errors of the predicted value sequence and the actual value sequence is $E = \sum\limits_{i=1}^{k} \left| V_i^1 - V_i' \right|, \alpha = \alpha + 0.001$;

Step 5. Repeat steps 1–4 to $\alpha = 1$, and record the absolute error generated by each cycle as E_j ($j = 0, 1, \cdots, 1000$), and obtain min $\{E_j\}$ ($j = 0, 1, \cdots, 1000$). The corresponding α value is taken as the optimal dynamic solution of the static smoothing coefficient under the current sliding window, and is denoted as α_{best}.

Step 6. Let $t = k = m$, $\alpha = \alpha_{best}$, $T = 1$, and obtain the safety situation value of the $m + 1$th cycle from Eqs. (7)–(13).

5   Predictive Value Correction Sub-module

According to the theoretical analysis, it can be known that the initial predicted value of the network security situation in each period is in error with the actual value of the known security situation in the same window, and the error is related to the fluctuation of the security situation in the sliding window. In order to reduce the difference between the predicted value and the actual safety situation quantization value, this paper proposes a time-varying weighted Markov correction model based on error state.

5.1   Error State Division

As the new security situation is added, the sliding window moves, and the volatility of the network security situation sequence contained in the sliding window changes, and the distance between the upper and lower limits of the error also changes. Suppose there are k known safe situation values in the current sliding window, taking $V = \{V_i' \mid i = 1, 2, 3, \cdots, k\}$, and the corresponding safety situation prediction value is $V^1 = \{V_i^1 \mid i = 1, 2, 3, \cdots, k\}$, the lower limit of the error is $F^L = \min\{V_i^1 - V_i' \mid i = 1, 2, 3, \cdots, k\}$, and the upper limit of the error is $F^U = \max\{V_i^1 - V_i' \mid i = 1, 2, 3, \cdots, k\}$, the distance between the upper and lower limits of the error is denoted as $FL = F^U - F^L$. The process of dividing the error state is designed as follows:

Step 1. The upper and lower limits of the error are divided into n intervals, and the interval length is FL/n, and the interval range is $[F^L, F^L + FL/n), [F^L + FL/n, F^L + 2FL/n), \cdots, [F^L + (n - 1) \bullet FL/n, F^U]$.

Step 2. The sequence of error values in the current sliding window is $F = \{F_i = V_i^1 - V_i' \mid i = 2, 3, \cdots, k\}$. If $F_i \in [F^L + (j-1) \cdot FL/n, F^L + j \cdot FL/n)$, then the error F_i is in the error state j, where $j \in \{1, 2, \cdots, n\}$. In particular, when $F_i = F^U$, F_i is considered to be in state n.

Step 3. If the predicted value does not satisfy the threshold test requirement after the error correction, the number of error states needs to be increased, that is, $n = n + 1$ to refine the error correction result.

5.2 Time-Varying Weighted Markov Chain Based on Error State

Based on the sequence of error states in the current sliding window, this paper uses the time-varying weighted Markov chain to predict the error value. The error prediction process is designed as follows:

Step 1. Determine an error state transition probability matrix. There are currently n error states, the current period is t, the error state at the adjacent time is $f_{t-1}f_t$, and the error state after q cycles is recorded as f_{t+q}, then

$$p_{ijr} = p\{f_{t+q} = r \mid f_{t-1} = i, f_t = j\}, i, j, r \in 1, 2, \cdots, n$$

Wherein, p_{ijr} represents the probability that the error state of the period $t - 1$ is i, and the error state of the period t is j, and the error state is r after q cycles, and the probability is obtained by a statistical method. When the initial value of the error state number n is 3, the q-order error state transition probability matrix is

$$P_{(n\times n)\times n}^q = \begin{pmatrix} p_{111} & p_{112} & \cdots & p_{11n} \\ p_{121} & p_{122} & \cdots & p_{12n} \\ \vdots & \vdots & \ddots & \vdots \\ p_{nn1} & p_{nn2} & \cdots & p_{nnn} \end{pmatrix} \qquad (14)$$

$q = 1, 2, \cdots, \beta$. In this paper, $\beta_0 = [L/3]$, L is the sliding window width, and the X value adjustment is determined by step 3.

Step 2: Calculate the weight of each order error state transition probability matrix. First calculate the correlation coefficient η_q between $f_{t-1}f_t$ and f_{t+q}

$$\eta_q = \frac{\sum_{t=1}^{n-q} (y_{t-1} + y_t - 2\bar{y})(y_{t+q} - \bar{y})}{\sqrt{\sum_{t=1}^{n-q} (y_{t-1} + y_t - 2\bar{y})^2 \sum_{t=1}^{n-q} (y_{t+q} - \bar{y})^2}} \qquad (15)$$

Where $q = 1, 2, \cdots, \beta$; y_{t-1}, y_t, y_{t+q} respectively represent the error value of period $t - 1$, period t and period $t + q$ in the original error sequence in the current window. $\bar{y}$ represents the average of the original error sequence in the current window. Then the q-order error state transition probability matrix weight q is

$$\omega_q = \frac{|\eta_q|}{\sum_{q=1}^{\beta} |\eta_q|}, \quad q = 1, 2, \ldots, \beta \qquad (16)$$

Step 3. Adjust the value of β. Check the value of ω_β, and set the weight threshold of the error state transition probability matrix to 0.05 [11]. If $\omega_\beta < 0.05$, it indicates that the β-order error state transition probability matrix can ignore the error, and discard the matrix, $\beta = \beta - 1$, recalculate the value of ω_β until $\omega_\beta \geq 0.05$, at this time $q_{max} = \beta$.

Step 4. The error of the predicted value of the security situation in the current window is predicted. The probability that the error value of the period $t + 1$ is in the error state r $(r = 1, 2, ..., n)$ is $P_{r(t+1)}$

$$P_{r(t+1)} = \sum_{q=1}^{\beta} p_{ijr}^{(q)} \cdot w_q \tag{17}$$

Where $q = 1, 2, ..., \beta$; $i, j \in 1, 2, ..., n$, $p_{ijr}^{(q)}$ is taken from the q-order error state transition probability matrix P^q, which represents probability of the adjacent error state $f_{t-q} = i, f_{t-q+1} = j$ steering error state $f_{t+1} = r$. q is the q-order error state transition probability matrix weight, then the error state probability distribution vector of period $t + 1$ is $P_{r(t+1)} = \{p_{1(t+1)}, p_{2(t+1)}, ..., p_{n(t+1)}\}$.

Let the error median vector composed of the median values of each error interval be

$$F_{mid} = \{[F^L + (F^L + FL/n)]/2, [F^L + FL/n + (F^L + 2FL/n)]/2, ..., [F^L + (n-1) \cdot FL/n + F^U]/2\},$$

Then the error prediction value operator at time $t + 1$ is

$$F'_{t+1} = P_{r(t+1)} \cdot F_{mid} \tag{18}$$

The $t + 1$ time prediction value correction result is

$$V_{c(t+1)} = V_{(t+1)}^1 - F'_{(t+1)} \tag{19}$$

Among them, V_{t+1}^1 is the uncorrected security situation prediction value based on the network security situation prediction sub-module.

5.3 Threshold Test

The proximity of the corrected safety situation predictor value to the actual value is analyzed to determine whether the error state division number n is sufficient. It is known that the sequence of corrected safety situation predictors and the actual value sequence in a window are as shown in Table 4.

Table 4. Sequence table of predicted and actual values

Correction value	$V_{c(2)}$	$V_{c(3)}$	$\cdots$	$V_{c(k)}$
Actual value	V_2	V_3	$\cdots$	V_k

The method for judging the accuracy of prediction in this paper is:

(1) Post-test difference test: the difference between the actual value and the predicted correction value is the residual, which is denoted as $R_i = V_i - V_{c(i)}, i = 2, 3, \ldots, k$. The safety situation value variance S_1^2 in the current safety situation sequence segment is

$$S_1^2 = \frac{1}{k} \sum_{i=2}^{n} (V_i - \frac{1}{k} \sum_{i=2}^{k} V_i)^2 \tag{20}$$

The residual sequence variance S_2^2 is calculated by Eq. (21):

$$S_2^2 = \frac{1}{k} \sum_{i=2}^{k} (R_i - \frac{1}{k} \sum_{i=2}^{k} R_i)^2 \tag{21}$$

Then, the posterior difference ratio $c = S_2/S_1$, the smaller the value, the better the prediction accuracy.

(2) Small probability test:

$$P = P(|R_i - \frac{1}{k} \sum_{i=2}^{k} R_i| < 0.6745 S_1) \tag{22}$$

A small probability test result P is obtained from Eq. (22), and the larger the value, the better the prediction accuracy.

According to the c value and the P value, according to the prediction accuracy level table (as shown in Table 5), it is judged whether it is necessary to increase the number of error state divisions. If the model prediction result is the first-level prediction accuracy or the second-level prediction accuracy, it is not necessary to increase the number of error state divisions, otherwise the number of error state divisions is $n + 1$.

Table 5. Rank table of prediction accuracy

Prediction accuracy level	c	P
First level	<0.35	>0.95
Second level	<0.50	>0.80
Third level	<0.65	>0.70
Fourth level	≥0.70	≥0.65

6 Experimental Results and Analysis

The predictive validity of the model is verified using Lincoln Laboratory's standard dataset LL_DOS_1.0.

6.1 Experimental Data Processing

Under the Ubuntu 16.04 operating system, the LL_DOS_1.0 packet is replayed using the Tcpreplay technology, and the Snort intrusion detection system is used to generate an alarm log for the replay traffic under the Windows 10 operating system.

Based on the entropy correlation method described in Sect. 2, the network security situation is quantified, and the quantization period $T = 4$ min is set, and 90 security situation values in the interval [2800, 4000] are generated in 1–360 min. In this paper, 10 safe situation values within 1–40 min are taken as the actual safety situation sequence. Compare the actual values of 80 network security situations and the corresponding network security situation predictions within 41–360 min to test the prediction effect of the model.

6.2 Experimental Comparison and Analysis

The experimental data set is the LL_DOS_1.0 data set, which uses the model of this paper, the traditional Markov prediction model, and the improved Convolutional Neural Network (ICNN) prediction model [6] to obtain the network security situation prediction value and compare the predictions effect. Through experiments, the network security situation prediction value sequence of three methods (as shown in Fig. 2) and the security situation prediction value absolute error sequence (shown in Fig. 3) are obtained.

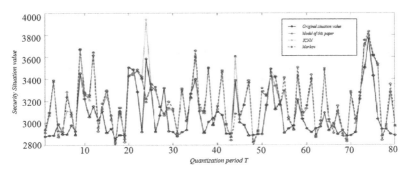

Fig. 2. Network security situation prediction value sequence diagram

It can be seen from Fig. 2 and Fig. 3 that the network security situation prediction result obtained by the model is better than the original state potential value. ICNN model fits better than Markov model. The reasons are as follows:

(1) The prediction of the traditional Markov model depends on the state transition probability matrix, and the state transition probability matrix lacks dynamic adjustment.

(2) The hyper-parameters in the neural network model training are set to be affected by prior experience, which causes the prediction results to deviate from the actual values;

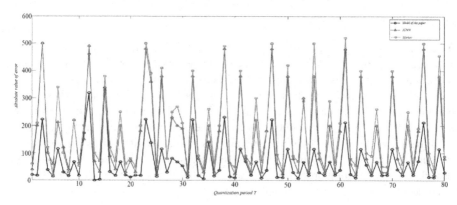

Fig. 3. Predictive value absolute error sequence diagram

(3) The model uses a sliding window mechanism to fragment a long nonlinear time series. The security posture value in the window is continuously updated, so that correlation coefficient can be adaptively and dynamically adjusted, and the network security situation prediction value with higher precision is corrected, and the accuracy of the network security situation prediction is improved.

7 Conclusion

This paper proposes an adaptive network security situation prediction model based on alarm elements. The model quantifies the network security situation values of several cycles by entropy correlation method, and segments the security situation values arranged in time series based on the sliding window mechanism. The adaptive three-dimensional exponential smoothing method is used to generate the initial safety situation prediction results, and the time-varying weighted Markov chain is used to predict the error and correct the safety situation prediction value.

Acknowledgements. This work was supported by the Civil Aviation Joint Research Fund Project of the National Natural Science Foundation of China under granted number U1833107.

References

1. Leau, Y.B.: Network security situation prediction: a review and discussion. Commun. Comput. Inf. Sci. **516**, 424–435 (2015)
2. Zhang, S.B.: Network security situation prediction model based on multi-swarm chaotic particle optimization and optimized grey neural network. In: IEEE International Conference on Software Engineering and Service Science, Piscataway. IEEE Press, Nanjing (2018)
3. Xiao, P.: Network security situation prediction method based on MEA-BP. In: IEEE International Conference on Computational Intelligence & Communication Technology. IEEE Press, Nanjing (2017)

4. Sun, S.X.: The research of the network security situation prediction mechanism based on the complex network. In: IEEE International Conference on Computational Intelligence and Communication Networks. IEEE Press, Nanjing (2015)
5. Zhou, X.W.: Multi node network security situation prediction model based on improved G-K algorithm. Sci. Technol. Eng. **18**(25), 72–77 (2018)
6. Zhang, R.C.: Network security situation prediction method using improved convolution neural network. Comput. Eng. Appl. **55**(6), 86–93 (2019)
7. Xi, R.R.: An improved quantitative evaluation method for network security. Chin. J. Comput. **38**(4), 749–758 (2015)
8. Debar, H., Wespi, A.: Aggregation and correlation of intrusion-detection alerts. In: Lee, W., Mé, L., Wespi, A. (eds.) RAID 2001. LNCS, vol. 2212, pp. 85–103. Springer, Heidelberg (2001). https://doi.org/10.1007/3-540-45474-8_6
9. Zhao, D.M.: Fuzzy risk assessment of entropy-weight coefficient method applied in network security. Comput. Eng. **30**(18), 21–23 (2004)
10. Fu, Y.: An approach for information systems security risk assessment on fuzzy set and entropy-weight. Chin. J. Electron. **38**(7), 1489–1494 (2010)
11. Wang, X.: Network anomaly detection model based on time-varying weighted Markov Chain. Comput. Sci. **44**(9), 136–141+161 (2017)

Watermark Based Tor Cross-Domain Tracking System for Tor Network Traceback

Jianwei Ding[✉] and Zhouguo Chen

30th Research Institute of China Electronics Technology Group Corporation,
No. 8 Adventure Road, High-tech Zone, Chengdu, China
mathe_007@163.com, czgexcel@163.com

Abstract. Anonymous network is widely used to access the Internet, causing varieties of cyber security incidents because of its anonymity, which increasingly affects the security of cyberspace. How to detect anonymous network flow to position the anonymous users, is becoming to a research hotspot. However, with rapid development of the encryption and network technology, it is a nontrivial task to detect and position the anonymous user in such a complex network environment.

In this paper, we design a prototype system called Watermark based Tor Cross-domain Tracking System that is effectively detects and determine the sender and the receiver on the real Tor network to testify its function. Moreover, instead of conventional passive network flow analysis, this paper learns from active network flow analysis to design three digital watermark models to implement the embedding, extracting and matching of watermark information, and meanwhile it will not affect the network flow's content and transmission. Experimental results on the real data sets show that when embedding the three watermark models on the sender, watermark based Tor cross-domain tracking system indeed yields the positioning function.

Keywords: The router onion · Watermark model · Tracking system

1 Introduction

In recent years, more and more anonymous networks appears with the rapid development of the Internet and encryption technology. After the emergence of anonymous networks such as the router onion (Tor), users can communicate anonymously on the Internet through anonymous networks, which also brings network security issues [13,25]. Using anonymous communication systems, cyber attackers can hide their identities. Cyber attackers usually join multiple intermediate springboard hosts into anonymous networks and use these springboards

Supported by: National Key R&D Program of China (No. 2016YFE0206700); Sichuan Science and Technology Program, NO. 2018HHO115.

to attack, intended to make network tracing and network supervision more difficult, which not only threatens the privacy of users, but also causes users to suffer economic losses.

The traffic of anonymous network is encrypted, which make it difficult to analyze. In this case, traditional intrusion detection technology has obvious shortcomings. Conventional passive network traffic analysis [4,6,19] can not confirm the communication relationship between two parties in communication, track anonymous attackers or find intermediate proxy hosts, which is difficult to be applied in a large-scale, high-bandwidth network environment, such as Tor. The controlled environment communication entity correlation technology that incorporates the idea of digital watermarking into network traffic analysis [17] can be applied to a variety of network environments, and has the advantage of high detection rate, low false alarm rate, strong concealment and short detection time.

This paper applies correlation positioning technology to track the illegal users in the Tor, which applies actively traffic analysis technology to track users who attack a certain party through a series of intermediate springboard hosts or users who communicate illegally through anonymous channels. It draws on the idea of digital watermark technology [8–10,24], and embeds watermark information by actively changing certain characteristics of the network flow (such as packet length, network flow rate, or packet sending time interval, etc.) generated by suspicious senders. The watermark information extracted by the suspicious receiver is compared with the original watermark information. If a certain detection rate can be achieved, it can be determined that there is a communication relationship between the sender and the receiver. At the same time, this paper designs a cross-domain collaborative tracking architecture based on network watermarks, support hidden signal detection, tracking and positioning, and path construction, in response to the problem that watermark transmission may cross multiple autonomous domains (ASs), construct a cross-domain collaborative flow watermark tracking architecture.

The rest of this paper is organized as follows. Section 2 reviews related work. Section 3 gives the design details of our cross-domain collaborative tracking architecture based on network watermarks. In Sect. 4, we design three kinds of watermark models, including IPD model, IW model and IWG model. Section 5 presents experimental results to validate our approach. Section 6 gives the conclusions of the research.

2 Related Work

The onion router (Tor) [5] is a world-renowned anonymous network. The core of the Tor network's anonymity is rerouting technology. After multi-layer routing and forwarding, each layer of routing only knows the upper-level node, so it is difficult to trace the source of the traffic. In addition, the bridge protocol will disguise its source as other data traffic, intended to distinguish the third party from the content of the message.

Hence, there are many studies on the traffic analysis of Tor [2,3,22], whose idea mostly is that analyzing the Tor's protocol and then extracting characteristics from network flow to train a machine learning model like SVM for Tor traffic analysis [14,26]. However, with Tor protocol's upgrading, pluggable transmission protocol, such as Meek, obfsproxy and FTE, are introduced into Tor. With the introduction of protocol hiding technology, the previous fingerprint characteristics methods are no longer applicable. Based on the new generation of Tor protocol, there are also some research studies on the Tor traffic analysis [7,15,21], but the recognition rate is not high.

There are a few studies on traceability technology, which are also affected by the pluggable transmission protocol. At present, there is no research on the traffic tracking technology for the Tor protocol. Traditional traffic tracking technologies [27] are generally divided into active traffic analysis and passive traffic analysis. Passive traffic analysis [1] will not have any impact on the anonymous communication process. It only observes the communication process to infer the relationship between users. Active traffic analysis [11,12,16,18,20,28] is mainly to artificially interfere with the traffic, and to achieve neither the purpose of exposure nor the means of intervention. There are some typical digital watermark model, such as packet transmission rate based watermark, inter packet watermark and packet sending interval watermark.

3   Watermark Based Tor Cross-Domain Tracking System

Watermark based Tor Cross-domain Tracking System is illustrated in Fig. 1. The overall structure of the Tor tracking system is based on the idea of SDN's (software defined network) data and control separation. SDN switches do not run any protocol between them, which are only responsible for forwarding data packets, and it is controlled entirely by the upper controller. Developer is able to customize routing decisions and transmission rules in the controller.

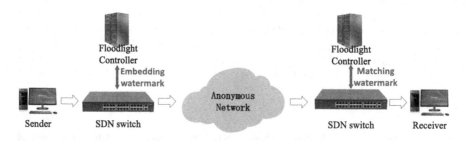

Fig. 1. Architecture of Watermark based Tor Cross-domain Tracking System.

Hence, this paper adopts the secondary development of the Floodlight controller, intended to analyze OpenFlow protocol and issue flow table. The SDN switch sends the matching data packet to the controller by matching the flow table. The controller embeds digital watermark into the target flow and then sends flow's data packets back to the SDN switch in the sender. The target flow passes through Tor network, and the controller matches digital watermark of target flow in the receiver. This architecture design only needs to modify the codes on the controller, and does not need to modify the OpenFlow protocol in the underlying switch, which make the implementation is more flexible.

The Floodlight controller communicates with the SDN switch through the OpenFlow protocol. The controller can add, update, and delete flow table. The switch selects and processes data packets according to flow table. OpenFlow protocol supports three message types in the workflow of the entire architecture, which include Flow-mod messages, Packet-in messages, and Packet-out messages. The controller uses the Flow-mod message to deliver flow table to the switch for updating flow table. If the data packet and flow table are inconsistent or the operation defined in the flow table is to forward the data packet to the controller, the switch sends a Packet-in message to hand the data packet processing right to the controller. The controller uses Packet-out messages to send data packets to the switch via the data channel, and hands the operation to the switch for execution.

All message processing modules in Floodlight controller need to listen to OpenFlow messages, and these modules must follow a certain calling sequence. After each module processes them in sequence, the packet-in message will eventually reach the forwarding module, which is the key to processing the forwarding of data packets between the sending device and the receiving device. Therefore, we modify the relevant code for processing data packets in the forwarding module to implement the digital watermark embedding and detection functions. The work process of watermark based Tor cross-domain tracking system is illustrated in Fig. 2.

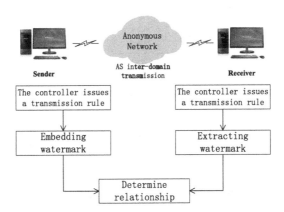

Fig. 2. Work Process of Watermark based Tor Cross-domain Tracking System.

4   Design of Digital Watermark Model

In this paper, we refer to the previous watermark model to design Tor's water-mark model in the watermark based Tor cross-domain tracking system. The design of watermark model need to follow the principle of not affecting the contents of the packet. Hence, we design three watermark models, that is inter packet delay based watermark model, interval based watermark model and interval gravity based watermark model.

4.1   Inter Packet Delay Based Watermark Design

The Inter Packet Delay (IPD) based watermark model encode watermark by fine-tuning the timing of the selected packet. It is necessary to ensure that there are enough packets in the watermarked network flow, and the watermark is embedded only on the selected IPD. In order to make it difficult for an attacker to detect the existence of a watermark without knowing the IPD selection function and other watermark embedding parameters, it is necessary to obtain an IPD on the basis of randomly selecting a packet set and randomly pairing.

Watermark Model. Suppose that network flow f has $n(n > 1)$ packets, and $<P_i, P_j>$ $(0 \leq i < j \leq n - 1)$ are two successive packet pair on the embedded side. The sending time of packet P_i and P_j is t_i and t_j respectively. The sending time interval $dipd_{i,j}$ of P_i and P_j is calculated as follows:

$$dipd_{i,j} = t_j - t_i \tag{1}$$

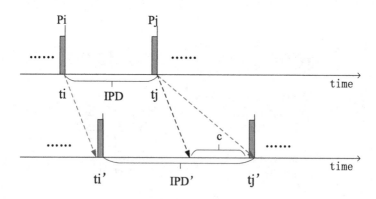

Fig. 3. Schematic diagram of embedded watermark information bits in one IPD.

For one inter-packet delay to be embedded one watermark bit, we need to add extra delay before sending the packet P_i. The watermark bit embed equation is shown as follows:

$$dipd'_{i,j} = dipd_{i,j} + c \tag{2}$$

where $dipd'_{i,j}$ is the time interval after embedding watermark, and the extra delay c is calculated as follows:

$$c = \left(round\left(\frac{dipd_{i,j}}{s} + \frac{1}{2} \right) + \left(w - round\left(\frac{dipd_{i,j}}{s} + \frac{1}{2} \right) \% \ 2 + 2 \right) \%2 \right) * s$$
(3)

where w is the binary watermark information bit '1' or '0', and s is the selected reference time length. The larger the value of s, the less the watermark bit is embedded, but the longer it takes. A schematic diagram of embedding watermark information bits in a single IPD is illustrated in Fig. 3.

Because of the randomness for embedding watermark bits in one inter-packet delay, the success rate of watermark embedding is not high, and it is easily affected by network disturbances. Therefore, the distribution of IPDs carrying watermarks in the longer duration of the network flow is considered. A single watermark bit is embedded in the average of multiple IPDs. As shown in formula (4), in addition to the reference time length s, the number of data packet pairs (redundancy amount) r needs to be selected.

$$dipd_{avg} = \frac{1}{r} \sum_{k=1}^{r} dipd_k$$
(4)

$<P_{i,k}, P_{j,k}>$ $(0 << j \le n - 1, 1 \le k \le r)$ is the k-th packet used to embed the watermark bit, and the packet $P_{i,k}$ is sent $t_{i,k}$, and packet $P_{j,k}$ is sent $t_{j,k}$, where $dipd_k = t_{j,k} - t_{i,k}$. A schematic diagram of embedding watermark information bits in multiple IPDs is shown in Fig. 4.

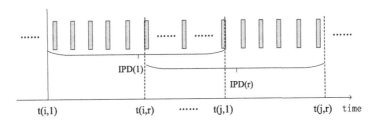

Fig. 4. Schematic diagram of embedded watermark information bits in multiple IPDs.

For the watermark information fw with a watermark bit number of l, $(l+1)*r$ random packets can be selected. Applying l times to embed l-bit watermark information in the selected network flow f.

Watermark Detection. Watermark detection is the process of determining whether a given watermark information is embedded in the IPD of the selected Flow. For the network flow f with l-bit watermark information arriving at the detection end, $<P_i, P_j>$ are two packets that have arrived one after another, and the arrival time of packet P_i is $t_i\prime$, and the arrival time of packet P_j is $t_j\prime$. The interval $aipd_{i,j}$ between the arrival of P_j and P_i is calculated as shown as follows:

$$aipd_{i,j} = t_j\prime - t_i\prime \tag{5}$$

The watermark information is extracted according to the watermark bit decoding function w' is calculated as follows:

$$w' = round\left(\frac{aipd_{i,j}}{s}\right)\%2 \tag{6}$$

Where w' is the extracted one-bit binary watermark information bit '1' or '0'.

Let the l-bit binary watermark information decoded from the watermark flow f be fw', and set the error threshold $h(1 \leq h \leq l)$. Compare fw' with the original watermark information fw. If fw' and fw have different digits less than or equal to h, then the watermark information fw can be considered to be detected in the flow f, so that it can be determined that there exists communication relationship between the receiver and the sender.

It should be noted that if the value of h is set smaller, the watermark detection rate is lower; if the value of h is set too large, even if the detection result is also with the error range, an un-watermarked network flow may be misunderstood that there exists communication relationship between the receiver and the sender. This situation is called system false positive. Hence, the error threshold h needs a tradeoff to satisfy both a higher watermark detection rate and a lower watermark false positive rate.

4.2   Interval Based Watermark Design

The Interval based Watermark (IW) model uses the characteristics of the invariable duration of the network flow to divide the duration of the selected network flow into fixed-length intervals, and adjusts the number of packets in a specific interval according to the watermark information bits to be embedded to achieve the embedded watermark information bit. In this design, the time interval is used as the carrier, so that the watermark is not affected by the change of the number of packets.

Watermark Model. Suppose that, a random time offset σ is set from the beginning of the selected network flow f, and it is watermarked after the time ? has elapsed. Divide the network flow into multiple time intervals I_i of length T. Each I_i contains $X_i(0 \leq i \leq n-1)$ consecutive packets. These packets X_i is independent and identically distributed with respect to the network flow f.

For the binary watermark information fw with a watermark bit number l, divide every three intervals of the flow f into one group. Randomly select n consecutive pairs of intervals, and randomly assign these consecutive pairs of pairs so that each r group of consecutive intervals is used in pairs. To encode a bit of watermark, where $r = \frac{n}{l}$.

The random allocation strategy for n sets of consecutive interval pairs is that: the $<3i, 3i+1, 3i+2>$, $<3(i+l), 3(i+l)+1, 3(i+l)+2>$, $<3(i+2l), 3(i+2l)+1, 3(i+2l)+2>$, ..., $<3[i+(r-1)l], 3[i+(r-1)l]+1, 3[i+(r-1)l]+2>$ consecutive interval pairs are used to encode the i-th watermark bit $(0 \leq i \leq l-1)$.

In order to avoid conflicts between consecutive interval pairs when embedding watermark information bits, every two pairs of interval pairs used to embed watermark information bits need to be inserted as a buffer without using watermark information bits. Therefore, in the selected n sets of consecutive intervals $<I_{i,j,1}, I_{i,j,2}, I_{i,j,3}>$ $(0 \leq i \leq l-1, 0 \leq j \leq r-1)$, $I_{i,j,1}$ is the buffer interval, and $<I_{i,j,2}, I_{i,j,3}>$ is used as the i-th watermark of the encoding the j-th group of bits embeds a space pair. Each interval contains $X_{i,j,k}(0 \leq i \leq l-1, 0 \leq j \leq r-1, k=1,2,3)$ consecutive data packets.

Without artificial interference, the arrival times of the data packets are evenly distributed in each time interval, so the exception u of the number of packets contained in each interval is the same. The model chooses to encode each watermark bit as $I_{i,j,2}$ and $I_{i,j,3}$ with a difference in the number of packets $Y_{i,j}(0 \leq i \leq l-1, 0 \leq j \leq r-1)$, as shown as follows:

$$Y_{i,j} = \frac{X_{i,j,2} - X_{i,j,3}}{2} \tag{7}$$

The average of all the r packet amount deviations $Y_{i,j}$ for encoding the i-th watermark bit $\bar{Y}_{i,r}$ is calculated as follows:

$$\bar{Y}_{i,r} = \frac{1}{r}\sum_{j=0}^{r-1} Y_{i,j} \tag{8}$$

The expectation of the number of packets in each interval is u, and the number of packets X_i is independent and identically distributed with respect to the network flow f, the expectation of $\bar{Y}_{i,r}$ is calculated to be 0. Therefore, the binary watermark information bit '0' or '1' can be encoded through increasing or decreasing $\bar{Y}_{i,r}$ by u, that is, by adjusting the number of packets $X_{i,j,2}$ and $X_{i,j,3}$ within $I_{i,j,2}$ and $I_{i,j,3}$ respectively to makes the distribution of $\bar{Y}_{i,r}$ shift by u to the right or the left.

When the watermark to be embedded is '0', increasing $\bar{Y}_{i,r}$ by u can be achieved by increasing $Y_{i,j}$ by u, that is, increasing r pieces of $X_{i,j,2}$ by u and decreasing r pieces of $X_{i,j,3}$ by u. The former is accomplished by adding packets to the interval $I_{i,j,2}$: adding a delay T to all packets $X_{i,j,1}$ in the buffer interval $I_{i,j,1}$ of the current interval pair, and moving all packets in the interval $I_{i,j,1}$ to the current interval $I_{i,j,2}$. The latter is accomplished by clearing all packets in the interval $I_{i,j,3}$: adding a delay T to all packets $X_{i,j,3}$ in the current interval $I_{i,j,3}$, and moving all packets in the interval $I_{i,j,3}$ to the buffer interval $I_{i,j,1}$ of the next interval pair. When the watermark to be embedded is '1', $\bar{Y}_{i,r}$ is reduced by u, the method is the opposite of the above. The schematic diagram of the embedded watermark information bit is shown in Fig. 5.

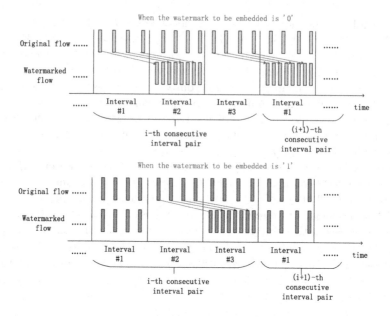

Fig. 5. Schematic diagram for embedding watermark in the IW model.

Watermark Detection. For the watermark flow f embedded with l-bit watermark information arriving at the detection end, by recording the number of packets $X_{i,j,2}$ and $X_{i,j,3}$ $1, 0 \leq j \leq r - 1$) for the embedding interval pair $<I_{i,j,2}, I_{i,j,3}>$ $(0 \leq i \leq l - 1, 0 \leq j \leq r - 1)$, the i-th watermark bit is extracted according to the following strategy:

- Step 1: Let $w_{i,0} = 0$, and calculate the average value of the number of packets $\bar{X}_{i,j,2}$ and $\bar{X}_{i,j,3}$ in the embedding interval pair as follows:

$$\bar{X}_{i,j,2} = (j+1)^{-1} \sum_{k=0}^{j} X_{i,k,2} \tag{9}$$

$$\bar{X}_{i,j,3} = (j+1)^{-1} \sum_{k=0}^{j} X_{i,k,3} \tag{10}$$

- Step 2: When $\bar{X}_{i,j,2} > \bar{X}_{i,j,3}$, let $w_{i,j} = 0$;
- Step 3: When $\bar{X}_{i,j,2} < \bar{X}_{i,j,3}$, let $w_{i,j} = 1$;
- Step 4: When $j > 1$ and $\bar{X}_{i,j,2} = \bar{X}_{i,j,3}$, let $w_{i,j} = w_{i,j-1}$;
- Step 5: Return to Step (1), repeat r times, and finally get $w_{i,r-1}$ is the i-th watermark bit obtained by decoding.

Repeat the above strategy l times to decode the complete l-bit binary watermark information fw' from the watermark flow f. Set the error threshold $h(1 \leq h \leq l)$ and compare fw' with the original watermark fw. If the number of different bits of fw' and fw is less than or equal to the error threshold h, then It is determined that there is a communication relationship between the receiver and the sender; otherwise, it can be considered that there is no communication relationship between the receiver and the sender.

4.3   Interval Gravity Based Watermark Design

According to the characteristics of the invariable constant duration of the network flow, the Interval Gravity-based Watermark (IGW) model divides the duration of the selected network flow into fixed-length intervals, and adds delay to change the time of the packet arrival interval, intended to make the time distribution gravity of the packet arrival time shift to achieve the purpose of embedding the watermark information bits.

Watermark Model. Suppose that there is a random time offset σ for a given network flow f, and the constant duration is defined as T_d after σ. There are X packets added watermark in this selected network flow, and he time stamp of the starting point of the watermark is t_0.

For binary watermark information fw with a watermark bit number of I, T_d is divided into $2n$ intervals with I_i of length T, and each I_i contains X_i consecutive packets. The sending timestamp of the data packet $P_{i,j}(0 \leq i \leq 2n - 1, 0 \leq j \leq X_{i-1})$ is $t_{i,j}$, which is time-lag relative to the start point of the interval I_i. The time offset of $P_{i,j}$ from the starting time in the time interval I_i is shown as follows:

$$\Delta t_{i,j} = \{t_{i,j} - t_0\}\%T \tag{11}$$

Choose n intervals from the $2n$ intervals in the I_i at random to construct a new interval named Interval Group A, denoted as $I_k^A(0 \leq k. \leq n - 1)$, and the other n intervals in the I_i form a new interval named Interval Group B, denoted as I_k^B. Then randomly assign intervals for groups A and B respectively, intended to make every $2r$ intervals to build a watermark bit, where $r = \frac{n}{l}$.

The randomly assignment strategy of the $2n$ intervals is that: set $x(0 \leq x \leq 2n - 1)$ to denoted as interval number, and then choose number $i?i + l?i + 2l????i + (2r - 1)l$ interval to encode the i-th $(0 \leq i \leq l - 1)$ watermark bit respectively. Assign the x-th interval to Interval Group A if $\frac{x-i}{l}\%2 = i\%2$; otherwise, assign the x-th interval to Interval Group B.

$I_{i,j}^A$ and $I_{i,j}^B$ are represented as the j-th $(i \leq j \leq r - 1)$ interval in the i-th $0 \leq i \leq l - 1$ encoded watermark bit for Interval Group A and B respectively. $X_{i,j}^A$ and $X_{i,j}^B$ are the packet amount for the interval $I_{i,j}^A$ and $I_{i,j}^B$ respectively, and X_i^A and X_i^B represent the packet amount in the i-th encoded watermark bit respectively.

$$X_i^A = \sum_{j=0}^{r-1} X_{i,j}^A \tag{12}$$

$$X_i^B = \sum_{j=0}^{r-1} X_{i,j}^B \tag{13}$$

According to the Eq. (11), we can calculate the time offset $\Delta t_{i,j,k}^A$ and $\Delta t_{i,j,k}^B (0 \le i \le l-1, 0 \le j \le r-1, 0 \le k \le X_{i,j}-1)$ for the k-th packet of the interval $I_{i,j}^A$ and $I_{i,j}^B$ respectively. Aggregate r timestamps in interval Group $A$and B respectively, and the time offset gravity of packet for Interval Group A and B is respectively calculated as follows:

$$A_i = \frac{\sum_{j=0}^{r-1} \sum_{k=0}^{X_{i,j}^A - 1} \Delta t_{i,j,k}^A}{X_i^A} \tag{14}$$

$$B_i = \frac{\sum_{j=0}^{r-1} \sum_{k=0}^{X_{i,j}^B - 1} \Delta t_{i,j,k}^B}{X_i^B} \tag{15}$$

As the time offset of the arrival interval for the packet $P_{i,j,k}$ evenly distributed over $[0, T)$, then we can calculate the time offset gravity of the packet arrival interval is $\frac{T}{2}$ for Group A and B respectively. Hence, the interval based watermark assignment chooses the time offset gravity deviation of each pair encoded watermark bit A_i and B_i, denoted as follows:

$$Y_i = A_i - B_i \tag{16}$$

According to adjust the time offset gravity of Group A or B, we can make the distribution of Y_i pan right or pan left to embed binary watermark information bit 1 or 0. Suppose that the maximum manual adding delay $c(c < c < T)$, when the encoded binary watermark bit is 1, then we can make the distribution of Y_i pan right $\frac{c}{2}$ by adding A_i, which means that manual adding extra delay $c_k(0 < c_k < c)$ for each packet $P_{i,j,k}(0 \le i \le l-1, 0 \le j \le r-1, 0 \le k \le X_{i,j}^A - 1)$ in the r intervals $I_{i,j}^A$. The packet $P_{i,j,k}$ delay strategy is follow the equation shown as follows:

$$\Delta t_{i,j,k}^A{}' = c + \frac{T}{T-c} \Delta t_{i,j,k}^A \tag{17}$$

where $\Delta t_{i,j,k}^A{}'$ represents the time offset of the corresponding interval after adding delay, and its calculation is shown as follows:

$$\Delta t_{i,j,k}^A{}' = (\Delta t_{i,j,k}^A - t_0 + c_k)\%T \tag{18}$$

Then the adding extra delay c_k for each packet $P_{i,j,k}$ is calculated as follows:

$$c_k = c - \frac{c}{T}[(\Delta t_{i,j,k}^A - t_0)\%T] \tag{19}$$

After adding delay, the time offset of the arrival interval for the packet $P_{i,j,k}$ evenly distributed from $[0,T)$ to $[c,T)$, illustrated as Fig. 6. The time offset gravity of the arrival interval in Group A_i is $\frac{T+c}{2}$, and the distribution of Y_i pans right $\frac{c}{2}$, intended to embed the binary watermark information bit '1'.

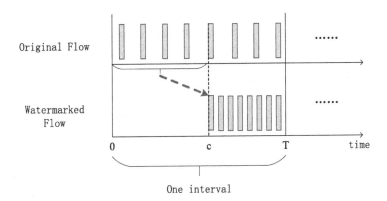

Fig. 6. Schematic diagram for altering the distribution of the packet arrival time.

Similarly, when the watermark bit to be encoded is '0', then we can make the distribution of Y_i pan left $\frac{c}{2}$ by adding B_i, which means that manual adding extra delay $c_k (0 < c_k < c)$ for each packet $P_{i,j,k} (0 \leq i \leq l-1, 0 \leq j \leq r-1, 0 \leq k \leq X_{i,j}^B - 1)$ in the r intervals $I_{i,j}^B$, intended to embed the binary watermark information bit '0'.

Watermark Detection. For the watermark flow f embedded with l-bit watermark information arriving at the detection end, we can record the arrival time offset of all the packets in every interval to calculate the arrival time offset gravity of Group A and B respectively, and then calculate the gravity deviation Y_i.

If Y_i is larger than 0, then the binary watermark information bit is determined to be '1'; otherwise, the binary watermark information bit is determined to be '0'.

Suppose that the complete l-bit binary watermark information decoded from the watermark flow f is fw', and at meanwhile the error threshold is set to $h(1 \leq h \leq l)$. Compare fw' with the original watermark information fw, if fw' and fw are not the same number of bits less than or equal to the error threshold h, it can be determined that there is a communication relationship between the receiver and the sender; otherwise, It can be considered that there is no communication relationship between the receiver and the sender.

5   Experimental Results

5.1   Experiment Design

The implementation of watermark based Tor cross-domain tracking system needs to reduce the possibility of watermark information being discovered, that is, to achieve high concealment, without affecting the service quality of the user as much as possible. Therefore, in the experiment, the controller of the watermark embedding end performs protocol analysis on the arriving data packets, and only the watermark embedding is performed on the TCP data packets. Depending on the selected watermark model, watermark the data packets returned from the server to adjust the server-side traffic, or add watermarks to the data packets sent from the client to adjust the client-side traffic. It ensures that the loading time of the webpage can be affected as little as possible when the watermark is embedded, thereby ensuring the user's access efficiency and improving the concealment of the watermark.

In the experiment, the three designed watermark models and detection schemes are respectively applied to watermark based Tor cross-domain tracking system, and the Eclipse software is used in the Floodlight controller to implement the programming in Java. In the experiment, a file with a size of 1G was placed in the Apache webpage set up by the server, and the client generated traffic by using the wget command on the terminal to download the file.

The experiment applies detection precision rate and false alarm rate [23] to estimate the results for the three watermark models. Meanwhile, the three watermark models are tested in an experimental environment without going through the intermediate springboard host, and the client directly accesses the server.

5.2   Results Analysis of IPD Model

Set the length l of the watermark information to 20 bits, the size of the watermark information bit interval to 2, the number of delayed packets embedded between different watermark bits to 5, and the error threshold h to 3. Adjust the value of the redundancy amount r and the additional delay time c to perform multiple experiments on the network flow. The shortest test duration is about 2 min.

First, test the effect of the additional delay c on the watermark detection precision rate and the system false alarm rate. Set the number of redundancy $r = 10$ and adjust the additional delay c. The test results are shown in Table 1.

Table 1. Effect of additional delay c on the watermarking system ($r = 10$).

Additional delay c(ms)	$c = 5$	$c = 10$	$c = 15$	$c = 25$	$c = 50$
Detection precision rate	84.10%	92.70%	93.20%	93.70%	94.00%
False alarm rate	9.10%	6.60%	7.30%	6.10%	4.10%

According to the experimental result shown in Table 1, when the redundancy amount r is Invariant, the detection precision rate will increase as the additional delay c increases, and the false alarm rate will decrease as the additional delay c increases. When error threshold h is set increased, both detection precision rate and false alarm rate will increase. Hence, the system need to set a reasonable error threshold to make that detection precision rate reach to 100% and false alarm rate can ben guaranteed to be below 5%.

Table 2 and 3 show the effect of redundancy amount r on the detection precision rate and the false alarm rate. The error threshold is set to $h = 0$, and additional delays $c = 5\,\text{ms}$ and $15\,\text{ms}$, respectively, and adjust the number of redundancy r.

Table 2. Effect of redundancy amount r on the watermarking system ($c = 5\,\text{ms}$)

Redundancy amount r	$r = 5$	$r = 10$	$r = 15$	$r = 20$	$r = 25$
Detection precision rate	80.50%	84.10%	90.00%	98.10%	99.00%
False alarm rate	7.20%	6.10%	4.30%	4.10%	3.90%

Table 3. Effect of redundancy amount r on the watermarking system ($c = 15\,\text{ms}$).

Redundancy amount r	$r = 5$	$r = 10$	$r = 15$	$r = 20$	$r = 25$
Detection precision rate	94.70%	93.20%	97.70%	99.30%	98.90%
False alarm rate	4.20%	3.30%	4.00%	2.00%	3.10%

According to the experimental result shown in Table 2 and 3, when the additional delay c is Invariant, as the redundancy amount r increases, the detection precision rate basically increases and the false alarm rate basically decreases. When $c = 5\,\text{ms}$ and the error threshold h is set to 11, the detection precision rate can reach to 100%, but the false alarm rate will also reach to 10%. When setting the error threshold to $h = 5$, the detection precision rate is above 90% and the false alarm rate is guaranteed to be below 5%. When $c = 15\,\text{ms}$ and the error threshold $h = 3$, detection precision rate reach to 100% and the false alarm rate is below 5%.

5.3   Results Analysis of IW Model

Set the length of the watermark information to $l = 24$ bits, the time offset to $\sigma = 6\,\text{s}$, the redundancy amount to $r = 12$, and the error threshold to $h = 3$. Adjust the value of the time interval length T to perform multiple experiments on the network flow, and test the effect of the length of the time interval T on the detection precision rate and the false alarm rate. The results are shown in

Table 4. As T increases, the test duration also increases, and the test takes about 15 min at one time.

Table 4. Effect of time interval T on the watermarking system.

Time interval T(ms)	$T = 10$	$T = 50$	$T = 200$	$T = 500$
Detection precision rate	65.20%	83.50%	100%	100%
False alarm rate	12%	3%	1%	1%

Analyzing the results in Table 4, it is known that with the increase of the time interval T, the detection precision rate is increasing, and the false alarm rate is decreasing. When the time interval T is set to 50 ms, setting the error threshold $h = 5$ can make the detection precision rate reach to 100% and the false alarm rate is lower than 1%.

The above test is performed without the system passing through the intermediate springboard host. The client directly accesses the server. Both are located on the same network segment, and the delay of network interference is very small. Therefore, in the experiment, a network interference delay D was artificially added to test the influence of the network interference delay on the detection precision rate and the false alarm rate. The fixed time interval is set to $T = 200$ ms.

Table 5. Effect of network interference delay D on the watermarking system.

Network interference delay D	$D = 0$	$D = 50$	$D = 100$	$D = 200$	$D = 300$
Detection precision rate	100%	100%	87.50%	58.30%	43.50%
False alarm rate	1%	3%	6%	15%	27%

Analysis of the results in Table 5 shows that when the time interval T is Invariant, as the network interference delay D increases, the detection precision rate decreases and the system false alarm rate increases. In particular, when the network interference delay D reaches the same as or exceeds the time interval T, the detection precision rate and false alarm rate of the system will be greatly affected.

The experiment also simply tested the effect of the length l and the redundancy amount r of the watermark information on the detection precision rate and the false alarm rate of the system. The results show that the larger the redundancy amount r, the higher the detection precision rate and the lower the false alarm rate. The length l of the watermark information has no significant effect on the system. But the larger both r and l, the longer the test duration of the experimental process.

5.4   Results Analysis of IWG Model

Set the length of the watermark information to $l = 32$ bits, the time offset to $\sigma = 10$ s, the redundancy amount to $r = 14$, and the error threshold to $h = 3$. Adjust the value of the time interval T and the artificially added maximum delay c to perform multiple experiments on the network flow. The test results are shown in Table 6. The test time will increase with the increase of T, and the test time is about 10 min.

Table 6. Effect of time interval T and maximum delay c on the watermarking system.

Interval T(ms)	$T = 10$		$T = 50$		$T = 200$		$T = 500$	
Max delay c(ms)	5	7	25	35	80	150	200	350
Precision rate	75.20%	79.50%	92.10%	94.30%	98.10%	100%	100%	100%
False alarm rate	4.80%	3.60%	3.10%	3.00%	1.60%	2%	0.70%	0.80%

The results in Table 6 shows that, as the time interval T increases, the detection precision rate increases, and the false alarm rate decreases. When the time interval T is invariant, the larger the ratio of the time interval T occupied by the maximum delay c, the higher the detection precision rate. When the time interval is set to $T = 50$ ms, setting the error threshold $h = 5$ can make the detection precision rate reach to 100% and the false alarm rate drops below 3%.

The experiment tested the effect of changing the value of the redundancy amount r on the system when the time interval length is set to $T = 50$ ms and the maximum delay is set to $c = 35$ ms. The test results are shown in Table 7.

Table 7. Effect of redundancy amount r on the watermark system ($T = 50$ ms, $c = 35$ ms)

Redundency amount r	$r = 14$	$r = 16$	$r = 18$	$r = 20$
Detection precision rate	94.30%	95.30%	94.60%	96.80%
False alarm rate	3.00%	2.80%	2.60%	2.70%

The results in Table 8 show that the larger the redundancy amount, the higher the detection precision rate and the lower the system false alarm rate.

The experiment also tested the effect of changing the value of the redundancy number r when the time interval length is set to $T = 500$ ms and the maximum additional delay is set to $c = 350$ ms. The test results were that the watermark detection rate reached to 100% and the false alarm rate is lower than 1%. It shows that when the time interval length T is sufficiently large, the effect of the redundancy number on the system is small. At the same time, the larger the number of redundancy r, the longer the test time required for the experiment.

Table 8. Comparison of three watermark models on the system.

Watermark model	IPD	IW	IWG
Min test time	2 min	15 min	10 min
Average of detection precision rate	about 80%	about 95%	about 95%
Average of false alarm rate	about 10%	about 3%	about 3%
Anti-interference ability	weak	strong	strong

5.5 Results Analysis

Compare the results of the three watermark models in Table 8, there are some conclusions:

- Among the three model, IPD-based watermark model has the simplest embedding and detection methods, the shortest test time, and the number of data packets in the selected network flow is not high. However, the anti-interference ability of the system is weak, and the watermark is not robust. Attackers can easily recover or modify the watermark information.
- Among the three model, IW model and IWG model's average detection precision rate is the highest and the false alarm rate is the lowest. When the time interval length T is longer, the anti-interference ability of the system is stronger, but at the same time, the service quality of the user is reduced. Both two models use a method to bind the watermark to a specific interval, so that the embedded watermark is not affected by changes in the number of data packets. However, these two solutions require a longer test time and require that the selected network flow has enough data packets.

6 Conclusion

In this paper we design watermark based Tor cross-domain tracking system and three watermark models on the Tor network tracking: IPD model, IW model and IWG model. The watermark model and detection system are designed separately. According to the test results, the impact of each watermark models on system performance is analyzed, and the availability of Tor tracking system is verified. The work of this paper is shown as follows:

- Design the watermark based Tor cross-domain tracking system for Tor's network communication entity based on the SDN, so that the control right is completely managed by the upper-level controller, without adding configuration information to the underlying network equipment, which is easy to operate and implement.
- Design three watermark models to design watermark schemes. The system can use different watermark schemes in different network situations, making the system more flexible.

The deficiencies in the watermark based Tor cross-domain tracking system designed in the paper are shown as follows:

- In the network transmission, there is only one intermediate springboard host between the client and the server. The network test environment is relatively simple, and the watermarking scheme is not tested in a more complicated network environment.
- The system uses the client to access the webpage set up by the server to generate traffic. In the three watermarking schemes that require artificial delay, the rate of the user accessing the webpage will be significantly slower, which affects the user's service quality.
- In the scheme that uses the time interval as the watermark carrier, the selected interval time is relatively long, so the number of data packets in the network flow is required, and the system does not have strong concealment.
- The offline detection method is used in watermark detection. Therefore, for a large number of network flows, the system cannot extract watermark information in time, and cannot quickly determine the communication relationship between the sender and receiver to locate the attacker.

In view of the shortcomings of the system designed in this paper, our future work contains the follows aspects: (1) Several intermediate springboard hosts are set up between the client and the server, so the network flow can go through a complex network environment after watermarking. Improve anti-interference ability of the system, the solution based on the test results, and enhance the robustness of the system. (2) To improve the method of generating traffic, we consider adding watermarks to the traffic generated by webpage advertisements, and try to optimize the active network watermark system without affecting users. (3) Improve the watermark detection method, so that the system can quickly determine the communication relationship between the sender and receiver to determine the location of the attacker, based on the large number of incoming network flows.

References

1. Agrawal, D., Kesdogan, D., Penz, S.: Probabilistic treatment of mixes to hamper traffic analysis. In: 2003 Symposium on Security and Privacy, pp. 16–27. IEEE (2003)
2. Cai, X., Zhang, X.C., Joshi, B., Johnson, R.: Touching from a distance: website fingerprinting attacks and defenses. In: Proceedings of the 2012 ACM Conference on Computer and Communications Security, pp. 605–616 (2012)
3. Cuzzocrea, A., Martinelli, F., Mercaldo, F., Vercelli, G.: Tor traffic analysis and detection via machine learning techniques. In: 2017 IEEE International Conference on Big Data (Big Data), pp. 4474–4480. IEEE (2017)
4. Das, A.K., Pathak, P.H., Chuah, C.-N., Mohapatra, P.: Contextual localization through network traffic analysis. In: IEEE Conference on Computer Communications, IEEE INFOCOM 2014, pp. 925–933. IEEE (2014)
5. Dingledine, R., Mathewson, N., Syverson, P.: Tor: the second-generation onion router. Technical report, Naval Research Lab Washington DC (2004)

6. He, G.-F., Yang, M., Luo, J.-Z., Zhang, L.: Online identification of tor anonymous communication traffic. Ruanjian Xuebao/J. Softw. **24**(3), 540–556 (2013)
7. He, Y.Z., Li, X., Chen, M.L., Wang, W.: Identification of tor anonymous communication with cloud traffic obfuscation. Adv. Eng. Sci. **49**(2), 121–132 (2017)
8. Hou, X., Chen, Y., Tian, H., Wang, T., Cai, Y.: Network watermarking location method based on discrete cosine transform. In: 3rd International Conference on Materials Engineering, Manufacturing Technology and Control. Atlantis Press (2016)
9. Houmansadr, A., Kiyavash, N., Borisov, N.: Rainbow: a robust and invisible non-blind watermark for network flows. In: NDSS (2009)
10. Iacovazzi, A., Elovici, Y.: Network flow watermarking: a survey. IEEE Commun. Surv. Tutor. **19**(1), 512–530 (2016)
11. Iacovazzi, A., Sarda, S., Elovici, Y.: Inflow: inverse network flow watermarking for detecting hidden servers. In: IEEE Conference on Computer Communications, IEEE INFOCOM 2018, pp. 747–755. IEEE (2018)
12. Iacovazzi, A., Sarda, S., Frassinelli, D., Elovici, Y.: DropWat: an invisible network flow watermark for data exfiltration traceback. IEEE Trans. Inf. Forensics Secur. **13**(5), 1139–1154 (2017)
13. Understanding Node Capture Attacks in User Authentication Schemes for Wireless Sensor Networks. Understanding node capture attacks in user authentication schemes for wireless sensor networks (2020)
14. Lashkari, A.H., Draper-Gil, G., Mamun, M.S.I., Ghorbani, A.A.: Characterization of tor traffic using time based features. In: ICISSP, pp. 253–262 (2017)
15. Lin, Z., Tong, L., Zhijie, M., Zhen, L.: Research on cyber crime threats and counter-measures about tor anonymous network based on meek confusion plug-in. In: 2017 International Conference on Robots & Intelligent System (ICRIS), pp. 246–249. IEEE (2017)
16. Liu, W., Liu, G., Xia, Y., Ji, X., Zhai, J., Dai, Y.: Using insider swapping of time intervals to perform highly invisible network flow watermarking. Security and Communication Networks (2018)
17. Tianbo, L., Guo, R., Zhao, L., Li, Y.: A systematic review of network flow watermarking in anonymity systems. Int. J. Secur. Appl. **10**(3), 129–138 (2016)
18. Luo, X., Zhou, P., Zhang, J., Perdisci, R., Lee, W., Chang, R.K.C.: Exposing invisible timing-based traffic watermarks with backlit. In: Proceedings of the 27th Annual Computer Security Applications Conference, pp. 197–206 (2011)
19. Panchenko, A., Niessen, L., Zinnen, A., Engel, T.: Website fingerprinting in onion routing based anonymization networks. In: Proceedings of the 10th Annual ACM Workshop on Privacy in the Electronic Society, pp. 103–114 (2011)
20. Peng, P., Ning, P., Reeves, D.S.: On the secrecy of timing-based active watermarking trace-back techniques. In: 2006 IEEE Symposium on Security and Privacy (S&P 2006), pp. 15-pp. IEEE (2006)
21. Pham, D.V., Kesdogan, D.: Towards a causality based analysis of anonymity protection in indeterministic mix systems. Comput. Secur. **67**, 350–368 (2017)
22. Saputra, F.A., Nadhori, I.U., Barry, B.F.: Detecting and blocking onion router traffic using deep packet inspection. In: 2016 International Electronics Symposium (IES), pp. 283–288. IEEE (2016)
23. Verde, M.F., Macmillan, N.A., Rotello, C.M.: Measures of sensitivity based on a single hit rate and false alarm rate: the accuracy, precision, and robustness of', A z, andA'? Percept. Psychophys. **68**(4), 643–654 (2006)
24. Wang, D., Cheng, H., Wang, P., Huang, X., Jian, G.: Zipf's law in passwords. IEEE Trans. Inf. Forensics Secur. **12**(11), 2776–2791 (2017)

25. Wang, D., Zhang, X., Zhang, Z., Wang, P.: Understanding security failures of multi-factor authentication schemes for multi-server environments. Comput. Secur. **88**, 101619 (2020)
26. Wang, T., Cai, X., Nithyanand, R., Johnson, R., Goldberg, I.: Effective attacks and provable defenses for website fingerprinting. In: 23rd USENIX Security Symposium (USENIX Security 2014), pp. 143–157 (2014)
27. Wang, X., Reeves, D.S.: Robust correlation of encrypted attack traffic through stepping stones by manipulation of interpacket delays. In: Proceedings of the 10th ACM Conference on Computer and Communications Security, pp. 20–29 (2003)
28. Flows By Non-Blind Watermarking. Enhancing invisibility in network flows by non-blind watermarking (2014)

System Security

Research on IoT Security Technology and Standardization in the 5G Era

Qin Qiu[1], Xuetao Du[2], Shengquan Yu[3], Chenyu Wang[4], Shenglan Liu[2(✉)], Bei Zhao[2], and Ling Chang[2]

[1] China Mobile Communications Group Co., Ltd., Beijing 100053, China
qiuqin@chinamobile.com
[2] China Mobile Group Design Institute Co., Ltd., Beijing 100080, China
{duxuetao,zhaobei,changling}@cmdi.chinamobile.com,
liushenglan94@163.com
[3] Beijing Normal University, Beijing 100875, China
yusq@bnu.edu.cn
[4] Beijing University of Post and Telecommunications, Beijing 100876, China
wangchenyu@bupt.edu.cn

Abstract. With the development of 5G technology, Internet of Things (IoT) is highly developing and deeply integrated with social life and industry productions, which brings about many security issues. In this paper, we first analyze the security risks for IoT in the 5G era, then summarize related security policies and standards. Furthermore, we propose security requirements and measures in aspects of sensor control equipment and IoT card, IoT network and transmission exchange, IoT business application and service, and IoT security management and operation. Finally, we put forward suggestions for promoting IoT security technology and the standardization work in the 5G era.

Keywords: 5G · IoT · Security · Privacy · Standardization

1 Introduction

5G communication technology supports high-speed information transmission and massive terminal connections, which accelerates the development of the IoT. In the 5G era, while IoT makes people 's life more convenient and intelligent, it faces many security risks.

1.1 5G Accelerates the Development of IoT

In recent years, new technologies such as 5G, big data, cloud computing, artificial intelligence have brought innovation vitality to IoT, making the intelligent connections of all things a reality. Main application scenarios of IoT are intelligent industry, intelligent agriculture, intelligent transportation, smart grid and so on. Therefore, the basic requirements of machine communication for 5G network are concentrated on massive terminal

D. Wang et al. (Eds.): SPNCE 2020, LNICST 344, pp. 77–90, 2021.
https://doi.org/10.1007/978-3-030-66922-5_5

access, ultra-low delay, efficient connectivity, low cost, low power consumption, and it has high level of reliability along with wide cover range [1]. The development trend of the IoT in the 5G era is mainly manifested in following aspects [2]:

From Narrowband to Broadband. With the development of UHD, VR, AR and other technologies, the IoT industry has a higher demand for the network bandwidth.

From Mixed Use to Exclusive Use. Traditional public network is difficult to meet the needs of industry applications. Customized and differentiated exclusive network services can meet the needs of vertical industry.

From Flat Coverage to Three-Dimensional Coverage. 5G provides all-round breadth and height coverage. Through network connection configuration and low altitude coverage optimization, it can meet the coverage of ATG, UAV and other scenes.

Improvement of Performance. The performance including time delay, reliability, security, positioning accuracy has been improved. Remote control services need high reliability and low time delay. UAV services need positioning accuracy and so on.

With the development of IoT in the 5G era, there are also corresponding industry application scenarios, mainly including three aspects: (1) 5G enabled industry private network. The integrated network slice service platform provides high reliable, strong performance, easy to deploy private network services for the vertical industry, to better meet the customized needs of industry users. (2) New intelligent network management. According to the requirements of industry customers for the stability and reliability of communication network, the intelligent network management of the IoT is built to realize the predictable failure, simple operation of the system, docking and expansion of the network management platform with other business platforms. (3) Diversified and customized terminals. In view of the diverse use scenarios and complex environment of industrial terminals, 5G communication module is integrated on the industrial terminals to achieve the diversification and customization of industrial terminals.

1.2 Security Risks for IoT in the 5G Era

5G introduces a new network structure, using SDN (Software Defined Network), NFV(network function virtualization), MEC(multi-access edge computing) and other technical means to meet the connection requirements of a large number of devices [3, 4]. 5G has penetrated into various industries, and various IoT devices have entered the communication network. At the same time, the network has become more vulnerable, bringing new risks and challenges.

Threats in Various Industries. 5G network can further strengthen the high-speed connection between not only people and things, but also things and things. It brings about Smart Healthcare, Internet of Vehicles, Intelligent Wear, etc., but it also increases the risk of network equipment being invaded, leading to further increase the threat to personal safety, industrial production, etc.

Risks Towards 5G Core Network. While the mobile edge computing technology is implemented as the enterprise network and 5G network are integrated, the 5G core network capability sinks to the network edge node, and there is an attack path from the edge equipment to the core network, which would introduce risks in enterprise networks into the core network.

Threats to SIM Based IoT Devices. Many IoT devices use SIM card within 5G network. According to security standards of SIM cards, it is possible to modify the content and function of the SIM card remotely by an invisible short message service (SMS) sent through OTA technology, which may be abused by attackers [5–7].

Complexity of Security Protection. 5G technology connects a large number of devices, and the network is becoming more complex. Massive equipment connection, and emerging new services and application scenarios would change the current network structure, and increase the difficulty of security protection.

2   IoT Security Policies and Standards in the 5G Era

With the arrival of 5G era, the application of IoT is gradually popularized. In order to create a favorable environment for industrial development, it is required to carry out security policies and standards for 5G and IoT. Many countries in the world are formulating 5G and IoT security policies and standards to varying degrees, so as to promote the development of IoT in 5G era in a multi-level and all-round way.

2.1   Security Policies

Focusing on promoting the security level of the IoT, many countries in the world have formulated strategies or policies to deal with risks of the IoT [8].

The United States has promoted the construction of the IoT security from the aspects of strategic formulation and policy implementation [8], issues strategic documents such as the *Strategic Principles for Securing the Internet of Things, Strengthening the Cybersecurity of Federal Networks and Critical Infrastructure*, and issues laws such as the *IoT in Government Network Security Solutions* and the *IoT Cybersecurity Improvement Act*. The EU focuses on the security baseline setting of the IoT and User data protection [9], the European Union security baseline guide for the IoT in critical information infrastructure environment comprehensively summarizes the security status of the IoT and security baseline recommendations, and the *General Data Protection Regulation* sets new regulatory requirements for the data protection of enterprises.

As early as 2013, the Chinese government has incorporated the IoT security into the work system in its policy planning and successively issued the guidance of the State Council on promoting the orderly and healthy development of the IoT, etc. In September 2019, the Ministry of industry and information technology issued the implementation opinions on promoting the quality of manufacturing products and services, which clearly defined 5G and the IoT related industries [10]. In terms of law, the Chinese government has issued laws and supported normative documents related to network security, including the *Cybersecurity Law of the people's Republic of China, the National Network Security Emergency Plan*, etc., which provides the legal system basis for the security supervision of the IoT industry.

2.2 Security Standards

5G and IoT security standards are constantly evolving and developing. It is a global common goal to accelerate the improvement of the IoT security standard system and the development of standards for key issues of IoT security in 5G era. At present, many standardization organizations at home and abroad have carried out research in the field of 5G and IoT security. The main research directions of the major standardization organizations are shown in Table 1.

In terms of 5G security standards, in 2016, 5G PPP released a white paper -*View on 5G Architecture* [11], which proposed the challenging problems to be solved to meet 5G requirements and discussed the impact of general 5G reference framework on 5G standardization. 3GPP has released the latest schedule of 5G global standards in June 2019, and planned to determine the second and third phase standards of 5G in March 2020 and June 2021 respectively. In addition, the NGMN organization jointly sponsored by multiple operators around the world has published a white paper on 5G, covering virtualization, privacy protection, IoT and other topics [12]. ETSI has studied NFV platform security specifications, MEC security standards and so on [13]. IETF's proposals on Internet protocol standards have also played an important reference role in the development of 5G standards [14]. In January 2020, CCSA held the release ceremony of the early 14 5G standards in China at the 5G Standard Release and Industry Promotion Conference. These 5G standards are fully in line with the global 5G standards, covering core network, wireless, antenna, terminal, security and other fields.

In terms of IoT security standards, at present, the international IoT security standards are mainly focused on the security system framework, network security, privacy protection, equipment security, etc. As 5G technology promotes the application of IoT to mature gradually, in recent years, the security standards tend to focus on the areas such as the application security and privacy protection. The industry alliance including 5GAA, IIC, GSMA, etc., has also opened in relevant key application fields development of security standards [8].The security standards of ISO/IEC for the Internet of Things are mainly focused on the architecture and security technology, etc., and ISO/IEC 30141:2018 [15], ISO/IEC 29192 [16] and so on have been issued. ITU-T has planned a series of security standards for the IoT [17], and actively carried out the security standardization of the Internet of Vehicles. ETSI has released the first consumer security standard for the IoT [18], which promoted the development of the future certification scheme for the IoT. IETF mainly studies IP network protocol the standard also proposes protocol optimization for the characteristics of IoT terminals [19]. China attaches great importance to the security supervision and technical support of the IoT. And China has carried out work in security reference model, perception and wireless security technology, key industry applications and other fields. At present, TC260 has issued a series of standards related to the IoT. *The Baseline for Classified Protection of Cybersecurity* [20] specifies the requirements for the security of the IoT. The *Security Reference Model and Generic Requirements*

Table 1. Main research directions of standardization organizations

Field	Standardization organizations	Main research directions
5G security	3GPP	Security architecture, RAN security, authentication mechanism, user privacy, network slicing [3]
	5GPPP	Security architecture, user privacy, authentication mechanism, etc. [11]
	NGMN	User privacy, network slicing, MEC security, etc. [12]
	ETSI	Security architecture NFV security, MEC security, privacy protection [13]
	IETF	Security solutions, user privacy, NSF, etc. for 5G mass IoT devices [14]
	CCSA	5G access network, core network, security and frequency, etc.
IoT security	5GAA	Security architecture of the Internet of Vehicles for all regions [8]
	IIC	Establish open interoperability standards and accelerate the implementation of industrial Internet [8]
	GSMA	Security Research of operators in the field of IoT [8]
	ISO/IEC	Architecture, security technology, including encryption, lightweight, authentication, privacy control, etc. [15, 16]
	ITU-T	Smart city and community, privacy protection, trust and identification, Internet of Vehicles, etc. [17]
	ETSI	Authentication authorization, quantum security threat assessment and analysis of the authentication security mechanism of the IoT group [18]
	IETF	Protocol of authorization, authentication and audit in IP network [19]
	TC260	Grade protection, IoT security reference model, IoT security standardization white paper, etc. [20, 21]
	CCSA	Communication network and system, security framework [22, 23]

for Internet of Things [21] clarify the reference framework for the security of the IoT. CCSA focuses on communication networks and systems, and has completed the industry standards for the IoT such as *the General Framework and Technical Requirements of IoT* [22], and *the General Requirements for Cellular Narrowband Radio Access for Internet of Things* (NB-IoT) [23], which provides a reference for the security construction of the IoT.

3   Security Requirements and Measures of IoT

According to the entity classification of ISO/IEC 30141:2018 [24] and the reference model of GB/T 37044-2018 [21], the security requirements of IoT are mainly located in sensing devices and cards, network and transmission exchange, business applications and services, and security management and operation [8], as shown in Fig. 1. There are four areas that work together as an organic whole to guarantee the security of IoT systems. The sensing security area focus on the security requirements of sensing devices and corresponding system. In the network security area, the security requirements major in the network and transmission exchange. The application security area is mainly to meet the service security requirements of the user, such as the identity authentication, access control and so on. As for the operation security area, it concentrates on the security requirements of the operation and maintenance management, and it is an indispensable part to ensure the security operation of the above three areas.

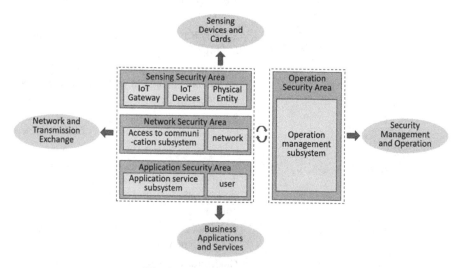

Fig. 1. Security requirements of IoT

Based on the Fig. 1, the existing security risks are considered, and the results are shown in Fig. 2. Combined with the risk points, the security requirements and key technologies are analyzed. In addition, security measures of IoT in the 5G era are summarized and shown in Fig. 3.

3.1 Sensing Devices and Cards

The main function of the sensing device is to realize the collection, identification and control of information. The smart IoT card refers to the smart card used in the field of IoT. Sensing devices and smart IoT cards have an important impact on the security of the IoT.

Sensing Devices. Sensing devices include sensing terminals and control devices. Sensing terminals are usually in a harsh environment without monitoring. Redundant deployment of key nodes is required to ensure that nodes can achieve network self-healing, so as to avoid work interruption in case of natural or man-made damage. Authentication between nodes before communication based on encryption algorithm is required to prevent attackers from illegally accessing the system by using the weakness of authentication mechanism [25–27]. It is necessary to limit the network sending speed and packet retransmission times to prevent the protocol vulnerability from being exploited [8].

Smart IoT Card. The smart IoT card embedded in the device in the form of software adopts the over-the-air card writing technology based on the public network, which may lead to eavesdropping, replay, denial of service, sensitive data disclosure, etc., so it is necessary to adopt the secure communication and storage encryption mechanism. In addition, the equipment is connected to the communication network based on the IoT network card as the identity, and the separation of the machine card and the card may also cause the problem of the card being misappropriated or abused, so it is essential to establish the corresponding security management and monitoring means.

Fig. 2. Risks faced by the IoT

3.2 Network and Transmission Exchange

The application of the IoT uses a complex network structure. During the process of data transmission and exchange in the network, a lot of risks will also be generated.

Wireless Communication. WIFI, Bluetooth, 2/3/4/5G and other wireless communication technologies have their own security problems, and the security problems of the IoT are also accompanied [28]. A wide range of sensing terminals and access devices are deployed in the unmonitored environment. The number of IoT nodes is huge and wireless radio frequency signals are used for data transmission. Attackers can send interference signals to interrupt communication, or hijack, eavesdrop, tamper with data in the process of signal transmission. Therefore, it is necessary to establish information transmission guarantee mechanism and improve data verification and encryption technology.

Transmission Switching. The transmission of IoT information will pass through different heterogeneous networks. When transmitting large amounts of data, it is easy to cause congestion in the core network. Therefore, multi-channel transmission is needed to alleviate the network pressure and resist the denial of service attack [8]. At the same time, the heterogeneous networks in the transport layer need to be interconnected, so it also faces problems such as cross-network authentication. The point-to-point and end-to-end encryption mechanism can be used to ensure the security of the transport layer. In addition, the data packets transmitted on IoT are not encrypted and signed, which is easy to be eavesdropped, tampered and forged. Thus PGP, SSL/TLS, IPsec and other protocols [29] are needed to provide communication encryption and authentication functions.

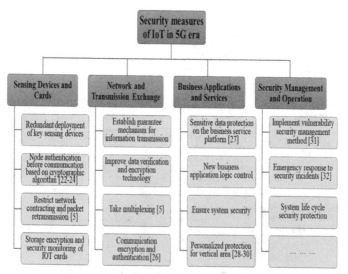

Fig. 3. Security measures of IoT in the 5G era

3.3 Business Application and Service

IoT business application and service security requirements mainly include business service platform security requirements and vertical industry security requirements.

Business Service Platform. The business service platform needs to avoid the risks of identity counterfeiting and unauthorized access due to the variety of access devices [30]. The business service platform needs to avoid stealing, tampering, forgery and so on when collecting, storing and processing a large number of sensitive data. As the emergence of various new business applications, attention should be paid to security threats in the realization of technology, logic, control and other aspects. It is also necessary to ensure the system security requirements.

Vertical Industry. Each vertical industry faces different security risks and needs. For example, in the smart city, its security requirements mainly lie in sensing devices protection and diversified network access [31]. In the industrial Internet, its security requirements mainly lie in the identification of industrial control equipment assets and network boundary, industrial network isolation measures, etc. [32]. As for the Internet of Vehicles, its security requirements mainly lie in ensuring the legitimacy of sensor data, the security of core control components and so on [33].

3.4 Security Management and Operation

IoT security management and operation security requirements mainly lie in IoT security management related requirements, system life cycle operation security requirements, etc.

IoT Security Management. Since IoT has different characteristics from the traditional Internet in many aspects such as terminal devices, firmwares and so on, it is important to research the vulnerability management methods in the IoT industry [30], including vulnerability library construction, vulnerability identification, vulnerability sharing, equipment reinforcement, etc. In addition, the emergency response of IoT security incidents is the last defense line of IoT security [34], and it is also crucial to ensure the normal operation of IoT business.

System Life Cycle. A complete life cycle of the IoT system includes planning and design, development and construction, operation management, and abandonment and exit. Each stage has different missions and security requirements, thus corresponding security protection measures shall be established in each stage.

4 Suggestions on the Development of IoT in the 5G Era

While 5G speeds up the rapid construction of the IoT security technology, in order to meet the current needs of the IoT industry, it is suggested to focus on the following aspects to promote the development of IoT security technology and standardization, and the key points of all suggestions are also extracted and shown in Fig. 4.

4.1 Sensing Devices and Cards

5G promotes the rapid development and application of Pan terminal, and the IoT sensing devices presents the characteristics of complexity and diversity. It is suggested to strengthen the security protection technology for the sensing devices in key application scenarios, and propose relevant standards, including UAV equipment security management requirements, intelligent medical equipment security technology and evaluation requirements.

Due to the limitation of energy, power consumption and storage space, the terminal sensing devices in the IoT system usually cannot work well, or needs too much cost to run complex cryptographic algorithms and security protocols. It is recommended to develop lightweight cryptographic algorithms [22] for resource constrained IoT equipment [28], and formulate relevant application implementation guidelines.

4.2 Network and Transmission Exchange

The IoT has penetrated into smart home, health care, public services and other scenarios, which will generate a large number of sensitive personal data, so it is necessary to strengthen the relevant security technology. In order to prevent risks from the identification and association of the device and identity, if necessary, temporary identification of the device can be used to replace the permanent identification, such as media access address, IPv6 address, etc. [35]. It is suggested to strengthen the research on key technologies of privacy protection and data security technology system, strengthen the application and implementation of data life cycle security management, personal information, data transaction and other relevant standards in key fields such as industrial Internet and smart city, and develop corresponding national standards such as data security implementation guides.

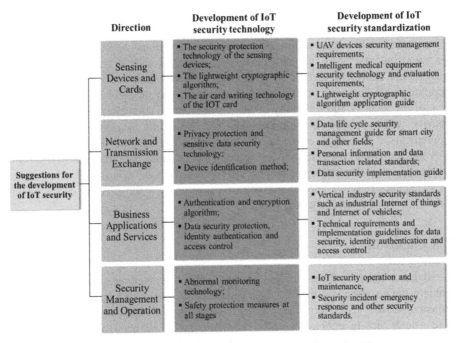

Fig. 4. Suggestions for the development of IoT security in the 5G era.

4.3 Business Application and Service

The 5G technology is widely used in IoT scenarios. The application security of the vertical industry of the IoT based on 5G shows the trend of expanding attack area, ubiquitous attack mode and blurring border. Traditional security mechanisms and defense means, such as authentication encryption algorithm, firewall, key management system, intrusion detection system, etc., need to be optimized in a targeted way [36]. Light-weight encryption algorithms are important methods to ensure the network security, and protect the data confidentiality along with the integrity [37–39]. Industrial Internet is an important field of national concern, and it is advised to speed up technologies research and standards promotion such as network facility security, platform and industrial application security, data security protection, test and experimental environment of industrial IoT. In addition, It is advised to promote the security of business platform and edge computing of the Internet of vehicles.

In the IoT ecosystem, there are some traditional industries, such as transportation, medical treatment, logistics, home furnishing, etc., which are not able to get through the IoT ecological chain, so the business operation needs to rely on the common business service platform. In order to cope with risks and challenges faced by platforms in providing IoT business assistance for enterprises, it is necessary to strengthen data security protection, access control, identity authentication [40] and other technologies, accelerate the construction of security capacity of the general business service platform of IoT, and support the development of relevant technical requirements, implementation guidelines and other specifications.

4.4 Security Management and Operation

IoT business involves responsibilities and interests of various parties, including users, equipment manufacturers, network operators, service providers, etc. When security issues arise, it is difficult to divide responsibilities. Therefore, in order to achieve secure operation and maintenance, on the basis of effective organization of all parties, we can enhance the security and reliability of the IoT system by strengthening the abnormal behavior implementation monitoring technology. According to missions and security requirements in each stage of the IoT business, establish corresponding security measures and improve the corresponding protection technologies. Simultaneously, it is proposed to speed up the development of security standards in the security operation of the IoT, emergency response and other aspects, so as to promote the effective coordination of the security ecology of the IoT.

5 Conclusion

The advancement of the 5G technology and IoT applications facilitates our lives unprecedentedly, but also brings some new risks and challenges. Focusing on the future development of the IoT, effectively guaranteeing the security of the IoT has become an urgent issue in the 5G era. In order to ensure the security of the IoT, we need to strengthen industry security management, improve security technologies and standards, build effective security protection systems, and explore new technologies. It is suggested that the industry regulatory authorities, technical research institutions and other relevant parties work together to actively promote the secure development of the IoT.

References

1. Schulz, P., Matthe, M., Klessig, H.: Latency critical IoT applications in 5G: perspective on the design of radio interface and network architecture. IEEE Commun. Mag. **55**(2), 70–78 (2017)
2. Skubic, B., Bottari, G., Rostami, A., et al.: Rethinking optical transport to pave the way for 5G and the networked society. Lightwave technol. **33**(5), 1084–1091 (2015)
3. Ahmad, I., Kumar, T., Liyanage, M., et al.: Overview of 5G security challenges and solutions. IEEE Commun. Stand. Mag. **2**(1), 36–43 (2018)
4. Ahmad, I., Kumar, T., Liyanage M., et al.: 5G security: analysis of threats and solutions. In: IEEE Conference on Standards for Communications and Networking (CSCN). IEEE Press (2017)
5. Huang, Q., Yang, C.: A lightweight RFID authenticate protocol based on smart SIM card. In: Proceedings of the 1st International Conference on Logistics, Informatics and Service Science, pp. 647–650. IEEE Press (2011)
6. He, R., Zhao, G., Chang, C., et al.: A PK-SIM card based end-to-end security framework for SMS. Comput. Stand. Interfaces **31**(4), 629–641 (2009)
7. Jia, F., Yang, Y., Peng, J.: Security mechanism for end to end SMS based on smart card. Appl. Res. Comput. **24**(5), 259–261 (2007)
8. TC 260, Communication security standards working group.: white paper on Internet of Things security standardization (2019)

9. Neisse, R., Steri, G., Baldini, G.: Enforcement of security policy rules for the Internet of Things. In: the 3rd International Workshop on Internet of Things Communications and Technologies (IoT-CT), IEEE Press (2014)
10. Ministry of industry and information technology: accelerate the development of 5G and Internet of Things related industries. http://www.sohu.com/a/339209778_166680
11. G PPP Architecture Working Group. View on 5G Architecture (2016)
12. Iwamura, M.: NGMN view on 5G architecture. In: Vehicular Technology Conference. IEEE Press (2015)
13. Jaeger, B.: Security orchestrator: introducing a security orchestrator in the context of the ETSI NFV reference architecture. In: IEEE Trustcom/ BigDataSE/ISPA, vol. 1, pp. 1255–1260. IEEE Press (2015)
14. Omheni, N., Bouabidi, I., Gharsallah, A., et al.: Smart mobility management in 5G heterogeneous networks. IET Netw. **7**(3), 119–128 (2018)
15. ISO/IEC 30141:2018 Information Technology - Internet of Things Reference Architecture (2018)
16. ISO/IEC 29192 Information Technology - Security Techniques - Lightweight Cryptography (2012)
17. Kafle, V., Fukushima, Y., Harai, H.: Internet of Things standardization in ITU and prospective networking technologies. IEEE Commun. Mag. **54**(9), 43–49 (2016)
18. ETSI, ETSI releases first globally applicable standard for consumer IoT security. China Standardization (2019)
19. Sheng, Z., Yang, S., Yu, Y., et al.: A survey on the IETF protocol suite for the Internet of Things: standards, challenges, and opportunities. IEEE Wirel. Commun. **20**(6), 91–98 (2016)
20. GB/T 22239-2019. Information Security Technology - Baseline for Classified Protection of Cybersecurity (2019)
21. GB/T 37044-2018. Information Security Technology - Security Reference Model and Generic Requirements for Internet of Things (2018)
22. YD/T 2437-2012. General Framework and Technical Requirements of IoT (Internet of Things) (2012)
23. YD/T 3331-2018. General Requirement for Cellular Narrowband Radio Access for Internet of Things (NB-IoT) (2018)
24. ISO/IEC 30141:2018. Internet of Things (IoT) - Reference Architecture (2018)
25. Wang, D., Wang, P., Wang C.: Efficient multi-factor user authentication protocol with forward secrecy for real-time data access in WSNs. ACM Trans. Cyber-Phys. Syst. (2019). https://doi.org/10.1145/3325130
26. Wang, C., Wang, D., Tu, Y., Xu, G., Wang, H.: Understanding node capture attacks in user authentication schemes for wireless sensor networks. IEEE Trans. Dependable Secure Comput. (2020). https://doi.org/10.1109/tdsc.2020.2974220
27. Wang, D., Li, W., Wang, P.: Measuring two-factor authentication schemes for real-time data access in industrial wireless sensor networks. IEEE Trans. Indu. Inf. **14**(9), 4081–4092 (2018)
28. Burg, A., Chattopadhyay, A., Lam, K.: Wireless communication and security issues for cyber-physical systems and the Internet-of-Things. Proc. IEEE **106**(1), 38–60 (2018)
29. Granjal, J., Monteiro, E., Silva, J.: Security for the Internet of Things: a survey of existing protocols and open research issues. IEEE Commun. Surv. Tutor. **17**(3), 1294–1312 (2015)
30. Cai, H., Xu, L., Xu, B., et al.: IoT-based configurable information service platform for product lifecycle management. IEEE Trans. Indu. Inf. **10**(2), 1558–1567 (2014)
31. Zhang, K., Ni, J., Yang, K., et al.: Security and privacy in smart city applications: challenges and solutions. IEEE Commun. Mag. **55**(1), 122–129 (2017)
32. Li, J., Yu, F., Deng, G., et al.: Industrial Internet: a survey on the enabling technologies, applications, and challenges. IEEE Commun. Surv. Tutor. **13**(3), 1504–1526 (2017)

33. Joy, J., Gerla, M.: Internet of vehicles and autonomous connected car - privacy and security issues. In: International Conference on Computer Communication and Networks. IEEE Press (2017)

34. Qiu, T., Lu, Y., Xia, F., et al.: ERGID: an efficient routing protocol for emergency response Internet of Things. J. Netw. Comput. Appl. **72**, 104–112 (2016)

35. Norrman, K., Dubrova, E.: Protecting IMSI and user privacy in 5G networks. In EAI International Conference on Mobile Multimedia Communications. ICST (2016)

36. Li, S., Xu, L., Zhao, S.: 5G Internet of Things: a survey. J. Indu. Inf. Integr. **10**, 1–9 (2018)

37. Singh, S., Sharma, P.K., Moon, S.Y., et al.: Advanced lightweight encryption algorithms for IoT devices: survey, challenges and solutions. J. Ambient Intell. Hum. Comput. 1–8 (2017)

38. Alizadeh, M., Hassan, W.H., Zamani, M., et al.: Implementation and evaluation of lightweight encryption algorithms suitable for RFID. J. Next Gener. Inf. Technol. **4**, 65–77 (2013)

39. An-Ping, L., Ji-Min, Y., Feng, L.I., et al.: A comparative study of several lightweight encryption algorithms. Modern Electronics Technique (2014)

40. Wang, D., Wang, P.: Two birds with one stone: two-factor authentication with security beyond conventional bound. IEEE Trans. Dependable Secur. Comput. **15**(4), 708–722 (2018)

MIMEC Based Information System Security Situation Assessment Model

Lixia Xie[1]([⊠]) [iD], Liping Yan[1] [iD], Xugao Zhang[1] [iD], Hongyu Yang[1] [iD],
Guangquan Xu[2] [iD], and Jiyong Zhang[3]

[1] Civil Aviation University of China, Tianjin 300300, China
lxxie@126.com
[2] Tianjin University, Tianjin 300350, China
[3] Swiss Federal Institute of Technology in Lausanne, Lausanne 1015, Switzerland

Abstract. The accuracy of existing information system security situation assessment methods is affected by expert evaluation preferences. This paper proposes an Information System Security Situation Assessment Model (ISSSAM), which is based on the Modified Interval Matrix-Entropy Weight based Cloud (MIMEC). Based on the system security situation assessment index system, the interval number judgment matrix reflecting the relative importance of different indicators is modified to improve the objectivity of the indicator layer weight vector. Then, the entropy weight based cloud is used to quantify the criterion layer and the target layer security situation index, and the security level of the system is graded. The feasibility and effectiveness of this model are verified by the security situation assessment of the Departure Control System (DCS). Through the comparison and analysis of the evaluation results based on entropy weight coefficient method and traditional AHP method, it is shown that the model we proposed has good stability and reliability.

Keywords: Security situation · Index system · Interval number judgment matrix · Entropy weight based cloud

1 Introduction

Security situation assessment refers to the process of predicting the security situation of the system based on the perception and acquisition of security elements in a certain time and space, and the integrated analysis of the acquired data information [1]. The security situation assessment model is necessary for information system security managers to obtain the dynamic security situation of the system, determine system abnormal events and make reasonable decisions.

Fu et al. [2] established a comprehensive assessment model based on the entropy weight coefficient method which reduces the subjective impact of expert experience. Luo et al. [3] proposed a two-stage information security risk assessment model based on grey comprehensive measures. This model reduces the ambiguity, uncertainty and subjectivity of information security risk assessment, but lacks management dimension

© ICST Institute for Computer Sciences, Social Informatics and Telecommunications Engineering 2021
Published by Springer Nature Switzerland AG 2021. All Rights Reserved
D. Wang et al. (Eds.): SPNCE 2020, LNICST 344, pp. 91–101, 2021.
https://doi.org/10.1007/978-3-030-66922-5_6

indicators. Xi et al. [4] proposed an improved quantitative assessment model. This model optimizes the quantified value of network security situation through game method, but the information source is too single. Zhou [5] proposed a network security assessment model based on neural networks. The model reduces the burden of manually determining the network security situation. However, the quantitative results of the network security situation need to be refined.

In summary, the information system security situation assessment indicators of existing research do not consider human factors. At the same time, the security situation assessment is greatly affected by subjectivity of experts, and the quantitative results are difficult to accurately reflect the system security situation. This paper proposes an Information System Security Situation Assessment Model (ISSSAM) based on a Modified Interval Matrix-Entropy Weight based Cloud (MIMEC). On the premise of retaining subjective evaluation by experts, the interval judgment matrix is modified and the index weight vector is optimized to make the quantitative assessment result of security situation more reasonable.

2   Security Situation Assessment Model

In this section, we establish an Information System Security Situation Assessment Model (ISSSAM) based on a Modified Interval Matrix-Entropy Weight based Cloud (MIMEC). The information system assessment process is designed as follows:

Firstly, based on the Analytic Hierarchy Process (AHP) [6], a three-layer index system for information system security situation assessment is established, and five evaluation dimensions are set : physical dimension (I_1), host system dimension (I_2), network dimension (I_3), data dimension (I_4), manager dimension (I_5).

Secondly, through the intrusion detection system to obtain system vulnerabilities and other methods to obtain a variety of data as the basis for expert scoring, and determine qualitative indicators and quantitative indicators for quantitative indicators.

Thirdly, the security situation is quantified by interval matrix correction module and the entropy weight based cloud module. The interval judgment matrix is given by experts, and the interval matrix correction module is used to obtain the optimal deterministic judgment matrix. Then the indicator layer is constructed according to the experts' evaluation results. Combined with the indicator layer weight vector, the criterion layer based cloud model is constructed and the entropy weight coefficient of the criterion layer cloud model is calculated. At last, the situation value of information system is obtained by the situation value operator.

Finally, according to the "Information security technology—classification guide for classified protection of information systems security" [7] and the comprehensive security situation value of information system, the security situation level of information system is determined.

3   Multi-source Data Standardization

Since the heterogeneity of multi-source data makes it difficult for experts to evaluate, this paper proposes a standardized method for qualitative and quantitative indicators as follows.

3.1 Standardization of Qualitative Indicators

Let a qualitative index comment be classified into m, and then set to β_1, β_2, ..., β_m, in order to define $\beta_i \sim \beta_j$ $(i, j \in 1, 2, ..., m)$, which means that the comment β_i is better comment on β_j, then $\beta_1 \sim \beta_2 \sim ... \sim \beta_m(i, j \in 1, 2, ..., m)$. It is assumed that there is a variable θ that reflects the score of this indicator, and $\theta \sim N(0, 1)$. Suppose the variable corresponding to comment β_i which reflects the expert score is t_i and t_i is the quantile of $N(0, 1)$, then

$$P(\theta < t_i) = \frac{i}{m}(i = 1, 2, \cdots, m - 1) \tag{1}$$

Let the expert score be V_e, and $V_e = \mu t_i$, where μ is the correction factor (This paper takes $\mu = 100$).

3.2 Standardization of Quantitative Indicators

Let the quantitative interval of the index X be $[X_a, X_b]$, the standardization process for the quantitative indicators of different dimensions is as follows:

1. Positive indicator

$$X_1 = \frac{x - X_a}{X_b - X_a}, \ (X_b > X_a) \tag{2}$$

2. Reverse indicator

$$X_2 = \frac{X_b - x}{X_b - X_a}, \ (X_b > X_a) \tag{3}$$

4 Interval Matrix Correction Module

The interval matrix correction module is divided into three sub-modules. They are interval matrix consistency degree judgment sub-module (Interval_matrix_identify), interval matrix element adjustment sub-module (Interval_matrix_adopt) and best deterministic matrix acquisition sub-module (Best_interval_matrix).

The workflow design of the interval matrix correction module is as follows

Step 1. Calculate the consistency degree of a given interval matrix by the interval_matrix_identify module, and obtain the consistency degree value (consis_value).
Step 2. If the consis_value > threshold, go to Step 3. Otherwise, adjust the interval number elements in the interval matrix by interacting with experts and go to Step 1.
Step 3. Based on the optimized interval matrix, the Best_interval_matrix is calculated and obtained. This matrix is used as the basis for solving the weight vector.

The processing method and process of each sub-module are explained in detail below.

4.1 Interval Matrix Consistency Degree Judgment Sub-module

The interval judgment matrix given by the expert generates Q random matrices according to the uniform distribution probability, and sequentially calculates the consistency ratio CR_k $(k = 1, 2, ..., Q)$ of the generated random matrix. Let the number of random matrices with satisfactory degree of consistency be p, then the degree of consistency of the interval matrices is $\gamma = p/Q$. The larger γ, the better the consistency of the interval matrix; the smaller γ, the worse the consistency of the interval matrix. This paper takes $Q = 100$.

4.2 Interval Matrix Element Adjustment Sub-module

When the consistency degree γ o is less than a certain threshold, some elements in the interval matrix need to be adjusted [8]. The specific process is designed as follows.

Step 1. Delete the elements of the hth row and hth column in the interval matrix to form n sub-intervals of order $n-1$. Then calculate the consistency degree γ_h $(h = 1, 2,..., n)$ of n sub-interval matrices $\bar{A}_h^{(n-1)}$.

Step 2. If the degree of consistency γ_{h1} and γ_{h2} of the sub-interval judgment matrix $\bar{A}_{h1}^{(n-1)}$ and $\bar{A}_{h2}^{(n-1)}$ is greater than other sub-interval judgment matrices. The common interval elements $\left[ah1h_2^L, ah1h_2^U\right]$, $\left[ah2h_1^L, ah2h_1^U\right]$ of the original interval judgment matrix have a great impact on the consistency of $\bar{A}$, so interact with experts and adjust the interval elements $\left[ah1h_2^L, ah1h_2^U\right]$, $\left[ah2h_1^L, ah2h_1^U\right]$.

Step 3. Turn to the interval matrix consistency degree judgment module, and continue to calculate the consistency degree of the adjusted interval judgment matrix.

4.3 The Best Deterministic Matrix Acquisition Sub-module

This sub-module consists of two processes: interval matrix convergence and optimal deterministic matrix optimization. The specific process is designed as follows:

1. Interval matrix convergence
   Step 1. Based on the optimized interval judgment matrix, R deterministic matrices are randomly generated according to the uniform distribution probability.
   Step 2. The consistency ratios $CR_i (i = 1, 2, ..., R)$ of the R deterministic matrices are respectively calculated, and the deterministic matrices with the first ω consistency ratios are obtained by sorting to form the tth matrix cluster, which is denoted as Cluster_matrix_t.
   Step 3. Integrate the new interval by using the same position elements of different matrices in matrix clusters, and then obtain the upper and lower limits of the elements of each interval in the new interval judgment matrix. When $i = j$, $a_{ij} = 1$; when $i \neq j$, $a_{ij}^L = \min\left\{a_{ij1}^L, a_{ij2}^L \ldots a_{ij\omega}^L\right\}$, $a_{ij}^U = \min\left\{a_{ij1}^U, a_{ij2}^U \ldots a_{ij\omega}^U\right\}$, and $\bar{a}_{ji} = \left[1 \middle/ a_{ij}^U, 1 \middle/ a_{ij}^L\right]$.
   Step 4. The newly generated interval judgment matrix is transferred to Step 1, until the sum of $\left|a_{ij}^U - a_{ij}^L\right| (i, j, \in 1, 2, ..., n)$ (the lengths of the element lengths of

the interval matrix) is not more than 10% of the sum of the lengths of the elements of the original interval judgment matrix, then the loop stops.

In Step 1, the proportion of each determined number of the randomly generated deterministic matrix in the left half interval of each interval element of the original interval matrix is α, and $0.5-\eta < \alpha < 0.5 + \eta$ (This paper takes $\eta = 0.05$).

2. Optimal deterministic matrix optimization
Step 1. Take the converged interval matrix as input and record it as *Input_matrix*. First, a deterministic matrix M_0 is initialized, the elements of the deterministic matrix are: $a_{ij}, i, j \in 1, 2, ..., n$. When $i = j, a_{ij} = 1$; when $1 < j \leq n, 1 \leq i < j, \left(a_{ij}^L + a_{ij}^U\right)\big/ 2$; when $1 < i \leq n, 1 \leq j < i, a_{ij} = 1/a_{ji}$. The CR_0 is calculated as the initial consistency ratio.
Step 2. Randomly generate a deterministic matrix from *Input_matrix* which is denoted as $M_i(i = 1, 2, ..., k)$, calculate its consistency ratio CR_i, and compare it with CR_0. If $CR_i < CR_0$, let $CR_0 = CR_i, M_0 = M_i$; otherwise, keep CR_0 and M_0 unchanged.
Step 3. Adjust each element in each deterministic matrix according to Eqs. (4) and (5):

$$v = w * v + c * rand * (pbest - present) \tag{4}$$

$$present = present + v \tag{5}$$

where v is the speed of optimization; w is used to adjust the speed of optimization; c is the cognitive factor and usually $c = 2$; *rand* is the random number between $(0, 1)$; *pbest* is the current the element in the deterministic matrix with the smallest consistency ratio; *present* represents the element in the current deterministic matrix.

Let $v_{max} = \min\left(a_{ij}^U - a_{ij}^L\right)$ $(i \neq j, i, j \in 1, 2, ..., n)$, where $a_{ij}^L - a_{ij}^U$ are the upper and lower limits of each element of the converged interval matrix. If $v > v_{max}$, then take $v = v_{max}$; if $v < -v_{max}$, take $v = -v_{max}$. If *present* $\in \left[a_{ij}^L - a_{ij}^U\right]$, *present* does not need to be adjusted; if present $< a_{ij}^L$, then take *present* $= a_{ij}^L$; if *present* $> a_{ij}^U$; take *present* $= a_{ij}^U$. The initial value of v is taken as 0, *pbest*$_0$ corresponds to each element in the initial deterministic matrix M_0 in Step 1, and *present*$_0$ corresponds to each element in the deterministic matrix randomly generated in Step 2 for the first time.
Step 4. Iterate k times according to Step 2 to Step 3. On this basis, the eigenvector method can be used to solve weight vectors.

5 Entropy Weight Based Cloud Module

5.1 Expert Evaluation of Membership Cloud

n experts are invited to conduct the evaluation of a certain index, and the evaluation results are converted into a percentage form according to Sect. 3.2. The membership clouds [9] represents the evaluation results of the n experts. First, the three digital features (E_x, E_n, H_e) of the cloud model are calculated by the reverse cloud generator. Then the expert evaluation results are restored by the forward cloud generator. Finally, if the cloud drops are too discrete, it indicates that the expert evaluation opinions differ greatly, then we can apply for re-evaluation.

1. Reverse cloud generator

$$E_x = \frac{E_{x1} + E_{x2} + \cdots + E_{xn}}{n} \tag{6}$$

$$E_n = \frac{\max\{E_{x1}, E_{x2}, \cdots E_{xn}\} - \min\{E_{x1}, E_{x2}, \cdots E_{xn}\}}{n} \tag{7}$$

$$S^2 = \frac{1}{n} \sum_{i=1}^{n} (x_i - E_x)^2 \tag{8}$$

$$H_e = \sqrt{S^2 - E_n^2} \tag{9}$$

among them: E_{xi} indicates the percentage result of the ith expert evaluation, and n indicates the number of experts. The digital features of the membership cloud (E_x, E_n, H_e) are calculated by the above formulas.

2. Forward cloud generator.

Step 1. E_{nn}= Randn(E_n, H_e), which takes E_n as the expectation and produces a normally distributed random number E_{nn} with H_e as the standard deviation.
Step 2. x_i = Randn(E_x, E_{nn}), that is, taking E_x as an expectation, and generating a normal distribution random number x_i with E_{nn} as a standard deviation.
Step 3. $\xi_i = \exp\left[-(x_i - E_x)^2/(2E_{nn}^2)\right]$, the degree of membership is calculated according to the formula, and the pair (x_i, ξ_i) represents a cloud drop distributed over the domain U.
Step 4. Step 1 through Step 3 is performed cyclically until enough cloud drops are generated to restore the expert evaluation results in the form of a cloud model.

5.2 Membership Cloud Gravity Center

The result of the expert evaluation of f indicators subordinate to the criterion layer can be represented by f-dimensional membership clouds. The f-dimensional comprehensive membership cloud of the dimension can be formed by membership cloud gravity center. This paper uses the vector g to represent the gravity center vector of this cloud which is $g = (g_1, g_2, ..., g_f)$. Where $g_i = E_{xi} \cdot w_i (i = 1, 2, ..., f)$, E_{xi} represents the expected value of the ith membership cloud, and w_i represents the weight corresponding to the index which is calculated by the interval matrix correction module.

Assuming that the initial state of the system is ideal, the initial cloud center of gravity vector of the f-dimensional integrated membership cloud is $g^0 = \left(g_1^0, g_2^0, \ldots g_f^0\right)$.

The cloud gravity center vector representing the current expert evaluation result is $g' = \left(g_1', g_2', \ldots g_f'\right)$. Then normalize the changes in the gravity center vector of the f-dimensional integrated cloud is:

$$
g_i^G = \begin{cases} (g_i' - g_i^0)/g_i^0, & g_i' \leq g_i^0 \\ (g_i' - g_i^0)/g_i', & g_i' > g_i^0 \end{cases} \tag{10}
$$

among them, $i = 1, 2, \ldots, f$.

Calculate the weighted deviation δ from the weight vector $W = (w_1, w_2, \ldots, w_f)$:

$$
\delta = \sum_{i=1}^{f} g_i^G * w_i \tag{11}
$$

Enter δ into the evaluation cloud model to get the support level of this dimension index for different comments in the criterion layer [10].

The support matrix P of each dimension index in the criterion layer relative to each evaluation is determined by evaluating the activated comments in the cloud model and the support degree of each comment, as shown in Table 1.

Table 1. Support matrix.

Criteria layer	Worse	Bad	Average	Good	Excellent
X_1	P_{11}	P_{12}	P_{13}	P_{14}	P_{15}
X_2	P_{21}	P_{22}	P_{23}	P_{24}	P_{25}
X_3	P_{31}	P_{32}	P_{33}	P_{34}	P_{35}
X_4	P_{41}	P_{42}	P_{43}	P_{44}	P_{45}
X_5	P_{51}	P_{52}	P_{53}	P_{54}	P_{55}

X_1, X_2, X_3, X_4, X_5 in Table 1 correspond to the 5 dimensions of the criterion layer respectively, and p_{ij} indicates the degree of support of the ith index to the jth comment $(i, j \in 1, 2, 3, 4, 5)$.

Calculate the absolute entropy of each dimension indicator by using Eq. (12):

$$
\delta = \sum_{i=1}^{f} g_i^G * w_i \tag{12}
$$

when $p_{i1} = p_{i2} = \ldots = p_{in}$, there is $H_{max} = \ln n$. Calculate the relative entropy value of each dimension indicator by using Eq. (13)

$$
\mu_i = -\frac{1}{\ln n} \sum_{j=1}^{n} p_{ij} \ln p_{ij} \tag{13}
$$

The weight of the corresponding indicator is expressed by $(1-\mu_i)$, which is normalized:

$$\tau_i = \frac{1}{n - \sum\limits_{i=1}^{n} \mu_i}(1 - \mu_i) \tag{14}$$

where $\tau_i \in [0, 1]$ and $\tau_1 + ...+\tau_n = 1$, τ_i is the entropy weight coefficient of the subordinate cloud corresponding to X_i.

The weight vector corresponding to each comment in the given evaluation cloud model is set as $U = (u_{worse}, u_{bad}, u_{average}, u_{good}, u_{excellent}) = (1/15, 2/15, 1/5, 4/15, 1/3)$.

The information system comprehensive security situation value operator is Eq. (15):

$$V = 1 - \tau * P * U^T \tag{15}$$

Rating the comprehensive security situation of the information system based on the V. The security situation is divided into five levels in this paper: worse, bad, average, good and excellent, corresponding to the five probability intervals of V: 0.00–0.20, 0.21–0.40, 0.41–0.60, 0.61–0.80 and 0.81–1.00, respectively.

6 Experiment and Analysis

In this paper, the MIMEC based information system security situation assessment model is applied to a domestic Departure Control System (DCS). The system security situation assessment is conducted every Tuesday, from October 1 to December 23, 2018, for a total of 12 times. The following experiment uses the evaluation of the network dimension of the system criterion layer on October 9, 2018 as an example to illustrate the application process of the evaluation model.

For the four sub-indicators of the network dimension (I_3) $(I_{31}, I_{32}, I_{33}, I_{34}) = $ (network topology, network access control, security audit, network traffic), 10 experts are invited to evaluate each sub-indicator. According to the Sect. 3, the evaluation of qualitative and quantitative indexes was unified into the score under the percentage system.

The interval judgment matrix is given by experts on the relative importance of the four sub-indicators:

$$A^0 = \begin{pmatrix} 1 & [3,4] & [3,5] & [3,5] \\ [1/4, 1/3] & 1 & [1/2, 1] & [2, 5] \\ [1/5, 1/3] & [1/2] & 1 & [1/3, 1] \\ [1/5, 1/3] & [1/5, 1/2] & [1/3] & 1 \end{pmatrix}$$

According to Sect. 4, the best deterministic matrix is calculated:

$$A^{best} = \begin{pmatrix} 1 & 3.300\ 679 & 3.874\ 924 & 4.261\ 778 \\ 0.300\ 351 & 1 & 0.942\ 991 & 2.032\ 611 \\ 0.259\ 102 & 1.057\ 097 & 1 & 0.899\ 372 \\ 0.233\ 799 & 0.490\ 148 & 1.109\ 949 & 1 \end{pmatrix}$$

The consistency ratio is $CR = 0.022\,591 < 0.1$. The weight vector obtained is $w = (0.555\,665, 0.178\,134, 0.144\,072, 0.122\,129)$. The expected value vectors of the four sub-indexes of network dimension based on the graph is $(92.96, 87.50.92.22, 94.72)$.

According to Eqs. (10)–(11), the weighted deviation degree is $\delta = -0.079\,134$, and the security dimension value of the network dimension is $0.920\,866$. Inputting δ into the evaluation cloud model indicates that the network dimension is in "excellent" state, as shown in Fig. 1.

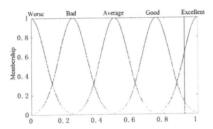

Fig. 1. Evaluation cloud activation

According to Sect. 5, the network dimension support vector is calculated as $(p_{31}, p_{32}, p_{33}, p_{34}, p_{35}) = (0.052\,86, 0.072\,56, 0.115\,66, 0.122\,04, 0.636\,88)$.

The evaluation support vector of the criterion layer is the same as the calculation process of the network dimension comment support vector. The obtained security level value vector of each dimension of the criterion layer is $(0.677\,2, 0.731\,4, 0.920\,9, 0.522\,5, 0.643\,4)$, and the comment support matrix P is shown in Table 2.

Table 2. Comment support

Criteria layer	Worse	Bad	Average	Good	Excellent
X_1	0.045 93	0.072 19	0.104 12	0.682 61	0.095 54
X_2	0.000 67	0.001 02	0.201 11	0.975 38	0.001 82
X_3	0.052 86	0.072 56	0.115 66	0.122 04	0.636 88
X_4	0.002 94	0.056 40	0.964 18	0.024 02	0.032 19
X_5	0.074 71	0.122 19	0.227 29	0.441 00	0.134 80

According to formula (12)–(15), the comprehensive security situation value of DCS is 0.752. So the security situation of the information system is in "excellent" state, which is consistent with the actual situation.

The security situation assessment method in this paper, the entropy weight coefficient method [2] and the AHP method [6] are applied to the evaluation of DCS. The criterion layer security situation and total security situation are evaluated, as shown in Fig. 2, 3.

As can be seen from Fig. 2 and Fig. 3 the fluctuation of the situation assessment value of the model in this paper is obviously smaller than that obtained by entropy weight coefficient method and AHP method. There are two reasons:

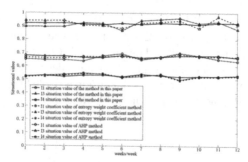

Fig. 2. Criterion layer security situation

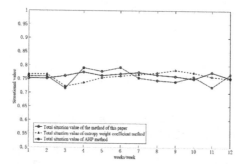

Fig. 3. Total security situation

First, the model in this paper improves the objective degree of the weight vector by correcting the interval matrix and overcomes the shortcoming of strong subjectivity of the traditional AHP method. At the same time, by judging the dispersion degree of the subordinate cloud droplets of the experts' evaluation results, abnormal indicator values can be found and reevaluation. Compared with entropy weight coefficient method, unreasonable indicator weighting can be avoided. Therefore, the quantitative result of the model in this paper is more appropriate to the actual DCS security situation, which improves the reliability of information system security situation evaluation.

Second, due to the difference of experts' personal ability, it is difficult to judge the relative importance of each dimension indicator in the criterion layer uniformly. On the basis of multi-source data standardization, the entropy weight coefficient of each cloud model corresponding to the criterion layer indicator is used to avoid weighting directly for the criterion layer indicator. Therefore, the total situation value of the DCS system can avoid large fluctuation and improve the stability of information system security situation assessment.

7   Conclusion

This paper proposes an Information System Security Situation Assessment Model (ISS-SAM) based on Modified Interval Matrix-Entropy Weight based Cloud (MIMEC). This model corrects the interval judgment matrix, finds the optimal deterministic matrix to determine the index layer weight vector, and combines the entropy weight membership cloud to quantify and grading the security situation. Compared with the evaluation results based on the entropy weight coefficient method and the traditional AHP method, it shows that the model has good reliability and stability.

Acknowledgements. This work was supported by the Civil Aviation Joint Research Fund Project of the National Natural Science Foundation of China under granted number U1833107.

References

1. Qu, X.H.F.: Research of network security situation assessment based on AHP. Tech. Autom. Appl. **37**(11), 43–45 (2018)
2. Fu, Y., Wu, X.P., Ye, Q., Peng, X.: An approach for information systems security risk assessment on fuzzy set and entropy-weight. Acta Electron. Sin. **38**(7), 1489–1494 (2010)
3. Luo, H., Shen, Y., Zhang, G., Huang, L.: Information security risk assessment based on two stages decision model with grey synthetic measure. In: IEEE International Conference on Software Engineering & Service Science. IEEE Press, Beijing (2015)
4. Xi, R.R., Yun, X.C., Zhang, Y.Z., Hao, Z.Y.: An improved quantitative evaluation method for network security. Chin. J. Comput. **38**(4), 749–758 (2015)
5. Zhou, L.B., F.: Study on applying the neural network in computer network security assessment. In: Eighth International Conference on Measuring Technology & Mechatronics Automation. IEEE Press, Macau (2016)
6. Cheng, X.F.: Information system security situation assessment and risk control method based on operation-flow. Civil Aviation University of China, Tianjin (2016)
7. Information security technology—classification guide for classified protection of information systems security: GB/T 22240—2008. Standards Press of China, Beijing (2008)
8. Zhu, J.J., Liu, S.X., Wang, M.G.: Novel weight approach for interval numbers comparison matrix in the analytic hierarchy process. Syst. Eng.-Theory Pract. **25**(4), 29–34 (2005)
9. Deyi, L., Haijun, M., Xuemei, S.: Membership clouds and membership cloud generators. J. Comput. Res. Dev. **32**(6), 15–20 (1995)
10. Feng, Z.H., Zhang, J.C., Zhang, K., Liu, W.: Techniques for battlefield situation assessment based on cloud-gravity-center assessing. Fire Control Command Control **36**(3), 13–15 (2011)

IoTFC: A Secure and Privacy Preserving Architecture for Smart Buildings

Amna Qureshi$^{(\boxtimes)}$ ⓘ, M. Shahwaiz Afaqui ⓘ, and Julián Salas ⓘ

Internet Interdisciplinary Institute (IN3), Universitat Oberta de Catalunya (UOC),
Barcelona, Spain
{aqureshi,mafaqui,jsalaspi}@uoc.edu

Abstract. In the pursuit of cities to be more efficient and responsive, various kind of Internet of Things (IoT) devices, such as actuators and sensors are used. This paper focuses on one specific IoT application - the smart building, and investigates the security and privacy issues in an integrated IoT-fog-cloud (IoTFC) smart building architecture. We consider the surveillance, maintenance, environment, and concierge use cases for smart building, in terms of their characteristics, compatible communication technology, and security and privacy requirements. IoTFC provides a comprehensive solution to the security and privacy challenges of authentication, access control, anomaly detection, data privacy and location privacy. To the best of our knowledge, IoTFC is a novel architecture, as it combines a complete set of light-weight security and privacy solutions suitable for smart buildings.

Keywords: Smart building · Fog computing · IoT solutions · Security · Privacy

1 Introduction

Despite the potential benefits of a smart building, the reliance on the collection of data by multiple entities contradicts the expectations of privacy. The raw information collected by different sensors such as temperature control, elevators, fire alarms, motion detectors or surveillance cameras, etc., can yield a tremendous amount of data that can be combined, analyzed and acted upon, all potentially without adequate accountability, security or meaningful consent e.g., the building manager can perform user profiling. Additionally, there are other privacy concerns [15], such as privacy-violating interactions, presentations, (e.g.,

This work was partly funded by the Spanish Government through grants INCIBEC-2015-02491 "Ayudas para la excelencia de los equipos de investigaciòn avanzada en ciberseguridad", RTI2018-095094-B-C22 "CONSENT". This research is partially funded by the SPOTS project (RTI2018-095438-A-I00) funded by the Spanish Ministry of Science, Innovation & Universities and 2017 SGR 60 Funded by the Generalitat de Catalunya.

D. Wang et al. (Eds.): SPNCE 2020, LNICST 344, pp. 102–119, 2021.
https://doi.org/10.1007/978-3-030-66922-5_7

answering a private query by displaying it on a screen that may be display of a private query accessible to others), transitions (when the IoT devices change of owner or location of devices), inventory attacks (the possibility to learn externally to a building what are its characteristics, equipment or personal devices) and the information linkage risk due to the collection of data from multiple sensors. Thus, privacy of a smart building requires the protection of occupant's private data, patterns, and interactions with other people or IoT devices. Unfortunately, such privacy challenges are only partially addressed in smart buildings.

Ubiquitous wireless communication technologies and protocols connecting the IoT devices, systems and application platform (building automation system) to each other and to the Internet are vulnerable to cyber attacks, e.g., if an automation system is compromised, a hacker can potentially take control of security functions such as locking doors to potentially gain access to a building, e.g., in Finland, a DDoS attack targeting the heating system left residents of two building apartments in the cold. In 2017, hackers reportedly took control of the electronic key system of a hotel in the Austrian Alps. Fully booked and with little alternative, the hotel paid the 1,600 euro ransom [5]. A recent Kaspersky report [6] on security of smart buildings revealed that out of the 26.5% protected smart building systems management computers, 12% were attacked with malware aimed at stealing account credentials and other valuable information. The sheer range of device types, conflicting and incompatible standards, and complex protocols makes securing these systems a daunting task for the security designers.

In the literature, there exists many solutions addressing security and privacy properties based on traditional cryptographic primitives, which are not compatible with the computational capacities of IoT devices. Thus, there is a need to design a reliable and light-weight security and privacy architecture for a smart building that can provide device authenticity, data confidentiality, location and data privacy, anomaly detection, secure data transmission and storage.

As an alternative, the paradigm of fog computing can be used as a viable solution to enhance the security within smart building systems. Fog computing extends the cloud computing platform with additional computational, storage and networking resources placed in the immediate vicinity of the end user devices. The proximity of the fog to IoT devices enables fog computing to deliver extremely low network latencies between the end user devices and the fog computing resources serving them, and to locally process transient data. In addition, decentralized and distributed data control makes fog computing a viable deployment method in a smart building to offer various features such as scalability, autonomy, manageability and security, etc. Fog nodes in the building can have security requirements as well as the connectivity and coordination to feed the larger cloud-based systems with the data needed to inform their difficult decisions.

Contributions and Plan of the Paper. This paper proposes a novel architecture based on fog computing and various IoT sensors considered for four use cases (surveillance, maintenance, environment, and concierge). IoTFC is a thorough architecture, which provides security (such as authentication, access control, confidentiality, etc.) and privacy (in terms of location and data) features through a combination of different network technologies, middleware transport protocols, and light weight cryptographic protocols implemented over open source solutions. As compared to other related works in literature, this architecture provides mechanisms to counter a complete set of security and privacy issues expected in smart buildings.

The remainder of the paper is organized as follows: In Sect. 2, we present the selected use cases of a smart building. Section 3 discusses the communication technologies suitable for smart buildings. In Sect. 4, we present the design details of the proposed IoTFC. Section 5 compares the relevant state-of-the-art smart building solutions with IoTFC. Finally, Sect. 6 concludes the paper.

2   Use Cases

This section describes four use cases for smart buildings spanning building and occupants' security, operational efficiency, energy conservation, efficient space utilization and occupant satisfaction. Also, a brief summary of the security and privacy threats in these cases is presented.

- **Smart surveillance:** It enables building managers to perform the following functions: 1) manage and control surveillance devices remotely; 2) make smarter decisions based on real-time security conditions; (3) detect false alarm without physical inspection; and 4) collect and analyze data to make important improvements to security processes of the building.
- **Smart maintenance:** It allows the building managers and operators to improve efficiency and optimize building operations in the following ways: 1) detect temperature changes, water pressure variations, or other abnormalities, and alert operational systems to prevent damage; 2) use predictive maintenance to avoid outages or significant downtime caused by maintenance; and 3) provide a swift response to any breakdown incident.
- **Smart environment:** It enables the building managers to perform the following functions: 1) monitor and adjust building systems, such as lighting, air quality, etc. to match comfort levels and reduce energy waste; 2) control and monitor entry and exit points of the building for better security decision-making and disaster prevention; and 3) automatic garbage collection from the dumpsters upon reaching the fill level.
- **Smart concierge:** It enables the building managers to support occupants in performing following predefined day-to-day tasks: 1) locate occurring events or available meeting spaces by date, time or room size; 2) book appointments or conference halls; 3) order a taxi or rental car/bike; and 4) registering visitors/guests without the inhabitant having to personally go to the reception.

Table 1 summarizes the possible security and privacy threats of each of the aforementioned use case. Many solutions have been proposed to either provide security or privacy against the attacks mentioned in Table 1, such as protection

Table 1. Security and privacy threats.

Threats	Use cases			
	Smart Surveillance	Smart Maintenance	Smart Environment	Smart Concierge
Security	• Remote tampering by hackers to use attacked cameras as spy tools • Manipulation or deletion of surveillance footage stored on local or global data centers • Collection of additional sensitive information (Wi-Fi or users' credentials) • DDoS attacks • Unauthorized access to the building's sensitive or critical information	• Create network congestion with false alarms generated by hacked IoT device • Access to other sensitive building systems (e.g., payment) by third party vendors responsible of remotely maintaining IoT devices (HVAC, lighting, etc.) • Physical intrusions into the network • Eavesdropping to obtain information without authorization, or man-in-the-middle (MITM) attack	• Bypass authentication attacks to gain full control of the devices and carry out malicious actions such as injecting malware, spyware, etc. into the building system • Stealthy Deception attacks • Jamming attacks to disrupt the functionality of devices • Rogue devices connecting to the network as legitimate devices	• Vectoring and sniffing attacks for data theft • Distribution of active X scripts by authorized but malicious entities to disrupt the whole system • Malware infiltrated in the building by the infected occupants' devices • Privilege escalation to gain access to occupant's data
Privacy	• Linking data from users to their identities (e.g., surveillance cameras and image recognition can be used to identify users) • Tracking and profiling all users' activities • Misuse of microphones or cameras for invading private spaces without consent	• Indirectly tracking and profiling the activities of a user (e.g., inferring the time of cooking, showering, or watching TV by the power consumption meter)	• Inferring the indoor activities by sensor measurements (e.g., inferring sleeping patterns from the change in temperature or humidity measurements in a bedroom) • Information disclosure	• Profiling users' preferences and behaviours • Inferring users' locations and interests through their location-based queries • Inferring social relationships between building occupants • Privilege abuse of continuous monitoring for malicious activities

of data from illegal disclosure or malicious violation by using an access control mechanism (discretionary access control (DAC), mandatory access control (MaC), role-based access control (RBAC), etc.), confirming one's identity and limiting unauthorized access to the system by providing authentication (elliptic-curve or public key cryptography based techniques), identifying malicious events or jamming attacks (machine learning-based anomaly detection), providing end-to-end secure communication (Zigbee with custom security, MQTT, etc.), preventing user profiling, tracking and re-identification (privacy-preserving data mining, location privacy protection techniques, and statistical disclosure control, etc.).

Although there exists a few solutions to counter some of the aforementioned security and privacy threats, IoTFC addresses a complete set of these threats and integrates lightweight security and privacy solutions that have not been integrated in a single architecture for a smart building. The proposed security and privacy solutions in IoTFC are provided in detail in Sect. 4.3.

3 Communication Technologies

The most essential part of IoT infrastructure is the communication system that acts as a bridge for the delivery of data and control messages between leave stations and a central processing unit (for our case, between a fog node and an end node). Since the amount of information generated over smart building scenario is assumed to be huge and increasing (due to the increase of end devices such as actuators or sensors embedded in home appliances along with the already existing in-building sensors such as smoke alarm and security window sensors, etc.), there is a need to adopt universally accepted, cost effective and scalable communication technology within IoT frame work. In this section, we provide a brief overview of the technologies and protocols that are currently being used within IoT smart building environments.

3.1 Available IoT Smart Building Systems

There are numerous smart building standards available, where some of them are proprietary solutions (e.g., Hubitat, Insteon, etc.), others are partially open standards (e.g. Zwave, EnOceon, etc.), and some of them are open source solutions (e.g., KNX, LonWorks, etc.). However, most of the open source standards support expensive devices as compared to the ones used in conventional installations. Also, these open standards utilize topologies and architectures that are predefined and are difficult to accommodate within fog computing paradigm. Moreover, these solutions are vulnerable to security attacks [1]. In this paper, we intend to utilize the fog computing-based Home Automation System (HAS) proposed in [8], which uses a custom made solution. The authors designed a home gateway based on Raspberry Pi and OpenHAB to provide home automation fog services through local gateways. Though the system is efficient as compared to traditional cloud-based systems, it lacks security and privacy solutions so as to be implemented in a building. We aim to build IoTFC architecture and its security and privacy solution over the aforementioned solution. We intend to improve the system by incorporating Raspberry Pi 4, which includes Bluetooth LE, Wi-Fi, Ethernet and increased computational capacity (required to support the lightweight security and privacy enabling solutions).

3.2 Networking Standards/technologies for Smart Buildings

From technological point of view, an architecture of a smart building can be divided into following four layers: 1) sensor, 2) network, 3) middleware, and 4) application.

Table 2. Comparison of indoor connectivity standards.

	Wi-Fi	ZigBee	Zwave	BLE	NFC
Operation range in building (m)	60	30	30	60	0.1
Maximum Data rate (kbps)	54000	250	100	1000	424
Frequency of operation (GHz)	2.4	2.4	0.86842	2.4	0.01356
Network topology	Star	Star, Mesh	Partial Mesh	Star, Mesh	Peer-to-Peer
IP layer at end devices	Yes	No	No	Yes	No
Security features	AES	AES	AES	AES	AES, RSA
Privacy features	×	×	×	Address Randomization	×

Sensor Layer. This layer detects and collects useful real-world data from the environment (i.e., temperature, humidity, etc.) or things (i.e., motion, vibration, etc.). Moreover, this layer processes information into digital form and then transmits it to the network layer. The main security issues includes DDoS attacks through malicious node placement that can result in battery depletion and physical attack on hardware component through tampering of sensors and devices.

Network Layer. This layer provides the means through which data is transferred among IoT hubs and devices to realize the integration of communication network and the perceptions. It constitutes the communication software and hardware components (i.e., topologies, network nodes, and gateways). Different aspects of these technologies are highlighted in Table 2. In terms of security, this layer is highly susceptible to DoS attacks, confidentiality and privacy attacks through eavesdropping and passive monitoring, MITM attack, illegal access, asynchronous and conspiracy attacks.

Below, we describe the details of the access methods commonly used in smart buildings and highlight the security measures of each to counter different security challenges.

- **Wi-Fi:** IEEE 802,11 based Wi-Fi is the most popular wireless technologies that is ubiquitously available at a global scale. The new Wi-Fi standards (IEEE 802.11ah and IEEE 802.11ax) have been particularly designed for IoT use cases. For security, Wi-Fi uses Wireless Equivalent Privacy (WEP) and Wi-Fi Protected Access version 1 or 2 (WPA, or WPA2) protocols. WEP uses a 64 or 128-bit encryption key that enforces confidentiality, access control and data integrity with the goal to protect the privacy of user data from eavesdropping. WEP, due to a brute force attack vulnerability, was superseded by WPA and WPA2. WPA supports Temporal Key Integrity Protocol (TKIP) that operates by performing per-packet key mixing with re-keying and WPA2 uses Advanced Encryption Standard (AES) block cipher TKIP protocol called Counter Mode Cipher Block Chaining MAC Protocol (CCMP) that provides stronger encryption.
- **Zigbee:** Zigbee is a low power, low data rate and low cost radio network standard which operated on top of the MAC and PHY layer of IEEE 802.15.4 standard for personal area standard. The Medium Access Control (MAC)

layer of IEEE 802.15.4 defines security services for access control based on MAC address, encryption using AES cryptography, frame integrity through detection and sequential freshness through MAC frame serialization. Apart from the MAC layer security features defined by IEEE 802.15.4 standard, Zigbee has its own security model which includes cryptographic key initiation, key transport, frame protection and device management. Some of the drawbacks are: assigned network key cannot be changed, eavesdropping or ejection of fake packets by adversary, and problem of inter-node coordination.

- **Z-wave:** It is on of the oldest low power wireless technology specifically designed for home automation. To enable faster and simpler development of application, Z-wave uses a simpler protocol architecture. The physical and data link layer is included as standard G.9959 by the International Telecommunication Union (ITU). In order to provide protection against authentication, confidentiality and replay attack, Z-wave defines a distinct security layer with two classes of security: Security 0 (S0) for lightweight, and Security 2 (S2) for stronger security. Both the classes provide confidentiality by encrypting information with AES-128. In S0 class, the network key is shared with all devices in a network. While in S2, each new network of subclass prevents a low-security class device from compromising higher-security device. A major security issue for Z-wave protocol is the requirement to support older devices that do not include encryption and authentication procedures. Also, Z-wave has been found to be susceptible to eavesdropping and spoofing attacks. Moreover, implementation of brute force resistant asymmetric cryptography at the end nodes can be challenging due to limited processing power and energy.

- **Bluetooth Low Energy (BLE):** Bluetooth, which is based on the IEEE 802.15.1 standard, is a low power, low cost wireless communication technology. BLE was introduced as version 4 of Bluetooth in 2011 and was designed for applications requiring periodic transmission of small amount of data. It provides confidentiality using encryption and uses Cyclic Redundancy Checks (CRC) with hashing and AES encryption to ensure integrity of data. In addition, BLE also aims to improve the availability by providing protection against MITM and DoS attacks through the use of secure connection pairing. To enable privacy, BLE protocol includes a privacy mode which uses random MAC addresses to help achieve anonymity. BLE has shown vulnerability to eavesdropping for nodes that lack display capability to present a six digit number and thus follow the just work association model.

- **Near Field Communication (NFC):** NFC, which is breakthrough of the Radio-frequency identification (RFID), is a short range and high frequency P2P (Peer-to-Peer) wireless technology used for wireless identification. It operates by storing information within tiny micro chips (tags) which is transmitted to readers within a certain physical range. In comparison to Bluetooth, NFC does not require pairing before sending the data. Although short-range, NFC technology is vulnerable to many security challenges such as eavesdropping, unauthorized manipulation of data and MITM attacks, which can be countered by using hardware secure elements. The communication between a tag and the reader is performed using AES-128 encryption. The tag is capable of producing random ID to improve privacy. Authentication is provided using Secure Unique NFC (SUN) procedure.

Middleware Layer IoT Network Technologies. This layer is responsible to provide abstraction to application layer. It receives data from network layer and stores it in a database. Also, it is responsible for ubiquitous computing and information processing. Due to wide the rapid and wide-spread evolution of IoT devices, different application layer protocols have been proposed. In this section, we discuss a few important application protocols that are being used within smart building paradigm.

- *Message Queue Telemetry Transport (MQTT):* It is the most popular light machine weight M2M protocol optimized for centralized data collection and analysis. MQTT uses a publish and subscribe architecture and is used for constrained devices operating with low-bandwidth, high-latency, and unreliable networks. It operates on top of TCP and uses the Transport Layer Security (TLS) protocol to provide encryption, authentication, and integrity. For security, each MQTT message contains a variable header with a user name and password for authentication support. The disadvantage of MQTT is the delay caused by TLS and the lack of support for priority messages. On the contrary, MQTT is the most widely used in IoT applications.
- *Advanced Message Queuing Protocol (AMQP):* It is a message-oriented light weight middle ware open standard that aims to create an open, asynchronous messaging protocol. Similar to MQTT, AMPQ uses a publish and subscribe architecture and the protocol is built on top of TCP. In terms of security, AMQP does provide higher security mechanisms (including Secure Sockets Layer (SSL) and Kerberos) over the cost of more computational power and resource. Therefore, it is difficult to implement AMQP on IoT devices with limited resources.
- *Extensible Messaging and Presence Protocol (XMPP):* It is a an IETF defined protocol developed for near real-time messaging and is based on extensible markup language that helps different entities within a network to communicate. It uses XML as data model and is build on top of TCP. XMPP supports both request/response and publish/subscribe models. The drawbacks of XMPP are: higher processing power and more bandwidth consumption, no QoS guarantee, restriction to simple data and lack of end-to-end encryption.
- *Constrained Application Protocol (CoAP):* This is also an IETF defined protocol developed for constrained devices that are capable of connecting to the Internet. It supports a variant of publish and subscribe and request and response architectures. It uses UDP (as opposed to TCP) with poor level of reliability and is assumed be more power efficient than MQTT. In terms of security, CoAP uses a lighter version of TLS. Key management and heavy cost of computation are considered as the main drawbacks of CoAP.

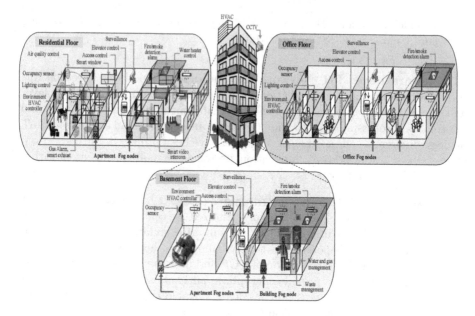

Fig. 1. Smart building architecture.

Application Layer IoT Network Technologies. This layer is used to provide application specific functionalities requested by customers. It provides an interface to lower layer protocols for end users to access data and to communicate with the IoT devices. The application layer typically includes Machine to Machine (M2M) communication protocols, cloud computing, a middle ware and a service support platform. The main security issues related to this layer are data leakage (stealing of data by attackers by knowing the vulnerabilities), malicious code injection (upload malicious codes in software), DDoS attack (to disrupt the availability), inability to receive security patches, and sniffing attack (corruption of the system through the injection of sniffer application).

4   System Architecture

The proposed IoTFC architecture is depicted in Fig. 1. As can be observed, the building is a mixed-use building that includes both offices and residential spaces. The residential portion takes up more square footage than the commercial part. The office floors inhabited by companies consist of open office areas with a few private office rooms, meeting rooms, reception area, open spaces for informal meetings, printing area, storage space, cafeteria, corridors and WCs. In each residential floor, there are 8–10 small to large-sized apartments that may consist of two to four bedrooms, living room, study/library, laundry area, storage space, kitchen, WC(s), terraces and balconies. Also, the building features a fitness centre, a heated indoor swimming pool, smart elevators and a large reception area. The lowest floor (basement) houses heating stations, parking, electric cars charging points, waste collection points, water tanks and pumps, and technical support area.

4.1 Entities

The proposed smart building architecture consists of the following stakeholders:

- Management staff: Their goal is to operate the building as smoothly and efficiently as possible. It can be further divided into three categories: (1) operational staff who bridge the technology-operations conversations with their technological counterpart; (2) technological staff who manage the network infrastructure and information backbone of the building; and (3) maintenance staff who performs day-to-day activities (e.g. repairs, inspections, etc.) related to electrical, mechanical equipment and civic duties.
- Occupants: These are the true users who utilize advanced technology of the building to achieve their goals of productivity, health, comfort, privacy, and well-being.
- Visitors: A non-resident who may be visiting the building for a brief time.

4.2 Functionality of Layers

The proposed IoTFC model is designed as a four-tier architecture as shown in Fig. 2. A brief functionality of each layer is presented below:

- **IoT Sensor layer:** Following are a few sensors used in the selected four use cases: smart surveillance (night vision security camera), smart maintenance and environment (digital temperature and humidity (DHT11), NFC reader (EZ430), motion detection (HC-SR 501), ultrasonic (HC-SR 04), gas leakage (MQ2), air quality control (MQ135), fire safety), and smart concierge (ultrasonic (HC-SR 04), motion detection (HC-SR 501), light sensing (LDR)).
- **Connectivity layer:** The aforementioned actuator nodes can relay the useful information through any of the communication technologies mentioned in Sect. 3.2. Apart from the default support of Wi-Fi and BLE, the 26 dedicated general-purpose input-output (GPIO) pins available in Raspberry Pi 4 can be used to connect a ZigBee coordinator (XeBee module). Also, Z-wave coordinator can be connected through the available USB port. Due to the comprehensive security and privacy features, in this work, we aim to utilize BLE. For this purpose, we utilize ESP32 wireless micro controller developed by Espressif, which contains an integrated Wi-Fi/BLE chip-set, RF core, amplifier, power management module and a builtin antenna. ESP32 can be programmed using Arduino IDE and can establish a secure connection to a MQTT broker.
- **Fog layer:** In IoTFC (Fig. 1), fog nodes are deployed at each apartment and office areas of the building closed to IoT devices for an efficient real-time processing and data analytics at the edges of the network. Each of these fog node will be connected to the sensors monitoring temperature, humidity,

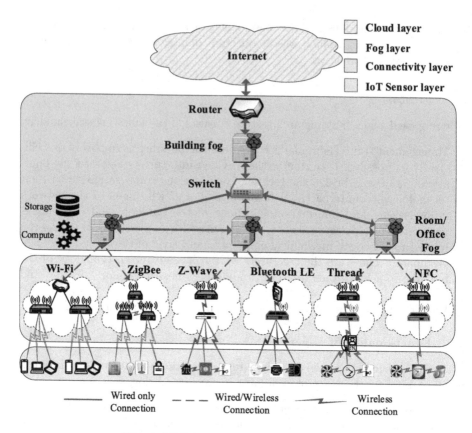

Fig. 2. IoTFC Smart building architecture.

air quality, occupancy, energy usage, and numerous other factors via Wi-Fi, Bluetooth or Zigbee. The fog nodes will have enough analytical power to perform real-time actions such as conserving plug load and HVAC usage when any room is unoccupied, to learn an occupant's preferred temperature and maintain the space at that comfort level, to perform access control, other security (cryptographic) operations, fault tolerance of a room, and privacy-preserving operations, and provide storage (for 1–2 h), and to make an escape route in an emergency situation. By deploying these services in the fog layer, lower latency and improved QoS can be achieved to deliver outstanding user experience. These nodes will be associated with the building's connectivity access infrastructure (wired or wireless), and will be able to perform continuous security scan of all devices connected to them. A building fog node communicates with all the apartment/office area fog nodes and takes slower, more deliberate actions, such as to set equipment maintenance schedules, optimize load balancing (also for connectivity layer) and fault tolerance of floor, move applications around if any apartment/office area fog node fails or becomes overloaded, and share with the cloud any information that requires non-critical decision making or very high storage space in tens of terabytes. The apartment/office area fog nodes and the building fog node is assumed to be Raspberry Pi 4, which runs a Linux distribution called Raspbian and utilizes OpenHAB (popular IoT integration platform) with additional bindings (to extend the functionality for parallel integration of different systems and devices) so as to create a distributed fog environment. Every fog node is assigned a static IP address and is connected to a common Ethernet switch, resulting in formation of a network. The building fog node will act as a master node connected to cloud and will be connected to the Internet via USB-Ethernet adaptor. Data from the IoT devices to the cloud are passed through the respective fog nodes using MQTT broker. It is pertinent to highlight here that the security and privacy solutions proposed in this paper can be added as additional bindings to OpenHAB.

- **Cloud layer:** Data from the fog layer that is less time sensitive is sent to the cloud for long-term storage, historical analysis, and big data analytics to determine operational aspects of the building. Over period of time, the cloud performs analysis on the received data from fog nodes to gain business insight, and based on these insights can send new or update application rules to the fog nodes to further optimize building operations.

4.3   Security and Privacy Solutions

There are mainly four kinds of data services in fog computing: data storage, data sharing, data query, and data computation. All these services demand different unique data security (access control, authorization revocation, availability, confidentiality of inputs, outputs and computing tasks) and privacy (identity, query, location, and preventing tracking and profiling) requirements. In IoTFC architecture, we propose light-weight solutions to mitigate the security and privacy issues mentioned in Table 1 for the four use cases.

Security in IoTFC. Three-tier light-weight security model is proposed in the following:

Access Control: In multi-users fog computing environment such as a smart building, permissibility of certain network or data resources must be allowed to only those users or devices, who possess certain rights to use the requested resources. The existing access control mechanisms (mentioned in Sect. 2) offer some advantages and a few limitations, e.g., RBAC model is more scalable than the DAC and MaC models. However, in RBAC, roles are assigned statically, whereas, in order to fulfill fog access control requirements (latency, efficiency, resource restriction, policy management), RBAC model can be evolved to support dynamic environment since the user's access privileges not only depend on "who the user is" but also on "where the user is" and "what is the user's state and the state of the user's environment". In IoTFC, we propose a dynamic RBAC (dRBAC) model in IoTFC architecture that dynamically grants and adapts permissions to users according to context. Since the fog devices are resourceful and are used closed to IoT devices, therefore, there is no need to deploy access control mechanism at IoT sensors layer.

dRBAC consist of the following elements: 1) subject: an entity such as user, fog node, etc. that accesses objects; 2) object: any computing or data resources; 3) roles: an entity associated with specific authority and responsibility, e.g., building management, occupants, visitors, etc.; 4) permissions: operations such as read, write, delete, etc. that bind objects and a set of actions that can be executed by a subject; 5) session: a set of interactions between subjects and objects; 6) context agent: an entity that collects context information such as domain, hierarchical role (mapping between inheritance relationships and roles, e.g., operational staff inherits the permissions of maintenance staff), and location; 7) user assignment list (*UA*): mapping between roles and subjects; 8) permission assignment list (*PA*): mapping between permissions and roles, and 9) conflict-of-interest (*COI*): a list mapping conflicting and non-conflicting permissions to roles.

In dRBAC, a policy maker center (PMC) is deployed at the building fog node that creates and maintains *UA*, *PA* and *COI*. The intermediate fog nodes (apartment/office) act as sub-coordinators and perform part of the policy decision-making tasks delegated by the building fog node. To successfully access building services or objects, a subject initially sends access right request to the PMC to get an access token. Given registered subject information established in profile database, the PMC evaluates the access request by enforcing defined authorization policy rules. If the access request is granted, the access token encapsulating access right is generated with issuer's signature and sent back to the subject. Otherwise, the access right request is rejected. Then, the subject can request access to other services or objects by presenting a valid access token to the service provider.

For each query, the coordinator or sub-coordinators would perform the following evaluation:

$$
\begin{aligned}
User(?user) \wedge Role(?role) \wedge Permission(?action) \wedge \\
Session(?activeorinactive) \wedge Context(?attributes) \wedge \\
Conflict(?action) \wedge Accessdecision(?decision)
\end{aligned}
\tag{1}
$$

For example, the subject (building resident) using his/her access token looks for an available meeting room by checking the meeting room schedule (that gets updated from the occupancy sensor). The coordinator of the office area verifies the access right policies and conditional constraints against the provided access token. If satisfied, it grants the access request to the resident, who can book the available time-slot of the meeting room. The building resident can update the schedule, but he/she is not allowed to delete or change any other information in the schedule such as deleting other booked time-slots.

Authentication: As services in IoTFC are offered to a large number of building users by fog nodes, authentication becomes a critical issue. Fog nodes need to authenticate at different levels so as to ensure security. Also, in an IoT multi-user fog environment, the IoT sensors layer has two types of devices, mobile and fixed IoT devices. The mobile IoT devices are carried by their owners (e.g., smart phones, tablet, NFC tags etc.), while the fixed IoT devices (e.g., occupancy sensors, NFC reader, etc.) are pre-deployed in specific areas to provide services. This mobility of users creates a problem in designing identity authentication with minimal latency. Hence, in IoTFC architecture, we need to design an identity control mechanism at three layers (IoT sensors, fog, cloud) with minimal latency.

The device level authentication is performed by running MQTT broker (open source Mosquitto) on the apartment/office fog node, which is a Raspberry Pi 4 and thus can act as both publisher and subscriber. A token authentication (*TA*) server is assumed to be deployed in IoTFC, which is responsible to issue a token and check its validity against any authentication request. The device first sends its credential i.e., valid ID (universal unique identifier (UUID) or a MAC address), and username to *TA* server for obtaining a token. Once the authentication server checks the requested credentials against its database, it then returns a valid token containing the header, payload and signature. Upon receiving the token, the device sends the "CONNECT" message to the broker by providing the access token as its username and the password. The broker can use the token to perform various validations, such as: check the validity of the signature, check the expiry date of the token, and check the token authentication server to see whether the token was revoked. If token is valid, the device is then allowed to publish the sensor data. MQTT payload encryption is performed between the publisher and the broker for secure communication.

In order to perform fog node authentication, first it needs to be registered in IoTFC. To perform registration, *TA* chooses a unique identity (ID_{FDi}) for each fog device (FD_i), and then calculates its pseudo identity $PID_{FDi} = H(K||ID_{FDi})$ using the its own secret key K, and generates corresponding temporary credentials $TC_{FDi} = H(K||TS||ID_{FDi}||nonce)$, where TS is the timestamp. *TA* stores these credentials in FD_i's database before it is deployed. Similar procedure is deployed for cloud server registration. Once registered, the fog nodes can communicate with other fog nodes or cloud through a secure key management protocol based on these temporary credentials to establish a secret key for secure communication.

Anomaly Detection: In order to identify events that appear to be anomalous in nature with respect to the normal behavior, IoTFC will utilize anomaly based intrusion detection system (AIDS) and network based intrusion detection system (NIDS). All sensor devices will be analyzed by AIDS of respective fog nodes (i.e., anomaly detection operates in a distributed manner). In the event of an unusual event, an alarm is sent by a fog node to the building fog node and to the IoT sensors. AIDS, operating at apartment/office fog node, comprises of a learning module (contains traffic patterns observed dynamically, such as, request count, failed authentication count, device usage at different time periods, bandwidth consumed, etc. and a known behavior models for different attacks, such as DDoS, TCP flooding, etc.), Classification module (Support Vector Machine (SVM) to

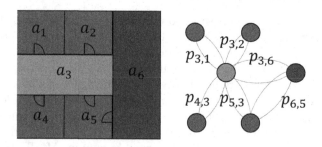

Fig. 3. Swap Areas and Markov Chain obtained from their measurements.

classify and detect anomaly) and flow management (methods to block a flow in the event of true detection and blacklisting of the node). NIDS, implemented at the building fog node, will monitor the entire network traffic and can centrally detect a malicious activity by using a light-weight SVM algorithm. The overall hierarchical approach will allow detection of both malicious end nodes as well as a malicious fog node. In addition, this procedure will ensure no access to a node black listed within the network.

Privacy in IoTFC. To guarantee privacy in case data stored for smart surveillance, "pixelating technique" will be used to automatically obscure the identities of persons, while all of their movements and actions remain recognizable. This approach provides important information to support decisions, appropriately plan interventions, and minimize damage and costs; all without violating the privacy of individuals.

To protect device level data, we propose to aggregate room sensors' measurements, called *Area nodes* $(a_1, a_2, \ldots, a_6)$ as shown in Fig. 3, that will act as Mix Zones [2]. Additional privacy can be provided by increasing the size of the Mix Zones by joining adjacent locations (e.g., aggregating a_1 and a_2 nodes) in Fig. 3 to obtain a super node a_{12}, which may also guarantee k-anonymity. With IoTFC, we also provide privacy for location-based services, users are protected from re-identification by the servers and from inference of their interests and habits from the location information contained in their queries [13]. We adopt a solution from [12] that is suited for IoTFC, in which the operation is performed by a component called *SwapMob*, which is integrated to the area nodes. We also consider privacy profiles for each of the user roles used for role-based access control. To comply with consent acquisition (as required by GDPR), we modify the SwapMob collection mechanism to a role-based swapping in which the fog nodes will only swap the currentIDs depending some pre-specified rules on the roles.

For example, in a smart building, a technician that comes to the building for maintenance will not expect to have the same privacy as an inhabitant or permanent worker from the building, however if there is a team of technicians all of them would be expected to have the same level of privacy. Therefore, in this case the Fog node may swap their currentIDs but only among the technicians or inhabitants, not between a inhabitant and a technician.

The SwapMob mechanism will carry out the privacy protection for all mobile IoT devices. It preserves the sufficient statistics for discrete time Markov chains specified by time interval τ and spatial resolution χ, hence will preserve the Global Markov Model [4]. It will be used for predicting locations, and analyzing collective patterns of mobility inside the building. All of these tasks will be performed without the need to know the specific locations for particular individuals. In Fig. 3, the global mobility Markov chain is represented as a directed graph in which $p_{i,j}$ denotes the probability of a device that is in area a_i to move to area a_j.

5   Comparative Analysis

This section carries out a comparative analysis of the proposed architecture with the relevant state-of-the-art smart building solutions. The comparison, summarized in Table 3, considers the following properties: access control, authentication, anomaly detection, data and location privacy, and middleware.

Table 3. Comparison of IoTFC with the related state-of-the-art smart building solutions.

Properties	Lilis et al. [10]	Ramos et al. [9]	Wissam et al. [14]	Pappachan et al. [11]	Ferrandez et al. [7]	Boyer [3]	IoTFC
Access control	✓	✓	✓	✗	✗	✗	✓
Authentication	✓	✓	✓	✗	✗	✗	✓
Anomaly detection	✓	✗	✗	✗	✗	✗	✓
Data privacy	✗	✓	✓	✓	✓	✓	✓
Location privacy	✗	✗	✗	✓	✗	✓	✓
Middleware	✓	✓	✓	✗	✓	✗	✓

Table 3 shows that IoTFC offers a secure and privacy-preserving middleware architecture. In comparison, the architecture in [10] offers security properties but fails to provide data and location privacy. The secure middleware architectures proposed in [9,14] offer access control, authentication and data privacy, while fail to provide anomaly detection and location privacy. The systems proposed in [11] and [3] fail to provide the guaranteed security properties but offer both data and location privacy. In [7], the authors have proposed a middleware architecture that ensures data privacy while fail to provide any other security property.

6   Conclusions

In this paper, we have proposed IoTFC, a novel secure and privacy preserving architecture for smart building based on fog computing. This paper is a first step towards implementing both security and privacy by design in a smart city use case. Most of the available smart building solutions are proprietary based, but IoTFC has been designed with open source communication technologies and standards to provide interoperability. Light weight access control and authentication protocols have been proposed to provide resilience against known security IoT attacks. The privacy solution provides means of protecting occupants from profiling and tracking. In future, we aim to provide a complexity, security and privacy analysis of IoTFC.

References

1. Antonini, A., Maggi, F., Zanero, S.: A practical attack against a KNX-based building automation system. In: ICS and SCADA Cyber Security Research, pp. 53–60 (2014)
2. Beresford, A.R., Stajano, F.: Location privacy in pervasive computing. IEEE Pervasive Comput. **2**(1), 46–55 (2003)
3. Boyer, J.P., Tan, K., Gunter, C.A.: Privacy sensitive location information systems in smart buildings. In: Clark, J.A., Paige, R.F., Polack, F.A.C., Brooke, P.J. (eds.) SPC 2006. LNCS, vol. 3934, pp. 149–164. Springer, Heidelberg (2006). https://doi.org/10.1007/11734666_12
4. Chen, M., Liu, Y., Yu, X.: NLPMM: a next location predictor with Markov modeling. In: Tseng, V.S., Ho, T.B., Zhou, Z.-H., Chen, A.L.P., Kao, H.-Y. (eds.) PAKDD 2014. LNCS (LNAI), vol. 8444, pp. 186–197. Springer, Cham (2014). https://doi.org/10.1007/978-3-319-06605-9_16
5. Costante, E.: Gsmart buildings: trends and challenges for a secure future (2017). https://www.forescout.com/company/blog/smart-buildings-trends-and-challenges-for-a-secure-future/
6. Faro, C.: Nearly four in ten smart buildings targeted by malicious attacks in h1 (2019). https://usa.kaspersky.com/about/press-releases/2019_smart-buildings-threat-landscape
7. Ferrández-Pastor, F.J., Mora, H., Jimeno-Morenilla, A., Volckaert, B.: Deployment of IoT edge and fog computing technologies to develop smart building services. Sustainability **10**(11), 3832 (2018)
8. Froiz-Míguez, I., Fernández-Caramés, T.M., Fraga-Lamas, P., Castedo, L.: Design, implementation and practical evaluation of an IoT home automation system for fog computing applications based on mqtt and zigbee-wifi sensor nodes. Sensors **18**(8), 2660 (2018)
9. Hernández-Ramos, J.L., Moreno, M.V., Bernabé, J.B., Carrillo, D.G., Skarmeta, A.F.: SAFIR: secure access framework for IoT-enabled services on smart buildings. J. Comput. Syst. Sci. **81**(8), 1452–1463 (2015)
10. Lilis, G., Kayal, M.: A secure and distributed message oriented middleware for smart building applications. Autom. Constr. **86**, 163–175 (2018)
11. Pappachan, P., et al.: Towards privacy-aware smart buildings: capturing, communicating, and enforcing privacy policies and preferences. In: IEEE ICDCSW, pp. 193–198 (2017)
12. Salas, J., Megías, D., Torra, V.: SwapMob: swapping trajectories for mobility anonymization. In: Privacy in Statistical Databases, pp. 331–346 (2018)
13. Shin, K.G., Ju, X., Chen, Z., Hu, X.: Privacy protection for users of location-based services. IEEE Wirel. Commun. **19**(1), 30–39 (2012)

14. Wissam, R., Daniele, S., Kouichi, S.: A new security middleware architecture based on fog computing and cloud to support IoT constrained devices. In: IML 2017, pp. 1–8 (2017)
15. Ziegeldorf, J.H., Morchon, O.G., Wehrle, K.: Privacy in the Internet of Things: threats and challenges. Secur. Commun. Netw. **7**(12), 2728–2742 (2014)

A Secure Experimentation Sandbox for the Design and Execution of Trusted and Secure Analytics in the Aviation Domain

Dimitrios Miltiadou[1]($\boxtimes$) (iD), Stamatis Pitsios[1] (iD), Dimitrios Spyropoulos[1] (iD), Dimitrios Alexandrou[1] (iD), Fenareti Lampathaki[2] (iD), Domenico Messina[3] (iD), and Konstantinos Perakis[1] (iD)

[1] UBITECH, Thessalias 8 & Etolias, 15231 Chalandri, Greece
dmiltiadou@ubitech.eu
[2] SUITE5, Alexandreias 2, Bridge Tower, 3013 Limassol, Cyprus
[3] ENGINEERING Ingegneria Informatica S.p.A., Piazzale dell Agricoltura 24, 00144 Rome, Italy

Abstract. The undergoing digital transformation of the aviation industry is driven by the rise of cyber-physical systems and sensors and their massive deployment in airplanes, the proliferation of autonomous drones and next-level interfaces in the airports, connected aircrafts-airports-aviation ecosystems and is acknowledged as one of the most significant step-function changes in the aviation history. The aviation industry as well as the industries that benefit and are highly dependent or linked to it (e.g. tourism, health, security, transport, public administration) are ripe for innovation in the form of Big Data analytics. Leveraging Big Data requires the effective and efficient analysis of huge amounts of unstructured data that are harnessed and processed towards revealing trends, unseen patterns, hidden correlations, and new information, and towards immediately extracting knowledgeable information that can enable prediction and decision making. Conceptually, the big data lifecycle can be divided into three main phases: i) the data acquisition, ii) the data storage and iii) the data analytics. For each phase, the number of available big data technologies and tools that exploit these technologies is constantly growing, while at the same time the existing tools are rapidly evolving and empowered with new features. However, the Big Data era comes with new challenges and one of the crucial challenges faced nowadays is how to effectively handle information security while managing massive and rapidly evolving data from heterogeneous data sources. While multiple technologies and techniques have emerged, there is a need to find a balance between multiple security requirements, privacy obligations, system performance and rapid dynamic analysis on diverse large data sets. The current paper aims to introduce the ICARUS Secure Experimentation Sandbox of the ICARUS platform. The ICARUS platform aims to provide a big data-enabled platform that aspires to become an "one-stop shop" for aviation data and intelligence marketplace that provides a trusted and secure "sandboxed" analytics workspace, allowing the exploration, curation, integration and deep analysis of original, synthesized and derivative data characterized by different velocity, variety and volume in a trusted and fair manner. Towards this end, a Secure Experimentation Sandbox has been designed and integrated in the

D. Wang et al. (Eds.): SPNCE 2020, LNICST 344, pp. 120–134, 2021.
https://doi.org/10.1007/978-3-030-66922-5_8

holistic ICARUS platform offering, that enables the provisioning of a sophisticated environment that can completely guarantee the safety and confidentiality of data, allowing to any interested party to utilize the platform to conduct analytical experiments in closed-lab conditions.

Keywords: Big data · Security · Privacy · Cybersecurity · Data analytics

1 Introduction

The Aviation industry encapsulates the design, development, production, operation and management of aircrafts with a wide range of products ranging from aircraft, rotorcraft, engines, avionics and systems to leading operations and services. The undergoing digital transformation of the aviation industry is driven by the rise of cyber-physical systems and sensors and their massive deployment in airplanes, the proliferation of autonomous drones and next-level interfaces in the airports, connected aircrafts-airports-aviation ecosystems and is acknowledged as one of the most significant step-function changes in the aviation history. The number of data generating sensors fitted in new aircrafts increases in recent years as it obvious from the report from Airbus that the A350 aircraft model is equipped with 50000 sensors on board collecting 2.5 terabytes of data per day [1], while at the same time according to estimations the global fleet could generate up to 98,000,000 terabytes of data by 2026 [2]. The aviation industry as well as the industries that benefit and are highly dependent or linked to it (e.g. tourism, health, security, transport, public administration), are ripe for innovation in the form of Big Data analytics.

Leveraging Big Data requires the effective and efficient analysis of huge amounts of unstructured data that are harnessed and processed towards revealing trends, unseen patterns, hidden correlations, and new information, and towards immediately extracting knowledgeable information that can enable prediction and decision making [3, 4]. Big data technologies is a new generation of technologies that aims to add value to a massive volume of data with various formats by enabling high-velocity capture, discovery, and analysis [5]. Conceptually, the big data lifecycle can be divided into three main phases: i) the data acquisition, ii) the data storage and iii) the data analytics. For each phase, the number of available technologies and tools that exploit these technologies is constantly growing, while at the same time the existing tools are rapidly evolving and empowered with new features. However, the Big Data era, comes with new challenges and one of the crucial challenges faced nowadays is how to effectively handle information security while managing massive and rapidly evolving data from heterogeneous data sources. Inevitably, the data analytics and extraction of intelligence from them requires dynamic data sharing between different internal stakeholders of an organization or even between external stakeholders, as well as data access to all these stakeholders. However, this introduces multiple security threats such as the discovery of private or confidential information and unauthorized access to data at storage or data in motion. While multiple technologies and techniques have emerged, there is a need to find a balance between multiple security requirements, privacy obligations, system performance and rapid dynamic analysis on diverse large data sets [6].

The current paper aims to introduce the ICARUS Secure Experimentation Sandbox of the ICARUS platform. The ICARUS platform aims to provide a big data-enabled platform that aspires to become an "one-stop shop" for aviation data and intelligence marketplace that provides a trusted and secure "sandboxed" analytics workspace, allowing the exploration, curation, integration and deep analysis of original, synthesized and derivative data characterized by different velocity, variety and volume in a trusted and fair manner. Towards this end, a Secure Experimentation Sandbox has been designed and integrated in the holistic ICARUS platform offering, that enables the provisioning of a sophisticated environment that can completely guarantee the safety and confidentiality of data, allowing to any interested party to utilize the platform to conduct analytical experiments in closed-lab conditions. Therefore, the main contribution of our work is to provide a trusted and secure sandboxed analytics workspace that effectively and securely addresses the critical barriers for the adoption of Big Data in the aviation industry, and enables aviation-related big data scenarios for EU-based companies, organizations and scientists. We present a holistic security approach that capitalizes on the latest advancements and modern technological breakthroughs in the areas of Big Data, Data Analytics and Security in order to accommodate the needs of the data analytics stakeholders for secure access, processing and data analysis of big data.

2 Materials and Methods

In spite of the important developments in the big data technologies, an analysis of these technologies in respect to the adopted architectures and techniques revealed that many shortcomings in terms of security still exist [7]. Furthermore, the importance of security and privacy measures is increasing, along with the growth in the generation, access, and utilization of Big Data [8]. The typical characteristics of big data, namely velocity, volume and variety associated with large-scale cloud infrastructures and the Internet of Things (IoT) revealed the inadequacy of the traditional security and privacy mechanisms as they fail to cope with the rapid data explosion in such a complex distributed computing environment, as well as with the scalability, interoperability and adaptability of contemporary technologies that are required for big data [9]. At the same time, the number of malicious attacks against big data infrastructure is on the rise, as it was revealed by recent surveys focused on the security aspect of big data [10]. Gathering, storing, searching, sharing, transferring, analyzing and presenting data as per requirements are the major challenging task in big data [11]. In any big data platform, the strategic priorities related to security should be clearly defined and the guidelines for choosing the associated technologies in terms of reliability, performance, maturity, scalability and overall cost should be also clearly established to ensure that the design platform provide the necessary security mechanisms that include, among others, the anonymization of confidential or personal data, the data cryptography, the centralized security management, the data confidentiality and data access monitoring [6]. At the same time, it should ensure their future evolutions will be easily integrated in the existing solution.

Fortunately, the recent advancements and trends in the big data technologies and the adopted strategies provide a new compelling opportunity to design and build a big data platform that incorporates a holistic security approach capable of addressing the security and privacy challenges imposed by the nature of big data and the requirements of data analytics stakeholders.

2.1 The ICARUS Technical Solution

The main objective of the ICARUS platform [12] is to provide a multi-sided platform that will allow exploration, curation, integration and deep analysis of original, synthesized and derivative data characterized by different velocity, variety and volume in a trusted and fair manner. Furthermore, the platform builds a novel data value chain in the aviation-related sectors towards data-driven innovation and collaboration across currently diversified and fragmented industry players, acting as multiplier of the "combined" data value that can be accrued, shared and traded, and rejuvenating the existing, increasingly non-linear models/processes in aviation.

The key objectives of the platform is to offer a scalable and flexible big data-enabled environment that provides secure and trusted: a) data preparation and data upload, b) data exploration, data sharing and data brokerage and c) data analysis execution and visualization generation capabilities. Security and privacy were considered as crucial pillars of the ICARUS platform. To this end, a security and privacy by-design approach was adopted in order to effectively and efficiently cover all the aspects related to the information protection and secure data management over the entire data lifecycle. In this context, the ICARUS platform incorporates advanced security mechanisms that offer methods for increasing the security, the privacy and the data protection across all tiers of the architecture, taking into consideration the aviation industry's needs, requirements and peculiarities with regards to the security of information. The key points of this approach is the adoption of the end-to-end encryption imposed in all the datasets that are stored in the platform, as well as a secure decryption process for the effective data sharing of datasets within the scope of the platform.

The ICARUS platform's architecture is a modular architecture that provides enhanced flexibility and is composed of a set of key components that are built on top of efficient and state-of-the-art big data infrastructure, technologies and tools, maximizing the benefits of their effective combination. In detail, the platform architecture is composed by 22 key components that have been designed with the aim of delivering specific business services with a clear context, scope and set of features (see Fig. 1).

The components are conceptually organized in three main tiers, the **On Premise Environment**, the **Core Platform** and the **Secure and Private Space**. Each tier is undertaking a set of functionalities of the platform depending on the execution environment and context. The scope of the On Premise Environment is to provide the required services that will perform all the data preparation steps as instructed by the Core Platform. In this context, the On Premise Environment is composed by multiple components that are running on the data provider's environment with the main purpose to prepare the data provider's private or confidential datasets in order to be uploaded in the platform. The On Premise Environment undertakes the responsibility of performing the tasks according to the instructions provided by the Core Platform.

The scope of the Core Platform is to provide all the required components for the execution of the core operations of the platform, as well as compilation of the instructions that are executed by the On Premise Environment and the Secure and Private Space.

The Core Platform is composed of multiple interconnected components running on the platform's cloud infrastructure. It performs all core operations of the platform while also orchestrating and providing the instructions that are executed by the On Premise Environment and the Secure and Private Space. Furthermore, the Core Platform provides the only user interface of the platform as the rest of the tiers are incorporating only backend services.

The scope of the Secure and Private Space is to provide all the required components for the formulation of the trusted and secure advanced analytics execution environment of the platform. In order to cope with the emerging security and privacy requirements, the Secure and Private Space is providing a trusted and secure sandboxed analytics workspace supporting the existing rich stack of analytics tools and features, while at the same time providing strong security guarantees towards data confidentiality and data privacy. In this context, the Secure and Private Space contains a set of interconnected components that constitute the advanced analytics execution environment of the platform, whose management and orchestration is performed through the Core Platform. It provides the trusted and secure environment where the data analysis executed in accordance with the analytics workflow that is designed by the user within the Core Platform. The designed workflow is translated into a set of instructions which are executed by the responsible deployed components.

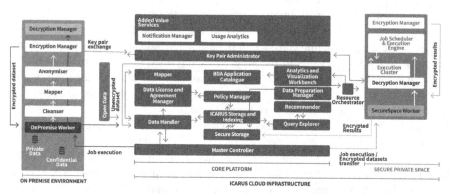

Fig. 1. The ICARUS platform conceptual architecture

The following section, focuses on the ICARUS Secure Experimentation Sandbox of the ICARUS platform's architecture that constitutes the trusted and secure sandboxed analytics workspace of the platform which facilitates the secure accessing and processing of big data towards the execution of advanced data analytics and visualizations over a modular and scalable architecture. The ICARUS Secure Experimentation Sandbox constitutes a cross-tier feature of the platform, that is orchestrated and controlled by the Core Platform tier and executed by the Secure and Private Space tier.

2.2 The ICARUS Secure Experimentation Sandbox

Data analytics rely on the effective aggregation and correlation of diverse multi-source data in order to generate new insights and knowledge. However, they are usually offered in conventional and in most cases through strictly defined fixed queries, constraining the fantasy of end-users and their experimentation potential which could lead to new knowledge insights in unexplored dimensions. At the same time, in the data storage and the data analytics phases of the big data lifecycle several security and privacy challenges arise as described in Sect. 2.1 which is imperative that they should be properly addressed so that data analytics will be embraced by the organizations.

The ICARUS platform is by design supporting data security and privacy principles, to safeguard personal data, but also business critical data. Towards to this end, one of the core features of the platform is the provisioning of a sophisticated environment that can completely guarantee the safety and confidentiality of data, allowing to any interested party to utilize the platform to conduct analytical experiments in closed-lab conditions with the offering of the ICARUS Secure Experimentation Sandbox. The ICARUS Secure Experimentation Sandbox architecture holds a key role in the platform's solution, covering the needs for data analysis execution and visualization generation in a secure and trusted manner, effectively addressing the security and privacy challenges imposed by the nature of the big data, capitalizing on the latest advancements in the techniques and technologies that enable easy, fast and secure deployment of containerized execution environments.

The design of the ICARUS Secure Experimentation Sandbox is composed by a set of components residing on the Core Platform tier, namely the Resource Orchestrator, the Secure Storage, the Analytics and Visualization Workbench and the Data Preparation Manager, as well as all the components that formulate the Secure and Private Space tier, namely the SecureSpace Worker, the Job Scheduler and Execution Engine, the Encryption Manager and the Decryption Manager. In the following paragraphs, all the core components that are utilized in the realization of the ICARUS Secure Experimentation Sandbox, as well as their interactions, are presented in detail.

Resource Orchestrator. The ICARUS Secure Experimentation Sandbox provides the deployment of a scalable isolated environment running on virtualized infrastructure and is based on technologies that enable easy, fast and secure deployment of containerized execution environments over virtualized infrastructure. The concept of containerized execution environments besides the portability and interoperability features, it also enables the monitoring, autoscaling and management of the deployed applications with the proper orchestration support. To this end, at the heart of the ICARUS Secure Experimentation Sandbox lays Docker, which is the container technology with the widest adoption by both the research and vendor communities. Docker containers are lightweight, standalone, self-contained systems that include everything that is needed for the proper execution of the system on a shared operating system such as code, runtime, system tools, system libraries and settings in an isolated manner. The container orchestration is handled by Kubernetes, which is also the most dominant open source platform for container orchestration, offering advanced monitoring capabilities, autoscaling and state-of-the-art management of deployed applications. Kubernetes encapsulates the concept of container virtualization thus enabling the deployment of containerized applications over virtualized infrastructure with extended orchestration support. Kubernetes is offering an additional layer of abstraction on top of the virtualized infrastructure providing out of the box service discovery and load balancing, storage orchestration, automated rollouts and rollbacks, automatic bin packing based on the available resources, health monitoring and self-healing mechanisms, as well as security and isolation between the deployed applications. The Resource Orchestrator is the component responsible for the provisioning of the Secure and Private Space, performing the deployment and monitoring of the required containerized services by exploiting the capabilities of Kubernetes and Docker.

Secure Storage. The Secure Storage undertakes the role of the private and secure storage of the ICARUS Secure Experimentation Sandbox. In this context, the Secure Storage component has a two-fold purpose: a) to store the data provider's private or confidential datasets, as well as the datasets acquired via the marketplace of the platform, prior to being used during the data analysis execution by the Secure and Private Space services and b) to store the data generated as a result of a data analysis performed by the Secure and Private Space services. In this process, the Secure Storage provides the private storage "spaces" with restricted access only to the owner of the "space". The design of the Secure Storage aims to better address the user requirements for the effective and efficient storage of datasets with high availability and high performance for the computationally-intense data analysis operations. Hence, the Secure Storage is composed by the Hadoop Distributed File System (HDFS), provisioned and monitored by the Apache Ambari, and the Apache Hive supported by the Presto query engine. The HDFS is utilized for the storage of the datasets that will be utilized in the data analysis. Apache Hive is operating on top of the HDFS as a data warehouse software that enables an abstraction layer on top of the datasets residing in HDFS, facilitating the complete data lifecycle management of the datasets, as well as providing an SQL-like interface enabling data query and analysis functionalities. Presto is offering the required high-performance query engine on top of the datasets by performing parallel query execution over a pure memory-based architecture running on top of Hive, enabling fast analytics queries against data of any size that are residing in Hive and consequently HDFS.

The SecureSpace Worker residing in the Secure and Private Space is interacting with the Core Platform via the Master Controller to fetch the required dataset, that are afterwards decrypted by the Decryption Manager and are finally stored in the Secure Storage. Thus, the Secure Storage is providing the means to the Analytics and Visualization Workbench component in order to access the transferred datasets and utilize them in the data analysis in an efficient and effective manner. Furthermore, the data generated in the Secure and Private Space upon the completion of a data analysis are encrypted by the Encryption Manager and provided through the SecureSpace Worker to the Secure Storage for storage. The encrypted results are fetched by the SecureSpace Worker and decrypted by the Decryption Manager in order to be provided as input for the visualization process that is executed by the Analytics and Visualization Workbench. Finally, the Secure Storage is enabling the data preparation operations performed by the Data Preparation Manager by enabling the access and manipulation of multiple datasets residing on the Secure Storage within the owner's private storage "space".

Analytics and Visualization Workbench. The Analytics and Visualization Workbench is the component enabling the design, execution and monitoring of the data analytics workflows within the platform and also where the visualization and dashboards are displayed. The users are able to select an algorithm from the extended list of supported algorithms and set the corresponding parameters according to their needs. Furthermore, the users are provided with scheduling capabilities for the execution of the designed analytics workflow in a selected date and time. In terms of design, the Analytics and Visualization Workbench consists of three main sub-components: (a) a very intuitive graphical user interface that allows users to exploit a set of machine learning algorithms to obtain meaningful information about his/her own data and display them with a set of built-in charts; (b) the repository for the implemented algorithms and (c) a microservice that exposes a RESTful API that enables the mediation between the frontend and the Job Scheduler and Execution Engine and the Resource Orchestrator backend components that the Analytics and Visualization Workbench interacts with. As the nature of the component includes both a graphical user interface and a set of backend functionalities, different technologies are utilized for each purpose adopting the backend-for-frontend design pattern. The graphical user interface is mainly based on TypeScript, JavaScript, CSS and HTML5, while for the backend functionalities of the service Node.js is used. Node.js is offering the mechanism for the implementation of frequent I/O bound operations, while also offering integration capabilities for a web client implementation.

While the design of the data analytics workflow is performed in the Analytics and Visualization Workbench, the execution of the analysis is performed within the Secure and Private Space by interacting with the Resource Orchestrator and the Job Scheduler backend components to ensure the existence of the Secure and Private Space and initiate the application execution (or schedule its execution via the respective interfaces offered by the Job Scheduler). During the execution of the analysis, it interacts via the Master Controller with the SecureSpace Worker and the Decryption Manager before the actual execution in order to ensure the transferring of the selected datasets to the Secure Space. Additionally, the user is able to create an application, which contains the list of datasets that were selected for analysis, the selected algorithm, as well as the selected visualization type, along with the corresponding parameters, and store it in the platform's application catalogue for later reuse. Furthermore, through the Analytics and Visualization Workbench and the interactions with the SecureSpace Worker and the Decryption Manager the advanced visualization capabilities of the platform are offered with a modern data visualization suite of charts and visualizations that span from basic charts to advanced multilevel visualizations.

Data Preparation Manager. The Data Preparation Manager is the component that offers the required data manipulation functionalities on top of the available datasets in order to make them suitable for consumption by the Analytics and Visualization Workbench. The Data Preparation Manager is taking the datasets of the user as input in order to apply a series of data manipulation steps based on the user needs and the results are fed in the Analytics and Visualization Workbench for data analysis or visualization. The Data Preparation Manager allows the user to create various data preparation jobs by defining on each of them the list of datasets that will be used as the basis for the data preparation and the sequence of data manipulation steps that should be followed in order to transform the original dataset into the one that the user needs.

The main data manipulation functionalities offered can be grouped into: a) column creation (timestamp-related, math-related, aggregation related, shift and conditional operations), b) column drop, c) row filtering, d) column renaming, e) dataset merging, f) null value fill-in operations and g) compute aggregations. The Data Preparation Manager performs all the designed data manipulation process within the Secure Storage as Presto queries and the newly created datasets are also stored with the Secure Storage with the appropriate data access control measures. Hence, the Data Preparation Manager that resides on the Core Platform directly interacts with the Secure Storage towards the preparation of the datasets that will be utilized as input for the data analysis or the visualization process as designed by the Analytics and Visualization Workbench.

SecureSpace Worker. The SecureSpace Worker resides at the Secure and Private Space and is responsible for the job or task execution that is related to the data analysis that is performed on the Secure and Private Space. The SecureSpace Worker is the component that is undertaking the local execution of the jobs as instructed by the Master Controller residing on the Core Platform, utilizing the deployed services on the local running environment of the Secure and Private Space. Hence, the SecureSpace Worker is tightly connected with the Master Controller component towards the realization of the Master/Worker paradigm that is adopted in the ICARUS platform architecture. The SecureSpace Worker is enabling the transfer of the selected encrypted datasets that will be utilized in the data analysis in the Secure Storage. Additionally, it receives a set of instructions that includes the decryption process of the selected datasets as performed by the Decryption Manager, the analytics job execution that is performed by the Jobs Scheduler and Execution Engine and the encryption of the produced results of the analytics as executed by the Encryption Manager. Moreover, the SecureSpace Worker is responsible for providing the encrypted results back to the Secure Storage. For the implementation of the SecureSpace Worker the Spring Roo framework, which provides the easy-to-use rapid development tool for building application in the Java programming language with a large range of features and integration capabilities, is leveraged.

Jobs Scheduler and Execution Engine. The Jobs Scheduler and Execution Engine is the component in charge of initiating, executing the analytics jobs as designed by the Analytics and Visualization Workbench, as well as of managing the resources available to the Execution Cluster nodes in the context of a Secure and Private Space. The main functionalities of the Job Scheduler and Execution Engine are: a) the deployment and management of the Execution Cluster by interacting the Resource Orchestrator, b) the execution (immediate or scheduled) of the data analysis on the Execution Engine, and c) the data handling operations related to the access or storage of the data assets that are utilized or produced in the process through the Secure Storage.

Under the hood, the analytics jobs are allocated to the Execution Cluster nodes, decoupling the invocation of a data analysis workflow coming from the Analytics and Visualization Workbench from its execution. The Jobs Scheduler and Execution Engine is designed as a multi-container service that consists of a job scheduler microservice which is responsible for scheduling the execution of the designed analytics workflows, and the execution engine which parses the definition of an analytic workflow and starts the respective computation. It deploys, scales and manages the nodes involved in analytics jobs execution and interacts with a set of local workers running on the Execution Cluster nodes for distributed computation. The Jobs Scheduler and Execution Engine is also responsible for monitoring the execution of the job and for reporting the execution status to the Analytics and Visualization Workbench. For the process of loading of datasets or storage of the produced results it is interacting with the Secure Storage, as well as the Encryption Manager for the encryption of the results prior to being stored.

The implementation of the Jobs Scheduler and Execution Engine is based on a customized version of the Spring Cloud Dataflow Server which provides an effective way to execute the designed analytics workflows. Following the micro-service approach, the component allows running algorithms via Spring Boot applications, as well as Apache Spark applications, using an intermediate microservice involved in a pipeline. This intermediate microservice implements a Spark client that interacts with a private Execution Cluster instance that is based on Apache Spark, that is utilized as the cluster computing framework of the platform.

Execution Cluster. The Execution Cluster, that is managed by the Jobs Scheduler and Execution Engine, is the cluster-computing framework of the platform and is deployed within the Secure and Private Space. The Jobs Scheduler and Execution Engine exploits the capabilities of the Execution Cluster in order to perform the actual analytics workflow execution within the context of a private Execution Cluster instance, guaranteeing the secure and isolated execution.

For the implementation of the Execution Cluster, the Apache Spark has been selected as the cluster-computing framework. Capitalizing on the rich set of features offered by Spark, the Execution Cluster offers the powerful processing engine that enables the data analysis execution across multiple datasets and support the extended list of data analysis algorithms that span from simple statistical analysis to more advance and complex machine learning and deep learning algorithms.

Encryption Manager. Within the Secure and Private Space, a running instance of the Encryption Manager is deployed. The role of the Encryption Manager in the Secure and Private Space is to encrypt the results of the analysis before they are securely transmitted and stored in the Secure Storage. The Encryption Manager provides the encryption cipher mechanism that generates the symmetric encryption key and the ciphertext that is produced by the encryption of the results. The encryption method that is adopted in the platform is based on a dual encryption approach that follows a symmetric key encryption, based on the AES256 symmetric key encryption algorithm, of the datasets or the results produced by the analysis execution and the secure exchange of a symmetric key between the involved parties upon their agreement for data sharing through secure SSL handshakes during the decryption process. To this end, the Encryption Manager instance of the Secure and Private Space is performing the symmetric encryption of the results interacting with the Job Scheduler and Execution Engine once the data analysis is finished. For the implementation of the Encryption Manager the Spring Boot Java framework is utilized, as well as the Bouncy Castle Java library that is offering a collection of open source lightweight cryptography APIs that complement the default Java Cryptographic Extension (JCE) with extended cryptography functionalities which are suitable for the implementation needs of the Encryption Manager.

Decryption Manager. In the same logic as with the Encryption Manager, within the Secure and Private Space a running instance of the Decryption Manager is deployed.

The role of the Decryption Manager in the Secure and Private Space is to enable the secure and effective decryption of the encrypted datasets on the data consumer side when legitimate access has been obtained without compromising the data privacy of the data provider. Hence, the Decryption Manager is facilitating the reception of the symmetric key that is utilized in the decryption process by interacting with the Encryption Manager through SSL-enabled connection and the decryption of the encrypted dataset with this symmetric key. To this end, the Decryption Manager is involved in the decryption of the selected dataset prior to the execution of the data analytics workflow and the decryption of the results prior to the execution of the visualization process. The implementation of the Decryption Manager is also based on the Spring Boot Java framework and the Bouncy Castle Java library, similar to the Encryption Manager implementation.

3   Results

The ICARUS platform, and the integrated Secure Experimentation Space that were presented in the previous section, have been thoroughly designed and were recently completed in terms of implementation, and thus no scientific results in terms of evaluation of the performance, efficiency, and of course security have yet been made available. Nevertheless, the concept, approach and technical solution will be verified, validated and evaluated through four core representative use cases of the overall aviation's value chain, which are briefly described in the forthcoming sections. The descriptions focus on the scope and expected results of each demonstrator, while a detailed description can be found on the project's deliverable [13].

3.1   Extra-Aviation Services in an Integrated Airport Environment

Airports constitute the most central point in the aviation services environment, as they are the hubs that interconnect passengers, airline companies, tourism organizations, commercial stores as well as city services. The capacity of the airport infrastructure (stands, gates vs planned aircraft arrivals) is generally adequate to meet the demands of the airport users at all times, but during a busy day and especially at peak hours, the demand exceeds the overall capacity of the airport and late arrivals or departures can create significant delays. So, the prime objective of the airport capacity planning is to ensure the most efficient use of the airport infrastructure, to achieve a sustained increase in throughput performance and to increase capacity in all weather conditions.

The aim of the demonstrator is to enable capacity enhancement decisions that are directly targeted to the needs of the airport, but shall also indirectly address interrelated problems that aviation stakeholders operating in the airport (such as airlines and ground handlers) currently face. Through the Secure Experimentation Sandbox of the ICARUS platform, various descriptive and predictive analytics will be conducted to address problems such as: Capacity Modelling, Airport Traffic Forecasting, Flight Delay Prediction, and Position and Slot Allocation/Scheduling, that are all interrelated to the core Airport capacity problem. The expected benefit for the airport is to optimize the Airport airside capacity (including Aircraft parking stands and Passengers Gates) and improve the utilization of all airport airside infrastructure and improve the Runway Operations Capacity. In the early demonstrator activities, the baseline activities for Capacity Modelling and Forecasting (across all business processes: (a) Improved planning of flight schedules per season, (b) Optimum coordination of ground services, (c) Optimization of airport operation services) will be performed. Essentially, they include data assets collection, exploration and experimentation with different analytics algorithms in the early ICARUS platform release. At the same time, the preliminary results for flight delay prediction to contribute to the optimum coordination of ground services will be showcased and evaluated.

3.2 Routes Analysis for Fuel Consumption Optimization and Pollution Awareness

Cutting operational expenses while reducing environmental impacts will certainly become among the top challenges of the aviation industry in the next decade. The success of this objective requires improvements not only in the use of resources and materials but also methods and tools. Flexible analysis options support assessing the economic viability of route network extensions or modifications and deliver reliable projections of operational key metrics such as block time, block fuel and payload capacity. In this context two distinct scenarios will be executed within the specific demonstrator: a) Pollution Data Analysis and b) Massive Route Network Analysis and Evaluation utilizing a tool for route analysis, aircraft performance and economic investigations, that will act as the data provider for both scenarios, and the ICARUS platform.

The scope of the first demonstrator scenario comprises a set of activities aiming to support a more accurate analysis of pollution data and aircraft emissions. Typical use cases in this field involve the modelling of pollution data and the prediction of aircraft performance in relation to the environmental impact. The scope of the second demonstrator scenario comprises a set of activities aiming to analyze pollution data on a larger scale, that of a massive route network. Typical use case examples in this field involve the statistical evaluation of weather data, the modelling of aircraft payload capacity scenarios and the prediction of aircraft performance in relation to the underlying route network. In both scenarios, adequate input data shall be compiled in order to conduct the required pre-processing calculations that they will then be processed with suitable ICARUS analytics and linked with other flight information data, if applicable. Finally, the produced results will be visualized with the suitable web dashboards/visualizations that will allow data consumers to review aircraft fuel burn and carbon emissions for defined flight legs or a massive route network. In the early demonstrator activities, the baseline activities for both scenarios for the concept evaluation, specification, prototyping and realization will be conducted, that includes the data sources collection and pre-processing, the candidate algorithms experimentation and evaluation, and finally the proper visualization definition.

3.3 Aviation-Related Disease Spreading

In the field of computational epidemiology, mathematical models are designed and used together with computational thinking to study the global spreading of epidemics in environments characterized by many degrees of complexity. The modelling tools aim at better understanding various phenomena related to the spread of infectious diseases, to analyze and forecast the evolution of specific epidemic outbreaks to assist policy making in case of public health emergencies. In this context, a meta-population model that uses a data-driven approach based on real-world data on populations and human mobility is currently available. Airline traffic data is a key component for the modelling of human mobility and the simulation of the global spread of an infectious disease and their effect on the population and the economy.

The aim of the demonstrator is to implement non-incremental improvements to the current model by integrating additional available aviation-related data, like travelers' age structure, gender and income data. As a consequence, the specific demonstrator aims at assessing both qualitatively and quantitatively the novel modelling capabilities, analyzing the accuracy of the predictions in historical and current epidemic forecasts. In the context of this demonstrator, updated datasets about population and airline traffic, together with official reports about passenger demographics coming from offices of statistics at the country level, will be leveraged to develop an upgraded version of its computational model. At a later stage, the demonstrator shall explore detailed passenger demographics originating from the airline booking systems, and use them to design the modelling framework with a full coupling between human mobility and intra-population interactions. In the early demonstrator activities, the baseline activities for the data exploration and collection, the data pre-processing and cleansing will be performed within the ICAURS platform towards the update of the design of the modelling approach based on the new available data, the simulation code adaptation and data importation in the updated model and finally the updated model validation.

3.4 Enhancing Passenger Experience with ICARUS Data

Passenger experience enhancement is a widely discussed topic in aviation, with all involved parties, from airlines and airports to ground handling companies and caterers, looking to optimize their services and product offerings for its achievement. Within the context of this demonstrator, two distinct scenarios will be executed utilizing the capabilities of the ICARUS platform: a) the reduce of cabin food waste towards the increase of revenue and b) the prediction of profitable discounts and offers to increase inflight sales.

The aim of the first scenario of the specific demonstrator is to enhance the existing analytics in the passenger experience by adding prediction capabilities for catering service companies and airlines in order to optimize the loading weight of the duty-free and catering trays on board, prior to the flight, while reducing the cabin food waste. Through this scenario, the implementation of predictive algorithms and methods that will suggest optimized tray loading is expected. The aim of the second scenario of the specific demonstrator is to enhance the existing analytics in the passenger experience by adding prediction capabilities for airlines and catering service companies, in order to suggest discounts and offers targeting to increase in-flight sales. Through this scenario, the targeted predictions and suggestions to airlines and caterers on products and bundles that can be offered on discount towards the increase of in-flight sales and the improvement of the passenger satisfaction and travel experience is expected. In the early demonstrator activities, the baseline activities for both scenarios include the data sources evaluation and availability, the experiment on the analytics algorithms and their evaluation, as well as the verification of the preliminary results from these activities towards their optimization in the subsequent versions of the demonstrator activities.

4   Conclusions

The scope of the current paper is to introduce the ICARUS Secure Experimentation Sandbox of the ICARUS platform. The ICARUS platform aims to provide a big data-enabled platform that aspires to become an "one-stop shop" for aviation data and intelligence marketplace that provides a trusted and secure "sandboxed" analytics workspace, allowing the exploration, curation, integration and deep analysis of original, synthesized and derivative data characterized by different velocity, variety and volume in a trusted and fair manner. To this end, the ICARUS platform offers the ICARUS Secure Experimentation Sandbox as one of its core features that enables the provisioning of a sophisticated environment that can completely guarantees the safety and confidentiality of data, allowing to any interested party to utilize the platform to conduct analytical experiments in closed-lab conditions in order to effectively and securely address the critical barriers for the adoption of Big Data in the aviation industry and enable aviation-related big data scenarios for EU-based companies, organizations and scientists. The concept, approach and technical solution will be verified, validated and evaluated through four core representative use cases of the overall aviation's value chain, as briefly presented.

Acknowledgement. ICARUS project is being funded by the European Commission under the Horizon 2020 Programme (Grant Agreement No 780792).

References

1. Airbus: Data revolution in aviation. https://www.airbus.com/public-affairs/brussels/our-top ics/innovation/data-revolution-in-aviation.html. Accessed 23 July 2020
2. Wyman, O.: Aviation's data science revolution (2017). https://www.oliverwyman.com/ content/dam/oliver-wyman/v2/publications/2017/jun/Aviations_Data_Science_Revolu tion_The_Connected_Aircraft_Final_web.pdf. Accessed 23 July 2020
3. Ajah, I.A., Nweke, H.F.: Big data and business analytics: trends, platforms, success factors and applications. Big Data Cogn. Comput. **3**(2), 32 (2019)
4. da Silva, T.L.C., et al.: Big data analytics technologies and platforms: a brief review. In: Latin America Data Science Workshop, 44th International Conference on Very Large Data Bases, Brazil (2018)
5. Amalina, F., et al.: Blending big data analytics: review on challenges and a recent study. IEEE Access **8**, 3629–3645 (2019)
6. Benjelloun, F.Z., Lahcen, A.A.: Big data security: challenges, recommendations and solutions. In: Web Services: Concepts, Methodologies, Tools, and Applications, pp. 25–38. IGI Global (2019)
7. Oussous, A., Benjelloun, F.Z., Lahcen, A.A.: Belfkih, S: Big data technologies: a survey. J. King Saud Univ.-Comput. Inf. Sci. **30**(4), 431–448 (2018)

8. NIST Big Data Public Working Group: NIST Big Data Interoperability Framework: Volume 4, Security and Privacy Version 2 (No. NIST Special Publication (SP) 1500-4r1), National Institute of Standards and Technology (2018)
9. Venkatraman, S., Venkatraman, R.: Big data security challenges and strategies. AIMS Math. **4**(3), 860–879 (2019)
10. Nelson, B., Olovsson, T.: Security and privacy for big data: a systematic literature review. In: 2016 IEEE International Conference on Big Data (Big Data), pp. 3693–3702 (2016)
11. Chidambararajan, B., Kumar, M.S., Susee, M.S.: Big data privacy and security challenges in industries. Int. Res. J. Eng. Technol. **6**(4), 1991 (2019)
12. ICARUS EC H2020 project Homepage. https://www.icarus2020.aero/. Accessed 20 June 2020
13. ICARUS: Demonstrators Execution Scenarios and Readiness Documentation. EC H2020 ICARUS project (2019)

Machine Learning

Research on a Hybrid EMD-SVR Model for Time Series Prediction

Qiangqiang Yang[1,2], Dandan Liu[2(✉)], Yong Fang[1], Dandan Yang[3], Yi Zhou[4], and Ziheng Sheng[5]

[1] School of Communication and Information Engineering,
Shanghai University, Shanghai, China
[2] College of Electronics and Information Engineering, Shanghai University
of Electric Power, Shanghai, China
liudandan@shiep.edu.cn
[3] Tianjin Navigation Instruments Research Institute, Tianjin, China
[4] MXSUN Software Company, Guangzhou, China
[5] School of Electrical Engineering and Telecommunications,
The University of New South Wales, Sydney, NSW 2052, Australia

Abstract. Time series prediction methods were widely used in various fields. The prediction method for non-stationary and nonlinear time series was studied in this paper. This method decomposed non-stationary time series into stationary sub-sequences using the Empirical Mode Decomposition method. And then an appropriate time-step was chosen and the Support Vector Regression algorithm was applied to predict each stationary sub-sequence. The sum of predicted values was the forecasting results of the original sequence. The method was applied to building energy consumption datasets, which were collected in some buildings. The experimental results showed that the hybrid algorithm of Support Vector Regression and Empirical Mode Decomposition had higher accuracy and was suitable for predicting non-linear and non-stationary time series. Moreover, this hybrid algorithm was used to predict the time series with outliers and to test its noise-resistant performance. The forecasting results also illustrated EMD-SVR algorithm was more robust than SVR algorithm.

Keywords: Time series · Empirical Mode Decomposition · Support Vector Regression · Building energy consumption · Prediction

1 Introduction

Time series is a set of data collected sequentially usually at fixed intervals of time. It is very common in real life. Basically, the goal of time series prediction is to estimate future values based on current and past data [1]. For example, forecasting models of building energy consumption can provide more reasonable building management solutions for building managers. Prediction models for financial time series will help people find out

D. Wang et al. (Eds.): SPNCE 2020, LNICST 344, pp. 137–150, 2021.
https://doi.org/10.1007/978-3-030-66922-5_9

the macro market operation rules. Therefore, many studies focused on how to develop accurate and effective time series prediction models.

Classic time series prediction models are based on stochastic process theory and mathematical statistics, which are divided into two categories: Auto-Regressive Moving Average Model (ARMA) and Autoregressive Integrated Moving Model (ARIMA). These classic time series prediction methods have achieved good prediction results in many fields [2, 3]. However, some literature also pointed out that it is difficult to develop accurate prediction models for nonlinear and non-stationary time series using the classic methods. Therefore, machine learning algorithms were applied to predicting nonlinear and non-stationary time series [4–6]. Among them, the support vector machine (SVM) algorithm is one of the most effective machine learning algorithms and has been widely verified in various fields [1].

However, it is still complicated to develop accurate prediction models for nonlinear and non-stationary time series even using SVM algorithm in some fields [7]. Then some hybrid algorithms that can decompose and predict non-stationary time series were discussed [8–11]. These algorithms turned non-stationary series into stationary subsequences and then analyzed decomposed subsequences to improve prediction accuracy.

In this paper, a time series prediction model for building energy consumption was established based on the EMD-SVM hybrid algorithm. In order to verify the robustness of the algorithm, it was applied to datasets with outliers, which proved that the algorithm had anti-noise ability and was suitable for the prediction of building energy consumption.

2   Methodology

2.1   Theory of Empirical Mode Decomposition (EMD)

Most of the natural phenomena are nonlinear and non-stationary systems. The Empirical Mode Decomposition (EMD) method has proposed by N.E. Huang et al. [12], which is effective in analyzing the non-linear and non-stationary time series. There are also some other transformation methods including Fourier transform, wavelet transform and so on. Fourier transform is always used to decompose linear and stationary signal and the Wavelet analysis depends on the Fourier transform even though it can be used to analyze non-stationary signals. Moreover, the basic wavelet function should be given before the wavelet analysis and it has a non-adaptive characteristic. Once the basic wavelet is selected, one will have to use it to analyze all the data. In order to solve these problems, EMD was adopted in lots of experiments and the results showed it has better performance.

The EMD method can be summarized as follows [12, 13]:

- Given a signal $cx(t)$, identify all extrema of $x(t)$.
- Interpolate between minima or maxima, ending up with some envelope $s_{min}(t)$ or $s_{max}(t)$ and compute the mean of envelope $m_1(t)$:

$$m_1(t) = (s_{max}(t) + s_{min}(t))/2 \tag{1}$$

- Extract the component $c_1(t) = x(t) - m_1(t)$, $c_1(t)$ is an Intrinsic Mode Function (IMF).
- Iterate on the residual $r_1(t) = x(t) - c_1(t)$.
- Repeat the steps until the decomposition results satisfy the stopping criterion. The original time series is decomposed into multiple IMFs and one corresponding residual:

$$x(t) = \sum_{i=1}^{N} c_i(t) + r_N(t) \tag{2}$$

where the residual $r_N(t)$ is computed as formula (3):

$$\begin{cases} r_1(t) - c_2(t) = r_2(t) \\ r_2(t) - c_3(t) = r_3(t) \\ \qquad \cdots \\ r_{N-1}(t) - c_N(t) = r_N(t) \end{cases} \tag{3}$$

2.2 Theory of Support Vector Regression (SVR)

The support vector machines algorithm is a machine learning method that was proposed by Vapnik [14]. SVM is called Support Vector Regression (SVR) when it is used to model and predict.

Actually, the purpose of the SVM algorithm is to classify the data points. If the data points are not linearly separable, they will be mapped into the N-dimensional feature space to make them linearly separable. That is to say, SVM will find an (N-1)-dimensional hyperplane in an N-dimensional space to classify the data points.

So kernel functions and optimizer algorithm are two parts of SVMs. The non-linear data is divided into high-dimensional space by kernel function and the optimizer algorithm is used to find the hyperplane in high dimensional space. SVR method minimizes the empirical risk and identifies an optimum hyperplane to maximize the distance separating the training data into subsets and minimize training error [15].

In this paper, the kernel function, radial basis function (RBF), was chosen and optimizer parameters were searched by the Particle swarm optimization (PSO) algorithm.

2.3 EMD-SVR Method for Time Series Analysis

EMD-SVR is a hybrid algorithm. The complete procedure of developing prediction model for time series using EMD-SVR algorithm can be described as Fig. 1.

1. Build the new time series for prediction models. Suppose that $y = \{y_1, \ldots, y_n\}$ is an original time series, then a new series $\bar{y}_i = \{(\bar{x}_i, \bar{z}_i)\}$ will be reconstructed, where $\bar{x}_i = \{y_i, y_{i+1}, \cdots y_{i+d-1}\}$, $\bar{z}_i = y_{i+d}$ and d is time-steps. $\bar{x}_i$ is the input of SVR model and $\bar{z}_i$ is the output of SVR model [16]. $\bar{x}$ is described by formula (4):

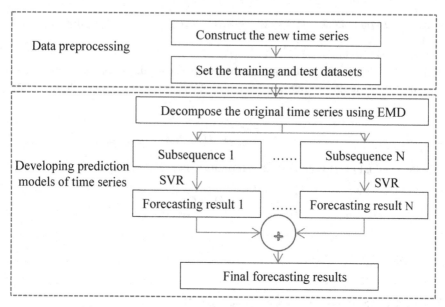

Fig. 1. The experimental procedure of SVR-EMD algorithm

$$\bar{x} = \begin{bmatrix} y_1\ y_2\ \cdots\ y_d \\ y_2\ y_3\ \cdots\ y_{d+1} \\ \cdots \\ y_{n-d}\ y_{n-d+1}\ \cdots\ y_{n-1} \end{bmatrix} \tag{4}$$

$\bar{z}$ is showed by formula (5):

$$\bar{z} = \begin{bmatrix} y_{d+1} \\ y_{d+2} \\ \cdots \\ y_n \end{bmatrix} \tag{5}$$

namely $\bar{x}_i$ is a $(n-d) \times d$ matrix and $\bar{z}_i$ is a $(n-d) \times 1$ matrix.

2. Set the training dataset and test dataset of prediction models. The number of data in the training set and test set can be specified according to the following formula (6).

$$\begin{cases} N = (n - d) \times 3/4 \\ M = (n - d) \times 1/4 \end{cases} \tag{6}$$

Then training dataset and test dataset were confirmed by formula (7):

$$\begin{cases} \bar{y}_{train} = \{(\bar{x}_i, \bar{z}_i)\}(i = 1, \cdots N) \\ \bar{y}_{test} = \{(\bar{x}_i, \bar{z}_i)\}(i = (N + 1), \cdots (N + M)) \end{cases} \tag{7}$$

3. Determine the evaluation criteria of prediction models. The two parameters, mean squared error (MSE) and squared correlation coefficient (R^2), were selected to evaluate the models. They are defined by formula (8) and (9), where R^2 is a number between 0 and 1. The lower MSE and higher R^2 would be expected.

$$\text{MSE} = \frac{1}{l} \sum_{i=1}^{l} (f(x_i) - y_i)^2 \tag{8}$$

$$R^2 = \frac{\left(l \sum_{i=1}^{l} f(x_i) y_i - \sum_{i=1}^{l} f(x_i) \sum_{i=1}^{l} y_i \right)^2}{(l \sum_{i=1}^{l} f(x_i)^2 - \left(\sum_{i=1}^{l} f(x_i) \right)^2)(l \sum_{i=1}^{l} y_i^2 - \left(\sum_{i=1}^{l} y_i \right)^2)} \tag{9}$$

4. Develop the forecasting model using training dataset. The steps of developing forecasting model for time series using EMD-SVR was shown as follows:

- Decompose time series into some subsequences.
- For every subsequence, the range of d was determined. For example, d can be chosen between 2 to 24 for the time series of building energy consumption.
  for d = 2 to 24
Apply the SVR algorithm and develop subsequence prediction models.
Compute the sum of prediction results to get the final results.
Compute MSE and R^2.
end for
- Choose the best time-steps according to MSE and R^2.
- Develop the forecasting model with the chosen time-steps for the original time series.

3 Experiments and Results

In an office building, the energy consumption data was collected. There are 120 data in one series because there are 120 working hours in one week.

After the ADF test, the non-stationary time series was selected to verify the validity of the EMD-SVR algorithm. Then the input and output datasets of SVM were illustrated by formula (10) and (11):

$$\bar{x} = \begin{bmatrix} y_1 \ y_2 \ \cdots \ y_{24} \\ y_2 \ y_3 \ \cdots \ y_{25} \\ \cdots \\ y_{96} \ y_{97} \ \cdots \ y_{119} \end{bmatrix} \tag{10}$$

$$\bar{z} = \begin{bmatrix} y_{25} \\ y_{26} \\ \cdots \\ y_{120} \end{bmatrix} \tag{11}$$

So training dataset and test dataset were confirmed by formula (12):

$$\begin{cases} \bar{y}_{train} = \{(\bar{x}_i, \bar{z}_i)\}(i = 1, \cdots 72) \\ \bar{y}_{test} = \{(\bar{x}_i, \bar{z}_i)\}(i = 73, \cdots 96) \end{cases} \tag{12}$$

3.1 The EMD-SVR Forecasting Model of Non-stationary Time Series for Building Energy Consumption

First, the non-stationary time series of energy consumption was decomposed into some subsequences as shown in Fig. 2. Then the prediction models were developed using SVR algorithm for every subsequence.

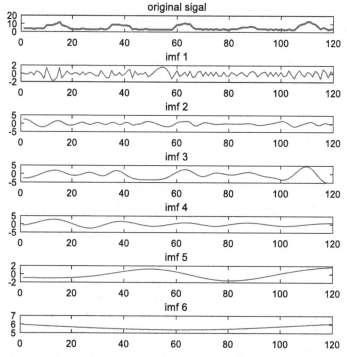

Fig. 2. The original time series and decomposed subsequences of building energy consumption

The comparison of prediction results between EMD-SVR and SVR was shown in Table 1. It illustrated that prediction results of EMD-SVR algorithm were better than SVR algorithm when different time-steps was chosen.

From Table 1 we knew the prediction performance was better when time-steps was 4. In this case, R^2 is 0.925743 and 0.943764 for the training dataset and the test dataset, respectively. As we knew the closer R^2 is to 1, the better the prediction results. Further, the value of R^2 for the test set is greater than the value of R^2 for the training set and they were all greater than 0.9, which indicated that the SVR predicting model was not

Table 1. The Comparison of Prediction performance between SVR and EMD-SVM models using different time-steps for building energy consumption

Time-steps	Algorithm	Training dataset		Testing dataset	
		MSE	R2	MSE	R2
2	EMD-SVR	0.571094	0.906797	1.196776	0.89809
	SVR	1.597639	0.750962	2.13689	0.803385
3	EMD-SVR	0.492307	0.922563	1.01905	0.918678
	SVR	0.836858	0.881241	1.807069	0.868493
4	EMD-SVR	0.482795	0.925743	0.861123	0.943764
	SVR	0.877369	0.874117	1.997539	0.884993
5	EMD-SVR	0.447164	0.931439	1.488371	0.90603
	SVR	1.404345	0.777708	1.728945	0.853921
6	EMD-SVR	0.458997	0.932865	1.27686	0.921485
	SVR	1.34057	0.791787	1.827347	0.851383
7	EMD-SVR	0.295075	0.965816	1.019978	0.929475
	SVR	1.343158	0.794188	1.724895	0.851919
8	EMD-SVR	0.28713	0.967779	1.12577	0.924394
	SVR	0.584819	0.919103	3.101792	0.73007
9	EMD-SVR	0.291151	0.970087	1.068173	0.924757
	SVR	1.378072	0.786893	1.786711	0.855755
10	EMD-SVR	0.28603	0.969325	1.307217	0.919837
	SVR	0.630584	0.932211	2.345651	0.87277
11	EMD-SVR	0.288488	0.968742	1.532112	0.894996
	SVR	0.501229	0.937123	2.324805	0.878927
12	EMD-SVR	0.295423	0.964094	1.948183	0.863141
	SVR	0.569224	0.934161	2.194049	0.887287
13	EMD-SVR	0.324873	0.955616	2.113517	0.834063
	SVR	0.984289	0.873426	2.512899	0.889048
14	EMD-SVR	0.390129	0.945416	1.692577	0.887154
	SVR	0.571432	0.942153	3.098651	0.855208
15	EMD-SVR	0.372134	0.941069	1.612551	0.892333
	SVR	0.448016	0.94655	3.103344	0.829688
16	EMD-SVR	0.39435	0.935235	2.180305	0.828951
	SVR	0.452569	0.952432	3.558848	0.837837
17	EMD-SVR	0.313769	0.948636	2.086048	0.830763
	SVR	0.451582	0.949555	3.592484	0.853068
18	EMD-SVR	0.26255	0.956753	1.460961	0.913228
	SVR	0.431549	0.948027	3.543799	0.843864
19	EMD-SVR	0.31237	0.951181	1.487005	0.905245
	SVR	0.476713	0.956992	5.24637	0.780169
20	EMD-SVR	0.459039	0.953733	5.217063	0.789639
	SVR	0.320427	0.959652	7.124436	0.371715

(continued)

Table 1. (*continued*)

Time-steps	Algorithm	Training dataset		Testing dataset	
		MSE	R2	MSE	R2
21	EMD-SVR	0.2697	0.957281	1.82646	0.885663
	SVR	0.922285	0.832968	3.082533	0.838191
22	EMD-SVR	1.044031	0.825763	3.315175	0.829745
	SVR	0.297115	0.962591	4.956615	0.584612
23	EMD-SVR	0.228852	0.964429	1.356089	0.913678
	SVR	0.905582	0.843615	2.261825	0.854461
24	EMD-SVR	0.293686	0.962969	3.510583	0.718327
	SVR	0.416496	0.943969	4.50047	0.765655

over-fitted. The prediction results of energy consumption models were seen in Fig. 3 and Fig. 4. It can also be seen that the forecasting curves were very similar to the actual curves.

3.2   Comparison of Noise-Resistant Capabilities for SVR and EMD-SVR

In fact, it is normal that forecasting results are disturbed by all kinds of noise, which is one of the mainly reasons that forecasting accuracy cannot be improved. Forecasting models were developed using SVR and EMD-SVR based on datasets with outliers in this section. The forecasting results responded the noise-resistant capability of EMD-SVR.

To learn the noise-resistant capability of different algorithms, the datasets that added outlier instead of original datasets were generated to simulate abnormal energy consumption. Suppose that the dataset contains A% outliers and there are N data in this dataset. The N*A% integers will be generated randomly between 1 and N and these integers are the serial number of outliers.

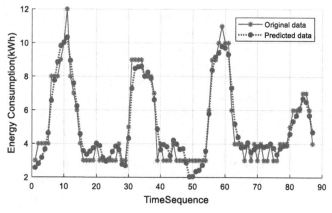

Fig. 3. Prediction results of building energy consumption using ESD-SVR method for the training dataset

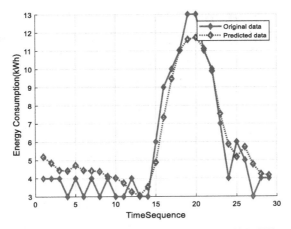

Fig. 4. Prediction results of building energy consumption using ESD-SVR method for the test dataset

$$j = randi([1, N], 1, N * A\%) \tag{13}$$

Then the dataset with outliers can be described by formula (14):

$$y(i) = y(i) + a\delta_i \cdot y(i)\, i = 1, 2 \cdots N \tag{14}$$

where a is a constant that can be set to 0.5 and δ_i is an unit function,

$$\delta_i = \begin{cases} 1, i = j \\ 0, i \neq j \end{cases} \tag{15}$$

An energy consumption time series with outliers was shown in Fig. 5.

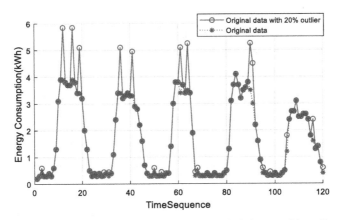

Fig. 5. Comparison of original dataset and original dataset with outliers

The SVR and EMD-SVR algorithms were applied to the datasets with 10%, 20%, 30% and 40% outliers, and the experimental results were shown in the Table 2, Table 3, Table 4 and Table 5. The comparisons among predicted values, original values and original values added to outliers were seen in Fig. 6, Fig. 7, Fig. 8 and Fig. 9. From these tables and figures we knew EMD-SVR algorithm was more robust than SVR algorithm. It meant that EMD-SVR algorithm had better noise-resistant capabilities.

Table 2. The Comparison of prediction performance between SVR and EMD-SVM models for building energy consumption dataset with 10% outliers

Time-steps	Algorithm	Training dataset		Testing dataset	
		MSE	R2	MSE	R2
2	EMD-SVR	1.049133	0.879419	0.998073	0.913055
	SVR	0.287353	0.899854	0.404267	0.657795
3	EMD-SVR	1.042892	0.894517	1.059364	0.909933
	SVR	0.282261	0.895995	0.392198	0.668
4	EMD-SVR	0.910117	0.900216	1.146224	0.901543
	SVR	0.250799	0.899624	0.398196	0.711845
5	EMD-SVR	0.869306	0.882654	1.554877	0.889715
	SVR	0.249604	0.898202	0.259128	0.787533

Table 3. The Comparison of prediction performance between SVR and EMD-SVM models for building energy consumption dataset with 20% outliers

Time-steps	Algorithm	Training dataset		Testing dataset	
		MSE	R2	MSE	R2
13	EMD-SVR	0.947983	0.891675	1.6985	0.865301
	SVR	1.34509	0.80497	3.979598	0.667633
14	EMD-SVR	1.115337	0.87121	1.662653	0.869568
	SVR	1.251211	0.818155	3.754576	0.676766
15	EMD-SVR	0.87896	0.892769	1.687701	0.865985
	SVR	0.901377	0.916191	4.943163	0.698825
16	EMD-SVR	0.7454	0.90294	2.266961	0.818882
	SVR	0.88956	0.887116	4.555295	0.75042

Table 4. The Comparison of prediction performance between SVR and EMD-SVM models for building energy consumption dataset with 30% outliers

Time-steps	Algorithm	Training dataset		Testing dataset	
		MSE	R2	MSE	R2
2	EMD-SVR	1.165155	0.901028	2.544622	0.927529
	SVR	2.206972	0.683425	3.55639	0.708367
3	EMD-SVR	1.096337	0.889661	2.460228	0.937368
	SVR	2.294979	0.693566	3.901687	0.694548
4	EMD-SVR	1.115835	0.882032	2.323355	0.92792
	SVR	2.316277	0.686902	3.950681	0.707086
5	EMD-SVR	1.334761	0.870688	2.62816	0.912298
	SVR	2.287445	0.719084	2.795342	0.79984

Table 5. The Comparison of prediction performance between SVR and EMD-SVM models for building energy consumption dataset with 40% outliers

Time-steps	Algorithm	Training dataset		Testing dataset	
		MSE	R2	MSE	R2
2	EMD-SVR	2.256755	0.773657	4.016026	0.840588
	SVR	2.304331	0.671405	2.667304	0.769721
3	EMD-SVR	2.042788	0.793416	2.935345	0.885079
	SVR	2.41279	0.677468	2.677986	0.776113
4	EMD-SVR	2.03426	0.793219	2.616165	0.886622
	SVR	2.471562	0.660606	2.563792	0.77963
5	EMD-SVR	2.093942	0.797823	2.7458	0.915917
	SVR	2.621712	0.695381	2.673072	0.818439

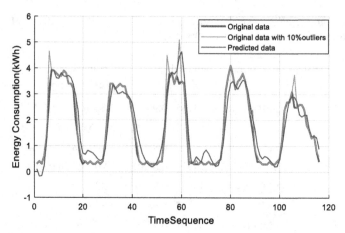

Fig. 6. Prediction results of building energy consumption using ESD-SVR method for the dataset with 10% outliers

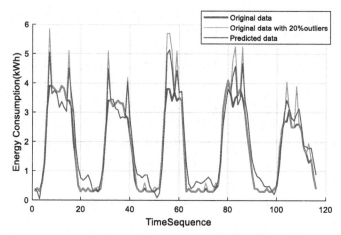

Fig. 7. Prediction results of building energy consumption using ESD-SVR method for the dataset with 20% outliers

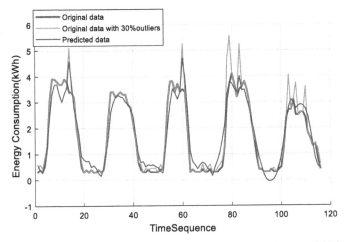

Fig. 8. Prediction results of building energy consumption using ESD-SVR method for the dataset with 30% outliers

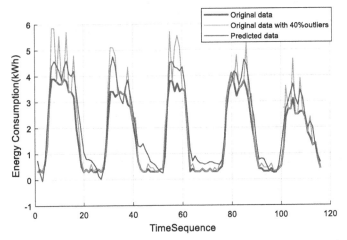

Fig. 9. Prediction results of building energy consumption using ESD-SVR method for the dataset with 40% outliers

4 Conclusion

In this paper, a method for predicting nonlinear and non-stationary time series was studied. The prediction models of time series were developed based on a hybrid EMD-SVR algorithm. Firstly, the nonlinear and non-stationary time series was decomposed into some subsequences using EMD. And then the prediction model for every subsequence was developed based on SVR. The sum of the prediction value of each subsequence is the final prediction result.

The forecasting model of time series for building energy consumption can be used to learn the law of building energy consumption and save energy. The EMD-SVR algorithm was applied to time series of building energy consumption. The results showed that the EMD-SVR algorithm was better than SVR algorithm. Moreover, the method was used to the dataset with outliers for verifying the robustness of the algorithm. The forecasting results of datasets with outliers illustrated EMD-SVR algorithm was more robust than SVR algorithm.

References

1. Sapankevych, N., Sankar, R.: Time series prediction using support vector machines: a survey. IEEE Comput. Intell. Mag. **4**(2), 24–38 (2009)
2. Wang, Q., Li, S.Y., Li, R.R., et al.: Forecasting U.S. shale gas monthly production using a hybrid ARIMA and metabolic nonlinear grey model. Energy **160**(10), 378–387 (2018)
3. Rounaghi, M.M., Zadeh, F.N.: Investigation of market efficiency and financial stability between S&P 500 and London stock exchange: monthly and yearly forecasting of time series stock returns using ARMA model. Phys. A Stat. Mech. Appl. **456**(15), 10–21 (2016)
4. Mohapatra, U.M., Majhi, B., Satapathy, S.C.: Financial time series prediction using distributed machine learning techniques. Neural Comput. Appl. **31**(8), 3369–3384 (2017). https://doi.org/10.1007/s00521-017-3283-2
5. Chen, S., Mihara, K., Wen, J.: Time series prediction of CO2, TVOC and HCHO based on machine learning at different sampling points. Build. Environ. **146**(12), 238–246 (2018)
6. Li, Z., Ye, L., Zhao, Y., Song, X., Teng, J., Jin, J.: Short-term wind power prediction based on extreme learning machine with error correction. Protect. Control Mod. Power Syst. **1**(1), 1–8 (2016). https://doi.org/10.1186/s41601-016-0016-y
7. Shiu, M.C., Wei, L.Y., Liu, J.W., et al.: A hybrid one-step-ahead time series model based on GA-SVR and EMD for forecasting electricity loads. J. Appl. Sci. Eng. **20**(4), 467–476 (2017)
8. Yuanfang, X., Yuanyuan, J., Xuemei, Z.: Gas outburst prediction model based on empirical mode decomposition and extreme learning machine. Recent Adv. Electr. Electron. Eng. **8**(1), 50–56 (2015)
9. Yaslan, Y., Bican, B.: Empirical mode decomposition based denoising method with support vector regression for time series prediction: a case study for electricity load forecasting. Measurement **103**(6), 52–61 (2017)
10. Wang, Z.Y., Qiu, J., Li, F.F.: Hybrid models combining EMD/EEMD and ARIMA for long-term streamflow forecasting. Water **10**(7), 1–12 (2018)
11. Yabing, J., Changwen, L., Fengrong, B., et al.: Sensitivity analysis on physical noise sources of small generator based on EMD-SVM. J. Tianjin Univ. **50**(10), 1077–1083 (2017)
12. Huang, N.E., Shen, Z., Long, S.R.: The empirical mode decomposition and Hilbert spectrum for nonlinear and nonstationary time series analysis. Proc. Roy. Soc. London A **454**(1), 903–995 (1998)
13. Rilling, G., Flandrin, P., Goncalves P.: On empirical mode decomposition and its algorithms. In: 2003 IEEE-EURASIP Workshop on Nonlinear Signal and Image Processing, pp. 8–11. IEEE, Grado (2003)
14. Vapnik, V.: The Nature of Statistical Learning Theory. Springer, New York (1995). https://doi.org/10.1007/978-1-4757-2440-0
15. Yu, J.Q., Nan, Y.L., Zhang, Y., Yang, X.: Research on fractal characteristics of building energy consumption time series. In: 4th International Conference on Energy Equipment Science and Engineering, pp. 1–8. IOP, Xi'an, China (2019)
16. Ghelardoni, L., Ghio, A., Anguita, D.: Energy load forecasting using empirical mode decomposition and support vector regression. IEEE Trans. Smart Grid **4**(1), 549–556 (2013)

Distant Supervision for Relations Extraction via Deep Residual Learning and Multi-instance Attention in Cybersecurity

Guowei Shen[1], Ya Qin[1], Wanling Wang[1], Miao Yu[2(✉)], and Chun Guo[1]

[1] College of Computer Science and Technology, Guizhou University, Guiyang 550025, China
gwshen@gzu.edu.cn, qyamail@163.com, 1733348173@qq.com,
gc_gzedu@163.com
[2] Institute of Information Engineering, Chinese Academy of Sciences, Beijing 100093, China
yumiao@iie.ac.cn

Abstract. A large number of open source threat intelligence resources provide regularly updated threat sources that can be applied to a variety of security analysis solutions. Fragmented security news, security forums, and vulnerability information are important sources of cyber threat intelligence, but it is difficult to correlate these multiple-source data. Cybersecurity knowledge graph is a powerful tool for data-driven thread intelligence computing. Relation extraction is a very important task in construction of cybersecurity knowledge graph from unstructured data. In order to reduce the influence of noisy data in deep learning model, we propose a distant supervised relation extraction model ResPCNN-ATT based on deep residual convolutional neural network and attention mechanism. This method takes word vector and position vector of the word as input of the model, extracts semantic features of texts through the piecewise convolutional neural network model PCNN, achieves the learning effect of less noisy data and better extracts deep semantic features in sentenses by using deep residuals Compared with other models, the model proposed in this paper achieves higher accuracy than other models.

Keywords: Thread intelligence · Cybersecurity knowledge graph · Relation extraction · Residual learning

1 Introduction

Currently, there are numerous open source threat intelligence sources providing periodically updated threat feeds fed into various analytical solutions. Security news, security forums, and vulnerability information are important data sources for cyber threat intelligence. However, the above data is fragmented and it is difficult to correlate such multi-source data.

Cybersecurity knowledge graph is a powerful tool for data-driven thread intelligence computing. Through cyber security knowledge graph, researchers can intuitively know network security entities and relations between the entities, such as utilization relation

D. Wang et al. (Eds.): SPNCE 2020, LNICST 344, pp. 151–161, 2021.
https://doi.org/10.1007/978-3-030-66922-5_10

between malware and vulnerabilities, employment relation between attackers and organizations, and ownership between software and vulnerabilities. Relation extraction is a very important task in construction of cybersecurity knowledge graph from unstructured data.

In relation extraction, the lack of labeled data for training is a challenge when constructing a network security knowledge graph. A common technique for coping with this difficulty is distant supervision in natural language processing. Distant supervision strategy is an effective method of automatically labeling training data. However, the assumption in distant supervision method is too strong, leading to the wrong label problem.

In this paper, we propose a novel cybersecurity relation extraction model ResPCNN-ATT combined **Res**idual Learning, **P**iecewise **C**onvolutional **N**eural **N**etworks (**PCNN**) and multi-instance **ATT**ention. The following list details the main contributions of the article:

- In order to reduce the impact of noise data in open source threat intelligence data sources, we propose a distant supervised cybersecurity relation extraction model based on ResPCNN-ATT. The model first uses the pre-trained word vector and the position vector between cybersecurity entity pairs as the model input, and then uses PCNN to extract the semantic features.
- Deep residual learning is used to solve the problem of gradient disappearance caused by noise data, so as to extract more effective semantic features.
- In order to better capture the more important semantic features in sentences, a multi-instance attention mechanism is used to calculate the correlation between instance and the corresponding relation to reduce the impact of noise data.

The rest of the paper is organized as follows. We describe related works in Sect. 2. The cybersecurity relation extraction model and details are shown in Sect. 3. Experiment is in Sect. 4. Section 5 draws conclusions.

2 Related Work

Data-driven cyber security event prediction and analysis are hot topics in current cyber security research [1]. Xiaokui Shu introduces a new methodology that models threat discovery as a graph computation problem for threat intelligence [2]. As a semantic knowledge base, knowledge graph is a powerful tool for managing large-scale knowledge consists with entities and relations between them. Haoze Yu proposed a relation extraction method for the construction of knowledge graph in food field [3].

Natural language processing technology [4–6] tends to only consider the domain name and IP address when analyzing the relation between malicious entities, both of which have very simple relation definitions. Aditya Pingle propose the RelExt [7] system, which strives to improve various cyber threat representation schemes, especially cybersecurity knowledge graphs (CKG), by predicting the relations between cybersecurity entities identified by cybersecurity named entity recognizer. VIEM [8] analyzes a large number of inconsistencies by extracting software names and software versions in Public Security Vulnerability Reports, so the extraction of relations is more complicated.

Relation Extraction (RE) is one of the most important topics in NLP. Many relation extraction methods have been proposed [9–11], such as bootstrapping, unsupervised relation discovery and supervised classification. Most existing supervised RE methods require a large amount of labelled relation-specific training data, which is very time consuming and labor intensive.

Distant supervision is proposed to automatically generate training data. Under the framework of distance supervised learning, some recent work [12–15] attempts to use deep neural networks in relation prediction. Although distant supervision is an effective strategy to automatically label training data, it always suffers from the wrong label problem.

3   The Proposed Model

In this section, we describe the overall architecture of our model. We introduce the components of our security entity relation extraction model one by one.

Under the framework of distant supervised learning, the problem of insufficient label data in deep learning can be solved, but at the same time it also brings some problems, such as the low-quality label data and the wrong label data. This would have a great impact on subsequent tasks of entity relation extraction. In view of the above problems, we propose a distant supervised relation extraction model ResPCNN-ATT based on deep residual neural network and attention mechanism. The framework is shown in Fig. 1. The model is mainly composed of a vector representation layer, a deep residual convolutional network layer, and a multi-instance attention layer.

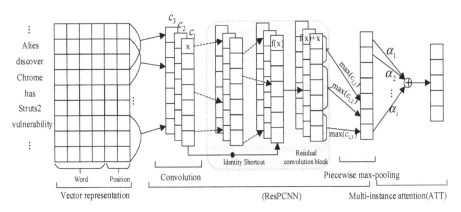

Fig. 1. Cybersecurity relation extraction model based on ResPCNN-ATT

The model first uses the pre-trained word vector and the position vector between entity pairs as input, which can highlight the role of the two entities, and then uses the piecewise convolutional neural networks to extract semantic features. At the same time, deep residual learning is introduced to solve the problem of gradient disappearance caused by noise data, so as to extract more effective semantic features. Finally, in order to better capture the more important semantic features in sentences, the multi-instance attention mechanism is used to calculate correlation between instances and corresponding relation, so as to reduce the impact of noise data and improve the performance of relation extraction.

3.1 Vector Representation

The vector representation layer in the model mainly includes word embedding and position embedding.

Word Embedding. Before training the relation extraction model, the text data needs to be vectorized so that the model can read the data. Compared with traditional one-hot coding, word vector mapping can represent more semantic and syntactic information. Word vector mapping is to map each word in the text to a k-dimensional real-valued vector. It is a distributed representation of words. When training a neural network model, the most common method is to randomly initialize all parameters and then use an optimization algorithm to optimize the parameters. Research shows that when a neural network is initialized with a pre-trained word vector, the parameters can be converged to a better local minimum.

For a given sentence $X = \{x_1, x_2, \ldots x_n\}$ consisting of n words, use word2vec to map each word to a low-dimensional real-valued vector space, then perform word vector processing on the sentence, and finally get a vector representation of each word in the sentence, to form a word vector query matrix D^c. Each input training sequence can be mapped by the word vector query matrix D^c to obtain the corresponding real-valued vector $x_t = \{w_1, w_2, \ldots w_n\}$.

Position Embedding. In the relation extraction task, we focus on finding the relation of entity pairs. Words that are often close to the entity are more able to highlight the relation between the two entities, such as some verbs: attack, use, etc. Therefore, in order to make full use of the information in the sentence, the position of each word in the sentence for two entities is an important feature in the relation extraction task. This paper uses the position vector (Position Embeddings, PE) mapping representation method proposed by zeng [14] et al., that is, the relative distance between the current word,entity e_1 and entity e_2 is stitched and converted into a vector representation through embedding. In sentence position vectorization, if the dimension of the word vector is d^c and the dimension of the position vector is d^p, then the dimension of the sentence vector is:

$$d^s = d^c + d^p * 2 \tag{1}$$

For example, the vectorized representation of "Alies discover Chrome has XSS vulnerabilities" is shown in Fig. 2, "Chrome" and "XSS" in the sentence correspond to entities e_1 and entities e_2 respectively. Then the distance from "Alies" to "Chrome" is 2, the distance from "Alies" to "XSS" is 4, and the distance from "vulnerability" to "Chrome" is -3, the distance from "vulnerability" to "XSS" is -1.

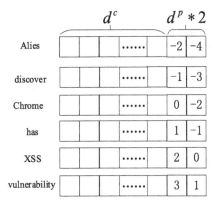

Fig. 2. Position embedding

3.2 Deep Residual Neural Network

In cyber security relation extraction tasks, the main challenge is that the length of the input sentence is variable and not fixed, and important feature information may appear in any area of the sentence. Therefore, in order to be able to use all local features and predict relations globally, this paper uses a piecewise convolutional neural network PCNN model to extract semantic features in sentences.

In this paper, a residual convolution block is designed for residual learning. Each residual convolution block is a sequence composed of two convolution layers. After each convolution layer, the activation function ReLU is used for nonlinear mapping, and features are then extracted using a local maximum pool. The kernel size of all convolution operations in the residual convolution module is w, and the newly generated features are guaranteed to be the same size as the original ones through the border padding operation. The convolution kernels of the two-layer convolution are $W_1, W_2 \in R^{w*1}$. The first layer of the residual convolution block is:

$$c_{i,1} = f(W_1 \bullet c_{i,i+w-1} + b_1) \tag{2}$$

The second layer is:

$$c_{i,2} = f(W_2 \bullet c_{i,i+w-1} + b_2) \tag{3}$$

Where b_1, b_2 are bias vector. In this paper, we optimize the residual learning to get the output vector c of the residual convolution block [16, 17].

After the semantic feature is acquired by convolution layer, the most representative local feature is further extracted by pooling layer. In order to capture characteristic information of different sentence structures, a Piecewise Max Pooling process is used.

3.3 Multi-instance Attention

In the relational extraction model, sentence-level attention is built on multiple instances, dynamically reducing the weight of noisy instances, and making full use of semantic information in these sentences to obtain final sentence vector representation.

For the instance set $S = (g_1, g_2, g_3, \cdots g_n)$ describing the same entity pair $< e_i, e_j >$, g_i is the instance vector output by the convolution layer, n is the number of instances contained in the set S. This paper will calculate the correlation degree between the instance vector g_i and the relation r. In order to reduce the impact of noise data and make full use of the semantic information contained in each instance in the set, the calculation of instance set vector S will depend on each instance g_i in the set:

$$S = \sum_i \alpha_i g_i \tag{4}$$

Where a_i is the weight of the input instance vector g_i, which measures the correlation of the corresponding relation r. The calculation formula of α_i is as follows:

$$\alpha_i = \frac{\exp(e_i)}{\sum_k \exp(e_k)} \tag{5}$$

e_i is a query-based function, which indicates the degree of matching between the input instance vector g_i and the prediction relation r.

Conditional probability of prediction relation $p(R|S)$ is calculated by softmax function:

$$p(R|S) = soft\max(\tilde{r}S + b) \tag{6}$$

Where $\tilde{r}$ is the relation matrix and b represents the bias vector. $p(R|S)$ is used to predict the relation between pairs of cyber security entities.

$$\tilde{R} = \arg\max p(R|S) \tag{7}$$

4 Performance Evaluation

In this section, we empirically demonstrate the performance of the proposed method on dataset CSER.

4.1 Datasets

In order to verify the performance of our proposed model, we build a Cyber Security Entity Relation dataset CSER. 10 types of relation were labeled. The dataset CSER is clawed from Freebuf website and wooyun vulnerability database, which includes network text data such as technology sharing, network security, and vulnerability information.

Commonly used Precision-Recall (P-R) curve, AUC value and average accuracy (P@N) are used to evaluate the model. The P-R curve is a curve drawn with the recall rate R as the abscissa and the accuracy rate P as the ordinate, using P and R at different confidence levels. The AUC value is the area included under the P-R curve. Generally, the larger the AUC value is, the better the model perform. P@N is the accuracy rate calculated by comparing the first N relation instances.

The set of dimensions of the word vector is {50, 60, $\cdots$, 300}. The set of dimensions of the position vector is {1, 2, $\cdots$, 10}. During the training process, the Adam optimizer performs optimization training. The value set of the learning rate is {0.01,0.001,0.0001}. The set of batch size processed in one iteration is {40, 160, 640, 1280}. In order to prevent the model from overfitting, the dropout method is used in CNN. Other parameters are shown in Table 1.

Table 1. Parameters

Parameters	Value
CNN window size	3
CNN hidden size	230
Learning rate	0.01
Batch size	160
Epoch	60
Dimension of the position vector	5
Dropout rate	0.5
Dimension of the word vector	50

4.2 Results

The experimental comparison in this paper mainly compares two aspects of the models.

On the one hand, it uses CNN algorithm with different performance to encode the training data and extract the semantic features in the sentence, mainly including the traditional models: CNN, PCNN, and ResPCNN.

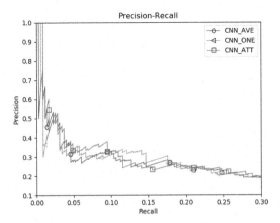

Fig. 3. The results of different bag methods AVE/ONE/ATT based on CNN

The second aspect is based on how CNN/PCNN/ResPCNN uses the information in the packaging bag for experimental comparison. Three different methods were used to process the information in the bag, namely AVE, ONE, and ATT. AVE assigns the same weight to all the sentences in the packet as the entity pair, that is $\alpha_i = 1/n$. ONE means to take the instance vector with the highest confidence, and find a sentence with the highest score from each bag to represent the entire bag. All models in this paper have been trained and tested on the dataset CSER. Figures 3, 4 and 5 show the P-R curves of the result on different bag models.

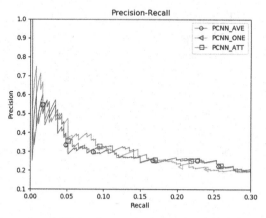

Fig. 4. The results of different bag methods AVE/ONE/ATT based on PCNN

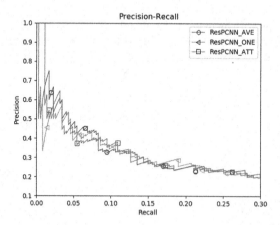

Fig. 5. The results of different bag methods AVE/ONE/ATT based on ResPCNN

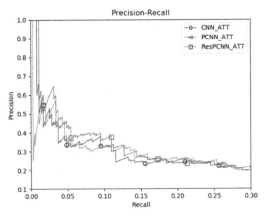

Fig. 6. The results of different sentence semantic feature extraction models CNN/PCNN/ResPCNN

From Fig. 6, the AUC value of the model ResPCNN-ATT is the highest value on the dataset CSER, which reachs 12.68%. The model ResPCNN-ATT proposed in this paper can better extract the deep semantic information of sentences, indicating that the introduction of the ATT method can effectively reduce the redundant data in distant supervised learning.

Table 2. Results for the first 100, 200, and 300 extracted relation instances upon manual evaluation.

Models	P@100	P@200	P@300	Mean	AUC
CNN+AVE	0.3267	0.2537	0.2452	0.2743	0.1062
CNN+ONE	0.2971	0.3035	0.2392	0.2799	0.1096
CNN+ATT	0.3267	0.2437	0.2425	0.2710	0.1121
PCNN+AVE	0.2971	0.2587	**0.2645**	0.2727	0.1096
PCNN+ONE	0.3168	0.2587	0.2358	0.2705	0.1109
PCNN+ATT	0.3267	0.2736	0.2525	0.2842	0.1121
ResPCNN+AVE	0.3267	0.2686	0.2458	0.2804	0.1205
ResPCNN+ONE	0.3564	0.2786	0.2558	0.2969	0.1184
ResPCNN+ATT	**0.4158**	**0.3084**	0.2558	**0.3267**	**0.1268**

As can be seen from Table 2, comparing the accuracy of the first 100, 200, and 300 relation instances on the dataset CSER, the relation extraction accuracy of ResPCNN-ATT is the highest, which reaches 32.67%. However, the accuracy of the CSER dataset is lower than other datasets. This is because the sentences in the CSER dataset are mixed with Chinese and English, the more complicated the sentence structure is, the less obvious the entity relation characteristics are, and the less the corpus data is.

In order to further analyze the relation extraction model proposed in this paper, by adding the depth of the ResPCNN-ATT model to verify the effectiveness of the introduction of residual learning, comparative experiments of convolutional layers with different depth is designed. In this paper, the number of convolutional layers is increased by increasing the number of residual convolution blocks, and the experimental comparison is performed on the CSER dataset. Figure 7 shows the P-R curves on models with different depth.

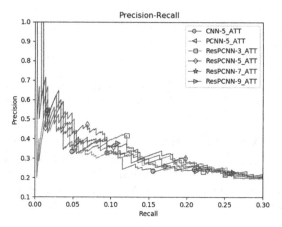

Fig. 7. The results on models with different depth

5   Conclusions

In this paper, we introduce a novel distant supervised cybersecurity relation extraction model ResPCNN-ATT. Algorithm ResPCNN is used to extract semantic features. Deep residual learning is introduced to solve the problem of gradient disappearance due to noise data. The mechanism calculates the correlation between the instance and the corresponding relation to reduce the impact of noisy data. The experimental results show that the model proposed in this paper has the highest accuracy of relation extraction compared with other model methods.

In the future, we intend to use reinforcement learning to further solve the problem of noise in the training data automatically generated by the distant supervised method.

Acknowledgement. Project supported by the National Natural Science Foundation Of China (No. 61802081).

References

1. Sun, N., Zhang, J., Rimba, P., et al.: Data-driven cybersecurity incident prediction: a survey. IEEE Commun. Surv. Tutor. **21**(2), 1744–1772 (2018)

2. Shu, X., Araujo, F., Schales, D.L., et al.: Threat intelligence computing. In: Proceedings of the 2018 ACM SIGSAC Conference on Computer and Communications Security, pp. 1883–1898 (2018)

3. Yu, H., Li, H., Mao, D., et al.: A relationship extraction method for domain knowledge graph construction. World Wide Web **23**, 1–19 (2020). https://doi.org/10.1007/s11280-019-00765-y

4. Liao, X., Yuan, K., Wang, X.F., et al.: Acing the IOC game: toward automatic discovery and analysis of open-source cyber threat intelligence. In: Proceedings of the 2016 ACM SIGSAC Conference on Computer and Communications Security, pp. 755–766 (2016)

5. Siracusano, G., Trevisan, M., Gonzalez, R., et al.: Poster: on the application of NLP to discover relationships between malicious network entities. In: Proceedings of the 2019 ACM SIGSAC Conference on Computer and Communications Security, pp. 2641–2643 (2019)

6. Zhu, Z., Dumitras, T.C.: Automatically learning the semantics of malicious campaigns by mining threat intelligence reports. In: 2018 IEEE European Symposium on Security and Privacy (EuroS&P), pp. 458–472. IEEE (2018)

7. Pingle, A., Piplai, A., Mittal, S., et al.: RelExt: relation extraction using deep learning approaches for cybersecurity knowledge graph improvement. In: Proceedings of the 2019 IEEE/ACM International Conference on Advances in Social Networks Analysis and Mining, pp. 879–886 (2019)

8. Dong, Y., Guo, W., Chen, Y., et al.: Towards the detection of inconsistencies in public security vulnerability reports. In: 28th {USENIX} Security Symposium ({USENIX} Security 19), pp. 869–885 (2019)

9. Socher, R., Huval, B., Manning, C.D., et al.: Semantic compositionality through recursive matrix-vector spaces. In: Joint Conference on Empirical Methods in Natural Language Processing & Computational Natural Language Learning, pp. 1201–1211 (2012)

10. Daojian, Z., Kang, L., Siwei, L., et al.: Relation classification via convolutional deep neural network. In: Proceedings of COLING, pp. 2335–2344 (2014)

11. Zhou, P., Shi, W., Tian, J., et al.: Attention-based bidirectional long short-term memory networks for relation classification. In: Proceedings of the 54th Annual Meeting of the Association for Computational Linguistics (Volume 2: Short Papers), vol. 2, pp. 207–212 (2016)

12. Santos, C.N.D., Xiang, B., Zhou, B.: Classifying relations by ranking with convolutional neural networks. Comput. Sci. **86**(86), 132–137 (2015)

13. Lin, Y., Shen, S., Liu, Z., et al.: Neural relation extraction with selective attention over instances. In: Proceedings of the 54th Annual Meeting of the Association for Computational Linguistics, vol. 1, pp. 2124–2133 (2016)

14. Zeng, D., Liu, K., Chen, Y., et al.: Distant supervision for relation extraction via piecewise convolutional neural networks. In: Proceedings of the 2015 Conference on Empirical Methods in Natural Language Processing, pp. 1753–1762 (2015)

15. Qin, P., Xu, W., Wang, W.Y.: Robust distant supervision relation extraction via deep reinforcement learning. arXiv preprint arXiv:1805.09927 (2018)

16. Huang, Y.Y., Wang, W.Y.: Deep residual learning for weakly-supervised relation extraction. arXiv preprint arXiv:1707.08866 (2017)

17. He, K., Zhang, X., Ren, S., et al.: Deep residual learning for image recognition. Comput. Vision Pattern Recognit. 770–778 (2015)

User Identity Linkage Across Social Networks Based on Neural Tensor Network

Xiaoyu Guo[1], Yan Liu[1(✉)], Xianmin Meng[2], and Lian Liu[2]

[1] PLA Strategic Support Force Information Engineering University, Zhengzhou 450001, China
hello_dreamer@126.com, ms.liuyan@foxmail.com
[2] Investigation Technology Center PLCMC, Beijing 100000, China
minmax-007@163.com, lianl11024@hotmail.com

Abstract. User Identity Linkage (UIL) across social networks refers to the recognition of the accounts belonging to the same individual among multiple social network platforms. The most existing methods usually apply network embedding to map the network structure space to the low-dimensional vector space and then use linear models or standard neural network layers to measure the correlations between users across social networks. However, they can hardly model the complicated interactions between users. In this paper, we propose a novel Neural Tensor Network-based model for UIL, called NUIL. Firstly, we use the Random Walks and Skip-gram model to learn the vector representations of users. Then, we apply the Neural Tensor Network, which has a stronger ability to express the interactions between entities, to mine relationships between users from a higher dimension. A series of experiments conducted on a real-world dataset show that NUIL outperforms the state-of-the-art network structure-based methods in terms of precision, recall, and F1-measure, specifically the F1-measure exceeds 0.66, with an increase of more than 20%.

Keywords: User identity linkage · Neural tensor network · Network embedding · Social network analysis

1 Introduction

With the rapid development of online social network, people usually join multiple social networks simultaneously according to their needs of work or life [1]. Each user often has multiple separate accounts in different social networks. However, these accounts belonging to the same user are mostly isolated without any connection or correspondence to each other [2].

The typical aim of User Identity Linkage (UIL) is to detect that users from different social platforms are actually one and the same individual, also known as Account Identification, Anchor Link Prediction, and Network Alignment [3]. UIL plays an important role in social network analysis, such as user behavior prediction, friend recommendation across platforms, and information dissemination across networks, etc.

© ICST Institute for Computer Sciences, Social Informatics and Telecommunications Engineering 2021
Published by Springer Nature Switzerland AG 2021. All Rights Reserved
D. Wang et al. (Eds.): SPNCE 2020, LNICST 344, pp. 162–171, 2021.
https://doi.org/10.1007/978-3-030-66922-5_11

Early research uses the public attributes and statistical features of users to solve the UIL problem [4, 5], such as username, user's hobbies, language patterns, etc. However, the correctness and richness of user's public attributes cannot be guaranteed. Compared with user's attributes, the connections between users are reliable and rich, and can also be directly used to solve the UIL problem. Therefore, the methods based on network structure are receiving more and more attention [6, 7].

With the development of network embedding (NE), many people use NE instead of traditional feature engineering to save the structural features of social network into low-dimensional vector space, which not only reduces the complexity of the algorithm, but also improves the accuracy of user identity linkage. For example, Man et al. [8] employed network embedding to capture the major and specific structural regularities and further learned a stable cross-network mapping for predicting anchor links. Zhou et al. [9] propose the FRUIP model which extracts the friend feature vector from the network neighborhood patterns and establishes a "one-to-one" mapping based on the similarity between users.

The existing methods generally use a linear model or standard neural network layers to measure the correlations between users after obtaining the low-dimensional vector space of social networks. However, the relationships between users across social networks are extremely complex, traditional methods can hardly model the complicated interactions between them. Inspired by the success of neural tensor network (NTN) for explicitly modeling multiple interactions of relational data [10], we propose a novel model NUIL: Neural Tensor Network for User Identity Linkage across Social Networks. The contributions of this manuscript are as follows:

- NUIL applies the Random Walks model and Skip-gram model to embed the network structure into a low-dimensional vector space to learn the effective vector representations of nodes and we also compare the performance of different Random Walk strategies in solving the UIL problem.
- NUIL replaces a standard neural network model with a neural tensor network model, which has a stronger ability to express the relationships between users, to relate two user vectors across multiple dimensions.
- We conduct a series of experiments on a real-world dataset consisting of two real social networks. The results show that NUIL can significantly improve the precision, recall, and F1-measure of user identity linkage compared to the state-of-the-art methods, e.g., more than 0.66 in terms of F1-measure.

2 Preliminaries

This section describes the terminologies used in this paper and then formally defines the problem of user identity linkage.

2.1 Terminology Definition

We consider a set of social networks as $G^1, G^2, \ldots, G^n$, each of which is represented as an undirected and unweighted graph. Let $G = (V, E)$ represent the network, where V

is the set of nodes, each representing a user, and E is the set of edges, each representing the connection between two users.

In this paper, we take two social networks as an example, which are treated as source network, $G^s = (V^s, E^s)$, and target network, $G^t = (V^t, E^t)$ respectively. For convenience, we have the following definitions.

Definition 1 (Anchor Link). *Link* (v_i^s, v_k^t) *is an anchor link between* G^s *and* G^t *iff.* $\left(v_i^s \in V^s\right)$ $\wedge$ $\left(v_k^t \in V^t\right) \wedge$ $\left(v_i^s \text{ and } v_k^t \text{ are accounts owned by the same user in } G^s \text{ and } G^t \text{ respectively}\right).$

Definition 2 (Anchor Users). *Users who are involved in two social networks simultaneously are defined as the anchor users (nodes) while the other users are non-anchor users (nodes).*

Definition 3 (Corrupted Anchor Link). *If a user from* G^s *and another user from* G^t *do not belong to the same natural person, we call that the two users form a corrupted anchor link.*

2.2 Problem Definition

Based on the definitions of the above terms, we formally define the problem of user identity linkage across social networks. Supposing we have two social networks, G^s and G^t, the UIL problem is to determine whether a pair of users, (v_i^s, v_k^t), $v_i^s \in V^s, v_k^t \in V^t$, corresponds to the same real natural person, which can be formally defined as:

$$\Phi\left(v_i^s, v_k^t\right) = \begin{cases} 1, v_i^s = v_k^t, \\ 0, otherwise. \end{cases} \tag{1}$$

where $\Phi\left(v_i^s, v_k^t\right) = 1$ means v_i^s and v_k^t belong to the same individual.

3 NUIL: The Proposed Model

The existing methods usually apply network embedding to map the social network structure space to the low-dimensional vector space, and then transform the UIL problem into a binary classification problem using a standard neural network layer, whose ability to express the relations between users is relatively weak. Inspired by the success of neural tensor network for explicitly modeling multiple interactions of relational data, we propose a neural tensor network-based framework for user identity linkage across social networks.

As shown in Fig. 1, the framework consists of three main components: Network Embedding, Neural Tensor Network, and Multi-layer Perceptron. Firstly, we generate multiple social sequences by applying Random Walks to sample the network, and then embed each user with a vector using the Skip-gram model. Secondly, we use the neural tensor network to mine the complicated interactions between users. Thirdly, the Multi-layer Perceptron model is used to transform the UIL problem into a binary classification problem. We will explain our model in detail later.

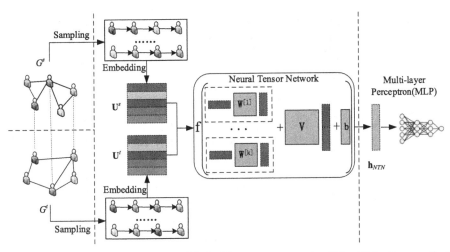

Fig. 1. The framework of NUIL: the red dashed lines represent anchor links. (Color figure online)

3.1 Network Embedding Based on Random Walks and Skip-Gram

To embed users into a latent space, we first generate multiple social sequences for each user in several rounds random walks, which encodes the social relationships among users in the social networks. All the sequences, called "corpus", are used in the Skip-gram model to learn the embedding vectors of users [11].

Network Structure Sampling Based on Random Walks. We conduct the network structure sampling as follows, taking the source network as an example. It starts at node v_m^s and proceeds along a randomly selected edge at each step, until length L is reached. The node sequence of v_m^s is written as $S_{v_m^s}^r$, where r is the rounds of sampling. Sampling by random walks can extract hidden structural social information, e.g., friendship and community in the social network.

User Latent Space Embedding. After getting the "corpus", we apply the Skip-gram model to generate the vector representations of nodes. Skip-gram model is originally used to predict the context of a word by maximizing the average log probability in the domain of word representation. We formally define the node sequence as $v_1, v_2, \ldots, v_L$, and maximize the log probability by the following equation:

$$\frac{1}{L} \sum\nolimits_{t=1}^{L} \sum\nolimits_{j=-w}^{w} \log p(v_{t+j}|v_t), j \neq 0 \tag{2}$$

where w is the size of the sliding window and L represents the length of the node sequence.

The conditional probability $p(v_{t+j}|v_t)$ is defined by a SoftMax function, which means the occurrence of the *j-hop* neighbor v_{t+j} given user v_t:

$$p(v_{t+j}|v_t) = \frac{\exp\left(u_{t+j}^T u_t^{'}\right)}{\sum_{i=1}^{L} \exp(u_i^T u_t^{'})} \tag{3}$$

where $\boldsymbol{u}_i$ and $\boldsymbol{u}'_i$ are, respectively, the input and output vector representations of user v_i.

But for a large-scale network, the calculation of $\sum_{i=1}^{L} \exp\left(\boldsymbol{u}_i^T \boldsymbol{u}'_t\right)$ is expensive. Therefore, the Negative Sampling [12] is adopted. The objective function is approximately converted as:

$$\log\left[\sigma\left(\boldsymbol{u}_{t+j}^T \boldsymbol{u}'_t\right)\right] + \sum_{i=1}^{K} \mathbb{E}_{v_i \sim p_n(v)}\left[\log\left(1 - \sigma\left(\boldsymbol{u}_i^T \boldsymbol{u}'_t\right)\right)\right] \tag{4}$$

where K is the number of negative examples. Empirically, each node is sampled with probability $p_n(v) \sim d_{v_i}^{3/4}$, where d_{v_i} is the degree of node v_i [13].

The objective function (2) is approximated by maximizing the objective function (4), and a vector representation of each node v_i is obtained by training using a stochastic gradient descent algorithm.

We apply the network embedding on the source and the target network respectively to obtain the corresponding vector spaces $\mathbf{U}^s$ and $\mathbf{U}^t$.

3.2   Modeling Relations Between Users Based on Neural Tensor Network

Neural Tensor Network. In deep learning literature, neural tensor network is originally proposed to reason the relationships between two entities in knowledge graph or used to classify relation between two entities [10]. It replaces a standard linear neural network layer with a bilinear tensor layer which relates two entity vectors across multiple dimensions.

Given two entities $(\mathbf{e}_1, \mathbf{e}_2)$ represented with d dimensional features, the goal of NTN is to predict whether they have a certain relationship R. Specifically, NTN computes a score of how likely it is that these two entities are in certain relationship R by the following function:

$$g(\mathbf{e}_1, R, \mathbf{e}_2) = \mu_R^T \tanh\left(\mathbf{e}_1^T \mathbf{W}_R^{[1:k]} \mathbf{e}_2 + \mathbf{V}_R \begin{bmatrix} \mathbf{e}_1 \\ \mathbf{e}_2 \end{bmatrix} + \mathbf{b}_R\right) \tag{5}$$

where $\mathbf{e}_1, \mathbf{e}_2 \in \mathbb{R}^d$ are the vector representations of two entities, $\mathbf{W}_R^{[1:k]} \in \mathbb{R}^{d \times d \times k}$ is a tensor and the bilinear tensor product $\mathbf{e}_1^T \mathbf{W}_R^{[1:k]} \mathbf{e}_2$ results in a vector $\mathbf{h} \in \mathbb{R}^k$, where each entry of $\mathbf{h}$ is computed by one slice $i(i = 1, \cdots, k)$ of the tensor: $h_i = \mathbf{e}_1^T \mathbf{W}_R^{[i]} \mathbf{e}_2$. The other parameters for relation R are the standard form of a neural network: $\mathbf{V}_R = \mathbb{R}^{k \times 2d}$ and $\mathbf{b}_R \in \mathbb{R}^k$. $\mu_R^T \in \mathbb{R}^k$ is used to convert the output of layer to a scalar as a score of the pair of entities for the specific relation, which is high when the entities contain the relation.

Modeling Relations Between Identities in NUIL. The neural tensor network models the relationships between two entities with a bilinear tensor product. We can naturally expand this idea: modeling relations between users, one from the source network G^s and another from the target network G^t, based on the NTN model.

Specifically, for any pair of users $(v_m^s, v_n^t), v_m^s \in G^s, v_n^t \in G^t$, we model the relationship between them according to the following equation:

$$\mathbf{h}_{NTN}\left(\boldsymbol{u}_m^s, \boldsymbol{u}_n^t\right) = f\left(\left(\boldsymbol{u}_m^s\right)^T \mathbf{W}^{[1:k]} \boldsymbol{u}_n^t + \mathbf{V}\begin{bmatrix} \boldsymbol{u}_m^s \\ \boldsymbol{u}_n^t \end{bmatrix} + \mathbf{b}\right) \tag{6}$$

where $\begin{bmatrix} \cdot \\ \cdot \end{bmatrix}$ is the concatenation operator on two column vectors.

Through the NTN model described above, we represent the relationship between a pair of nodes (v_m^s, v_n^t) as $\mathbf{h}_{NTN}\left(\boldsymbol{u}_m^s, \boldsymbol{u}_n^t\right)$. The difference from the original NTN model is that after obtaining the vector $\mathbf{h}_{NTN}$, we do not apply $\boldsymbol{\mu}^T$ to convert it to a scalar, but use it as the input of a Multi-Layer Perceptron.

3.3 Matching Identities Based on Multi-Layer Perceptron

The Multi-Layer Perceptron (MLP) is also known as Artificial Neural Network (ANN). There can be multiple hidden layers between the input layer and the output layer, and the layers are fully connected. In addition to the input layer, each node is a neuron with a nonlinear activation function.

The basic problem to be solved by neural networks is the classification problem, and the classic neural network model is MLP. In this paper, we apply MLP to transform the problem of node matching across social networks into a classification problem. Specifically, for any pair of nodes $(v_m^s, v_n^t), v_m^s \in G^s, v_n^t \in G^t$, with the ground-truth label g_{label}, we use NTN structure to model the complex interactions between them as a vector $\mathbf{h}_{NTN}\left(\boldsymbol{u}_m^s, \boldsymbol{u}_n^t\right)$. And then we input it to MLP, and output the predicted label p_{label} to achieve a binary classification.

$$g_{label}\left(v_m^s, v_n^t\right) = \begin{cases} 1, & v_m^s = v_n^t, \\ 0, & otherwise. \end{cases} \tag{7}$$

$$p_{label}\left(v_m^s, v_n^t\right) = MLP\left(\mathbf{h}_{NTN}\left(\boldsymbol{u}_m^s, \boldsymbol{u}_n^t\right)\right) \tag{8}$$

Therefore, combining the formulas (6) (7) (8), we use the cross entropy to construct the loss function of the entire model as:

$$Loss(\boldsymbol{\Omega}, \mathcal{D}, \mathcal{L}) = -\sum\nolimits_{(v_m^s, v_n^t) \in \mathcal{D}} [g_{label} \log p_{label} + (1 - g_{label}) \log(1 - p_{label})] \tag{9}$$

where $\mathcal{D}$ represents the set of node pairs used for model training, and $\mathcal{L}$ is the ground-truth labels corresponding to $\mathcal{D}$. $\boldsymbol{\Omega}$ is the set of parameters in the model. Please note that we abbreviate $g_{label}\left(v_m^s, v_n^t\right)$ and $p_{label}\left(v_m^s, v_n^t\right)$ as g_{label} and p_{label} respectively.

Assuming that the set of known anchor links is M, we construct the node pairs according to the ratio of positive and negative samples as $1 : C$. We apply the back-propagation algorithm and stochastic gradient descent algorithm to train the NUIL model in a supervised way. Finally, we can get a complete model for user identity linkage across social networks.

4 Experiments

In this section, we compare the proposed NUIL model with several state-of-the-art models on a real-world dataset consisting of two real social networks.

4.1 Dataset, Baselines and Parameter Setup, and Evaluation Metrics

Dataset. The real-world dataset is provided by [7], and it contains two social networks, Twitter and Foursquare. Table 1 summarizes the statistics of this dataset.

Table 1. Statistics of twitter-foursquare dataset.

Networks	#Users	#Relations	#Anchor users
Twitter	5120	164919	1609
Foursquare	5313	76792	

The number of anchor links is 1609, which can be seen as positive instances. Firstly, We set the ratio between positive instances (*Anchor Links*) and negative instances (*Corrupted Anchor Links*) to 1: 1. That is to say, for each anchor user v_i^s in Twitter, we randomly select one user from Foursquare, which is not corresponding to v_i^s, to construct a negative instance. Secondly, we set the ratio between training set, validation set, and test set to 8: 1: 1.

Baselines and Parameter Setup. The model we proposed in this paper is based on network structure, so we compare NUIL with several structure-based methods for UIL.

- **PALE:** Predicting Anchor Links via Embedding (PALE) [8] employs network embedding (such as DeepWalk) with awareness of observed anchor links as supervised information to capture the major and specific structural regularities and further learns a stable cross-network mapping for predicting anchor links.
- **FRUIP:** Structure Based User Identification across Social Networks [9], considers friends relationship with unsupervised learning. First, the friend feature vector is extracted with the network neighborhood patterns, and then compute their similarities for linkage prediction.

Parameter Setup. For the NUIL model proposed in this paper, the vector dimensionality d is 64, the number of tensor layers k in NTN is 8, and we set the MLP with two hidden layers: $32d$ (first hidden layer), $8d$ (second hidden layer) and $1d$ (output layer). The learning rate for training is 0.001 and the batch size is set to 8. The baselines are implemented according to the original papers.

Evaluation Metrics. Precision, Recall, and F1-measure are common metrics to evaluate the performance of a classifier. In this paper, we also evaluate all the methods in terms of Precision, Recall and F1-measure.

4.2   User Identity Linkage Performance

In this section, we present the performance of all methods on twitter-foursquare dataset.

Results Analysis. The Precision, Recall and F1 of all the algorithms are shown in Table 2. Based on the experimental results, we perform the following groups of comparisons to analyze the results.

Table 2. Comparisons of P, R and F1 on Twitter-Foursquare Dataset.

	Precision	Recall	F1
PALE [8]	0.5059	0.5342	0.5197
FRUIP [9]	0.5733	0.5342	0.5531
NUIL	**0.6437**	**0.6956**	**0.6686**

- **PALE - FRUIP.** As can be seen from Table 2, the performance on Precision of the two baseline methods exceeds 0.5. And, we can see that FRUIP, which considers the friend relationship in social networks, performs better than PALE which applies the traditional random walks-based network embedding.
- **PALE - NUIL.** The two methods both apply the traditional random walks-based network embedding. But we can find that NUIL, using the neural tensor model to mine the complicated interactions, is 27% higher than PALE on Precision. We can intuitively see the efficient performance of NTN model in solving UIL problem.
- **FRUIP - NUIL.** From Table 2, we can see that the three evaluation metrics of our method are all above 0.6. Specifically, NUIL is more than 12% higher than FRUIP on Precision.

Parameter Analysis. Through the above experimental results analyses, we can directly see the effectiveness of the NUIL model on user identity linkage across social networks. In this section, we analyze the influence of parameters on the problem of UIL, such as the percentage p of anchor nodes used for training, the vector dimensionality d.

We set the dimensionality to 16, 32, 64, and 128 respectively and the percentage to 0.2, 0.4, 0.6, and 0.8. Figure 2 shows the changes of F1-measure with percentage p and dimensionality d. On the whole, the F1-measure of NUIL model gradually increases and reaches convergence as the vector dimensionality and the percentage of anchor nodes used for training increase.

4.3   Discussions

According to our experiments, we have the following discussions.

- By comparing two pairs of models, PALE-FRUIP and FRUIP-NUIL, we can intuitively find that replacing a standard linear neural network with NTN model is very effective on the problem of user identity linkage across social networks.

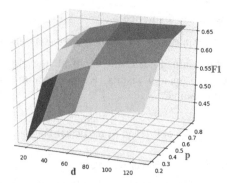

Fig. 2. Parameter Analysis on the vector dimensionality d and the percentage p of anchor nodes used for training.

- The NUIL model can not only be conveniently combined with the current popular network embedding methods, but also can easily be combined with user's attribute information in social networks, such as user's profiles or user's activity patterns.

5   Conclusion

In this paper, we studied the problem of use identity linkage across social networks and proposed a novel model, called NUIL. As the current mainstream methods did, we also applied the network embedding to map the network structure space to the low-dimensional vector space to capture the structural features of social network. Different from the traditional methods with the idea of nodes matching, we introduced the Neural Tensor Network into UIL to convert the node matching problem into a classification problem. The NTN model replaces a standard linear neural network layer with a bilinear tensor layer that directly relates the two entity vectors across multiple dimensions. Several experiments conducted on a real-world dataset indicate the effectiveness of NUIL. In the future work, we are committed to combining more comprehensive network embedding methods with neural tensor network model, such as global structure features of social network, community structure, and user attribute information.

Acknowledgements. This work was supported by the National Natural Science Foundation of China (U1636219, 61602508, 61772549, U1736214, 61572052, U1804263, 61872448) and Plan for Scientific Innovation Talent of Henan Province (No. 2018JR0018).

References

1. Kong, X., Zhang, J., Yu, P.: Inferring anchor links across multiple heterogeneous social networks. In: The 22nd International Conference on Information & Knowledge Management, pp. 179–188. ACM (2013)
2. Shu, K., Wang, S., Tang, J., Zafarani, R., Liu, H.: User identity linkage across online social networks: a review. In: SIGKDD Explorations Newsletter, pp. 5–17. ACM (2017)

3. Zhang, J., Yu, P.: Multiple anonymized social networks alignment. In: 2015 IEEE International Conference on Data Mining, pp. 599–608. IEEE (2015)
4. Goga, O., Lei, H., Parthasarathi, S., Friedland, G., Sommer, R., Teixeira, R.: Exploiting innocuous activity for correlating users across sites. In: The 22nd International Conference on World Wide Web, pp. 447–458. WWW (2013)
5. Zafarani, R., Liu, H.: Connecting users across social media sites: a behavioral-modeling approach. In: The 19th Knowledge Discovery and Data Mining, pp. 41–49. ACM (2013)
6. Wang, C., Zhao, Z., Wang, Y., Qin, D., Luo, X., Qin, T.: DeepMatching: a structural seed identification framework for social network alignment. In: The 38th International Conference on Distributed Computing Systems, pp. 600–610. IEEE (2018)
7. Liu, L., Cheung, W., Li, X., Liao, L.: Aligning users across social networks using network embedding. In: The 25th International Joint Conference on Artificial Intelligence, pp. 1774–1780. IJCAI (2016)
8. Man, T., Shen, H., Liu, S., Jin, X., Cheng, X.: Predict anchor links across social networks via an embedding approach. In: The 25th International Joint Conference on Artificial Intelligence, pp. 1823–1829. IJCAI (2016)
9. Zhou, X., Liang, X., Du, X., Zhao, X.: Structure based user identification across social networks. IEEE Trans. Knowl. Data Eng. **30**(6), 1178–1191 (2018)
10. Socher, R., Chen, D., Manning, C., Ng, A.: Reasoning with neural tensor networks for knowledge base completion. In: Advances in Neural Information Processing Systems, pp. 926–934. NIPS (2013)
11. Perozzi, B., Al-Rfou, R., Skiena, S.: Deepwalk: online learning of social representations. In: The 20th Knowledge Discovery and Data Mining, pp. 701–710. ACM (2014)
12. Mnih, A., Teh, Y.: A fast and simple algorithm for training neural probabilistic language models. In: The 29th International Conference on Machine Learning, pp. 1751–1758. ICML (2012)
13. Mikolov, T., Dean, J.: Distributed representations of words and phrases and their compositionality. In: Advances in Neural Information Processing Systems, pp. 3111–3119. NIPS (2013)

An Efficient and Privacy-Preserving Physiological Case Classification Scheme for E-healthcare System

Gang Shen[1], Yumin Gui[2(✉)], Mingwu Zhang[1,3,4], Yu Chen[1], Hanjun Gao[5], and Yixin Su[6]

[1] School of Computer Science, Hubei University of Technology, Wuhan 430068, China
shengang@hbut.edu.cn, csmwzhang@gmail.com, ychen@hbut.edu.cn
[2] Department of Ophthalmology, Wuhan Puren Hospital, Wuhan 430080, China
5537468@qq.com
[3] School of Computer Science and Information Security,
Guilin University of Electronic Technology, Guilin 541004, China
[4] Hubei Key Laboratory of Intelligent Geo-Information Processing,
China University of Geosciences, Wuhan 430074, China
[5] China Nuclear Power Operation Technology Corporation, LTD,
Wuhan 430070, China
gaohj@cnnp.com.cn
[6] School of Automation, Wuhan University of Technology,
Wuhan 430070, China
suyixin@whut.edu.cn

Abstract. In this work, an efficient and privacy-preserving physiological case classification scheme for e-healthcare system (EPPC) is proposed. Specifically, a homomorphic cryptosystem combined with a support vector machine (SVM) algorithm is applied to efficiently classify the physiological cases without compromising patients' privacy. In terms of the EPPC, it has the capability of diagnosing the patient's symptom in a timely manner. In addition, a signature authentication technology applied in EPPC can efficiently prevent data from being forged or modified. Security analysis result shows that the proposed EPPC scheme has the following advantages: protect the privacy of patients; ensure that the classification parameters of SVM are secured. Compared with the existing works, the proposed EPPC scheme shows significant advantages in terms of computational costs and communication overheads.

Keywords: E-healthcare system · Privacy protection · Physiological case classification · Homomorphic cryptosystem

1 Introduction

With the rapid increase of the aging population, the limited medical resources are far from satisfying the high requirement of medical services in quality [3,6,13].

D. Wang et al. (Eds.): SPNCE 2020, LNICST 344, pp. 172–185, 2021.
https://doi.org/10.1007/978-3-030-66922-5_12

Fortunately, e-healthcare, a new solution and one of the latest popular research fields, can effectively meet the demand for health monitoring and the limited medical resources. As a part of e-healthcare system, wireless body area network (WBAN), a popular technology, has the capability of diagnosing and monitoring of the patient's physical health in a real-time manner.

The e-healthcare system has following benefits [4,14]: i) provide remote health monitoring and real-time diagnosis for patients; ii) realize the online communication between patients and physicians; iii) improve the efficiency of medical treatment, and reduce the cost of medical treatment. However, security and privacy problems in e-healthcare system also pose enormous challenges [10,19]. For example, with these physiological data, a rogue is easier to deduce the physical condition of the patients, which may cause further psychological and physical harm to the patients. Therefore, protecting patients' privacy is imperative in the e-healthcare system.

Support vector machine (SVM), a machine learning tool, is commonly used in various fields, such as disease prediction [16], face recognition [8], text classification [11], handwriting recognition [12] and bioinformatics [2] etc. In general, training and testing are two parts of SVM [9]. The classification parameters of SVM classifier are obtained by creating the characteristics of different types of data sets in the training phase [5]. In the testing phase, all unlabeled data samples are classified by the classification parameters of SVM classifier and marked as a matching class. Therefore, classifiers with clinical data sets are used to identify patient data in medical diagnosis system in the testing phase. In this paper, the SVM is used to solve the problem of patient physiological case classification in the e-healthcare system.

In view of the sensitivity of information (e.g., patients' health information, medical institution information, etc.) in the process of e-health medical diagnosis, it is imperative to save the patients' physiological data and the classification parameters of the SVM classifier in healthcare centre (HC). In other words, the sensitive data in e-healthcare system shall be fully protected and shall not be compromised. Although numerous up-to-date clinical decisions based on classification schemes are proposed [15–18], the SVM tools are rarely used in the machine learning. Even if the SVM tools are used, these schemes have the disadvantages of large computation cost. What's worse, the method of obtaining the sign of an encrypted value is complicated, and the computational cost is also high. Therefore, the challenge in designing a secure and efficient scheme for physiological case classification is significant in the research field.

In this work, we propose an EPPC scheme for e-healthcare system, enabling to classify the patients' data without compromising their privacy, thereby protecting the security of the e-healthcare system. Specifically, the **main contributions** of this work are as fourfold.

(1) Firstly, we propose an EPPC scheme for e-healthcare system by combining Okamoto-Uchiyama (OU) homomorphic cryptosystem technology and SVM in machine learning. In the proposed EPPC scheme, the homomorphic properties of the OU encryption scheme can be directly used to implement

operations on the ciphertext data, and SVM algorithm is applied to obtain the accurate diagnostic results. Therefore, it protects the privacy of patient user (PU) and HC and also helps PU obtain the results of online diagnosis quickly.

(2) Secondly, with the method of scaling variables in scheme [9] for reference, we successfully solve the following problems: OU scheme only supports integers; system variables can only be continuous. In addition, the proposed EPPC scheme also applies BLS [1] signature technology to prevent the PU's physiological data from being forged or modified.

(3) Thirdly, we develop a novel method to obtain a sign of the ciphertext of classification label. Because it is a lightweight method, it can significantly decrease the computation cost of HC in our scheme.

(4) Finally, in order to demonstrate the efficiency of EPPC, we build a simulator in JAVA and compare the performance of EPPC with related scheme. The comparison results show that our scheme is more efficient.

The rest of the paper is structured as follows. In Sect. 2, we present the system model and security requirements of this paper. Then, we introduce the preliminary knowledge in Sect. 3. The concrete EPPC scheme is proposed in Sect. 4. In Sect. 5, the security analysis is described. Next, we illustrate the performance evaluation in Sect. 6. Finally, we summarize this paper in Sect. 7.

2   System Model and Security Requirements

In this section, we will present the system model, security requirements and design goals related to this paper.

2.1   System Model

The entities in system model include a healthcare centre and a patient user, as shown in Fig.1. The description is detailed as follows.

Healthcare centre (HC): HC is an incompletely trusted entity which can normally run the protocol as specified one but may try to learn about the maximum physiological data from the protocol under the influence of the adversaries. There are a large number of case training sets in the database of HC. HC is responsible for helping PU classify the physiological data, thus diagnosing his/her disease condition.

Patient user (PU): PU's physiological data and personal information can be sent to HC through the sensor devices. The physiological data is encrypted by user before it is sent to protect the patient's privacy.

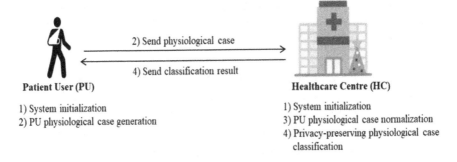

Fig. 1. System model of EPPC under consideration.

2.2 Security Requirements

The proposed EPPC scheme mainly meets the security requirements of the following three aspects.

(1) *Privacy.* The physiological data, reflecting PU's health status, involves PU's privacy, which should be protected. In other words, even if HC obtains PU's physiological case, it cannot identify PU's physiological data. Moreover, even if the adversary $\mathcal{A}$ obtains PU's physiological data from HC' databases, he/she still does not know about PU's health status.

(2) *Authentication.* The authentication of PU's physiological case enables HC to know whether the physiological case is sent by a valid PU, thereby preventing the physiological case from being modified or forged by adversary $\mathcal{A}$ during the transmission.

(3) *Confidentiality.* Since the training set data and the parameters of the SVM classification in HC are obtained by spending a lot of time and financial resources, these parameters are confidential data for HC. Even if PU receives the diagnosis result from HC, he/she will not learn any information of these parameters.

3 Preliminaries

3.1 Bilinear Pairing

Let $\mathbb{G}_1$, $\mathbb{G}_2$ be two cyclic groups of the same prime order q and $e : \mathbb{G}_1 \times \mathbb{G}_1 \to \mathbb{G}_2$ is a bilinear map that satisfies the properties as following:

(1) *Bilinearity.* $e(aP, bQ) = e(P, Q)^{ab}$ for all $a, b \in \mathbb{Z}_q^*$, and $P, Q \in \mathbb{G}_1$ are generators.
(2) *Non-degeneracy.* $e(P, Q) \neq 1$.
(3) *Computability.* $e(P, Q)$ can be computed by an efficient algorithm.

Definition 1. Let $\mathcal{G}en$ be a probabilistic algorithm, and input a security parameter κ, then output $(q, P, \mathbb{G}_1, \mathbb{G}_2, e)$, where q is a κ-bit prime number.

Definition 2. (Computational Diffie-Hellman (CDH) Problem). The CDH problem can be described as follows: For $a, b \in \mathbb{Z}_q^*$, given $(P, aP, bP) \in \mathbb{G}_1$, compute $abP \in \mathbb{G}_1$.

3.2   Support Vector Machine (SVM)

SVM, a two-class or multi-class classification model, is used to process data classification [9], which attracts wide attention and has been used for a long time since its birth. The liner classifier with the largest interval in the feature space can be used as the basic model of SVM. However, most of the data are not linearly divisible. In that way, it is necessary to map the sample from the original space to a higher dimensional feature space with the help of kernel function, so that the sample can be linearly separated in this feature space.

For linear classification problem, the training samples are linearly separable, so the decision function is

$$d(\mathbf{x}) = \mathbf{w}^T\mathbf{x} + b = \sum_{i \in S} \alpha_i y_i \mathbf{x}_i^T \mathbf{x} + b, \tag{1}$$

where α_i are Lagrangian variables and $\mathbf{x}_i \in S$ are support vectors for $i = 1, \cdots, |S|$, S is the set of support vectors. In addition, $\mathbf{w}$ and b are the classification parameters, and $y_i \in \{+1, -1\}$ is the classification label of sample $\tilde{\mathbf{x}}_i$ for $i = 1$ to $|S|$. For non-linear classification problem, the training samples are not linearly inseparable, then the dot product (i.e., $\mathbf{x}_i^T\mathbf{x}$) in Eq. 1 should be substituted by different kernel functions. So the decision function can be modified as

$$d(\mathbf{x}) = \sum_{i \in S} \alpha_i y_i k(\mathbf{x}_i, \mathbf{x}_j) + b. \tag{2}$$

In general, the samples with high feature dimension are linearly divisible, and the linear kernel function can be considered. Because the dimension of the sampled data in our scheme is high, we only consider a linear kernel in this work. Therefore, the decision function is

$$d(\mathbf{x}) = \sum_{i \in S} \alpha_i y_i (\mathbf{x}_i^T \mathbf{x}_j) + b. \tag{3}$$

3.3   Okamoto-Uchiyama (OU) Homomorphic Cryptosystem

The OU cryptosystem, proposed by Okamoto and Uchiyama [7], supports the additive homomorphism. In the OU cryptosystem, when the security parameter is 512 bits, the size of plaintext and ciphertext are approximately 512 bits and 1536 bits, respectively. The OU cryptosystem includes the following three algorithms:

(1) *Key generation.* Given the security parameter κ', select two large primes p' and q' with the same length $|p'| = |q'| = \kappa'$, and calculate $N = p'^2 q'$.

Then, choose $g \in \mathbb{Z}_N^*$ such that $g^{p'} \neq 1 \bmod p'^2$, and let $h = g^N \bmod N$. $pk = (N, g, h)$ and $sk = (p', q')$ are the public key and private key of the cryptosystem, respectively.

(2) *Message encryption.* Choose a random number $r \in \mathbb{Z}_N$, and calculate the ciphertext $C = E(m) = g^m \cdot h^r \bmod N$, where m is a message, $0 \leq m < 2^{\kappa'-1}$.

(3) *Message decryption.* Give the ciphertext $C \in \mathbb{Z}_N$, the corresponding message can be decrypted by calculating $m = D(C) = \frac{L(C^{p'-1} \bmod p'^2)}{L(g^{p'-1} \bmod p'^2)} \bmod p'$, where $L(x) = \frac{x-1}{p'}$.

In addition, the OU cryptosystem satisfies the additive homomorphism, and the specific form is as follow:

$$
\begin{aligned}
D(E(m_1) \cdot E(m_2) \bmod N) &= D(g^{m_1} h^{r_1} \cdot g^{m_2} h^{r_2} \bmod N) \\
&= D(g^{m_1+m_2} h^{r_1+r_2} \bmod N) \\
&= D(E(m_1 + m_2) \bmod N)
\end{aligned}
$$

where $0 \leq m_1 + m_2 < 2^{\kappa'-1}$.

4   The Proposed EPPC Scheme

In this section, we give the concrete EPPC scheme, which consists of five parts: system initialization, PU physiological case generation, PU physiological case normalization and privacy-preserving physiological case classification. In order to make reader have a better understanding, we only consider the classification of one PU's physiological case by HC in our work.

In EPPC scheme, we assume that HC owns the case training set of points $\tilde{\mathbf{t}}_i \in \mathbb{R}^n$, $i = 1, \cdots, n$, where $\tilde{\mathbf{t}}_i = (\tilde{t}_{i1}, \tilde{t}_{i2}, \cdots, \tilde{t}_{in})$ and each point $\tilde{\mathbf{t}}_i$ belongs to one of the two classes denoted by the label $y_i \in \{-1, +1\}$, $i = 1, \cdots, n$. Therefore, we can use these case samples to train SVM to classify the unlabeled test sample. In addition, the training case shall be normalized to keep their values on the same scale to prevent a large original samples from biasing the solution [10]. The normalized training case samples can be denoted as

$$
\mathbf{t}_i = \frac{\tilde{\mathbf{t}}_i - \bar{\mathbf{t}}}{\sigma^2}, \tag{4}
$$

where $\mathbf{t}_i \in \mathbb{R}^n$, $i = 1, \cdots, n$, σ and $\bar{\mathbf{t}}$ denote the standard deviation and the mean of the training case samples, respectively.

PU in our scheme means undiagnosed patient who has an n-dimensional physiological data $\tilde{\mathbf{d}} = (\tilde{d}_1, \tilde{d}_2, \cdots, \tilde{d}_n)$ received by sensors, where $\tilde{\mathbf{d}} \in \mathbb{R}^n$, $i = 1, \cdots, n$. These elements in the vector $\tilde{\mathbf{d}}$ can be expressed as different physiological indicators of PU. For example, d_1 represents blood sugar, d_2 denotes blood pressure, d_3 means blood lipids, etc. In this work, PU encrypts its physiological data and uploads them to HC's data server. In that case, the data normalization should be carried out in the form of ciphertext, and the specific process can be found in Sect. 4.3.

4.1 System Initialization

Given the security parameter κ, PU generates $(q, \mathbb{G}_1, \mathbb{G}_2, P, e)$ by running a key generation algorithm $\mathcal{G}en(\kappa)$, and then computes the OU cryptosystem's public key $pk = (N = p'^2q', g, h)$ and the corresponding private key $sk = (p', q')$, where p' and q' are two large primes with the same length $|p'| = |q'| = \kappa'$. PU also chooses a random number $x_u \in \mathbb{Z}_q^*$ as his/her signature private key, and computes the corresponding public key $Y = x_u P$. Additionally, PU chooses a cryptographic hash function H, where $H : \{0,1\}^* \to \mathbb{G}_1$.

Finally, PU exposes the common parameters $(q, \mathbb{G}_1, \mathbb{G}_2, P, e, N, g, h, Y, H)$, and keeps the private key (p', q', x_u) secretly.

HC randomly chooses a sufficiently large positive number ξ as a scaling factor to scale the decision function, which enables all the variables contained in the decision function to be quantized to the nearest integer value. HC also randomly selects two integers A and B to meet the following two conditions: i) $A > B$ and ii) $|A \cdot d(\mathbf{d}) + B| < 2^{\kappa'-2}$, where $|d(\mathbf{d})| < 2^l$ and $l \in \mathbb{Z}_N$. Since the data in our scheme is high dimensional, HC chooses a decision function with linear kernel function as

$$d(\mathbf{d}) = \sum_{i \in S} \alpha_i y_i \mathbf{t}_i^T \mathbf{d} + b, \tag{5}$$

where α_i are Lagrangian variables, $y_i \in \{+1, -1\}$ is the classification label of sample $\tilde{\mathbf{t}}_i$ for $i = 1$ to $|S|$, $\mathbf{t}_i$ and $\mathbf{d}$ are normalized training case samples for $i = 1, \cdots, |S|$ and normalized PU's physiological data, respectively. Here, $(\alpha_i, y_i, \mathbf{t}_i, b)$ are the classification parameters.

In the end, HC keeps $(\alpha_i, y_i, \mathbf{t}_i, b, \xi, A, B)$ secretly.

4.2 PU Physiological Case Generation

PU collects the physiological data $\tilde{\mathbf{d}} = (\tilde{d}_1, \tilde{d}_2, \cdots, \tilde{d}_n)$ through implantable, wearable or environmental sensor devices. The physiological data should be encrypted before they are sent to HC. Specifically, PU encrypts each element of the physiological data $\tilde{\mathbf{d}}$ individually using OU homomorphic cryptosystem to generate PU's physiological case $[\![\tilde{\mathbf{d}}]\!]$. The specific process is as follows:

(1) PU selects a random number $r_i \in \mathbb{Z}_N$, and calculates

$$[\![\tilde{d}_1]\!] = g^{\tilde{d}_1} \cdot h^{r_1} mod N,$$
$$[\![\tilde{d}_2]\!] = g^{\tilde{d}_2} \cdot h^{r_2} mod N,$$
$$\cdots \cdots$$
$$[\![\tilde{d}_n]\!] = g^{\tilde{d}_n} \cdot h^{r_n} mod N,$$

where $i = 1, \cdots, n$. Then, the PU's physiological case is generated as

$$\begin{aligned}
[\![\tilde{\mathbf{d}}]\!] &= (g^{\tilde{d}_1} \cdot h^{r_1} mod N, g^{\tilde{d}_2} \cdot h^{r_2} mod N, \cdots, g^{\tilde{d}_n} \cdot h^{r_n} mod N) \\
&= ([\![\tilde{d}_1]\!], [\![\tilde{d}_2]\!], \cdots, [\![\tilde{d}_n]\!]).
\end{aligned} \tag{6}$$

(2) Use the private key x_u to generate a signature δ as

$$\delta = x_u \cdot H([\![\tilde{\mathbf{d}}]\!] \| Y \| T), \tag{7}$$

where T denotes PU's current timestamp.

(3) Send PU's physiological case $[\![\tilde{\mathbf{d}}]\!] \| Y \| T \| \delta$ to HC.

4.3   PU Physiological Case Normalization

Because the OU cryptosystem does not support non-integers and the test samples and variables in SVM classification are continuous values [10], PU's physiological case should be quantized to the nearest integer value by HC before classification. The specific method is similar to Eq. 4.

(1) After receiving PU's physiological case $[\![\tilde{\mathbf{d}}]\!] \| Y \| T \| \delta$, HC first verifies the validity of the timestamp and signature. If $e(P, \delta) = e(Y, H([\![\tilde{\mathbf{d}}]\!] \| Y \| T))$ does hold, the signature is accepted. Since $Y = x_u P$ and $\delta = x_u \cdot H([\![\tilde{\mathbf{d}}]\!] \| Y \| T)$, $e(P, \delta) = e(P, x_u \cdot H([\![\tilde{\mathbf{d}}]\!] \| Y \| T)) = e(Y, H([\![\tilde{\mathbf{d}}]\!] \| Y \| T))$.

(2) Using the parameters such as mean $\bar{\mathbf{t}}$ and standard deviation σ of training case samples and scaling factor ξ, HC can compute the values $\{(p'-1)\frac{\xi \bar{t}_1}{\sigma^2}, \cdots , (p'-1)\frac{\xi \bar{t}_n}{\sigma^2}\}$.

(3) Since $(-1)\ mod\ p' = (p'-1)\ mod\ p'$, HC encrypts each of $\{(p'-1)\frac{\xi \bar{t}_1}{\sigma^2}, \cdots , (p'-1)\frac{\xi \bar{t}_n}{\sigma^2}\}$ by OU homomorphic cryptosystem as $\{[\![(-1)\frac{\xi \bar{t}_1}{\sigma^2}]\!], \cdots , [\![(-1)\frac{\xi \bar{t}_n}{\sigma^2}]\!]\}$.

(4) After verifying the signature, depending on the homomorphic property, HC computes the normalized value of each element in $\{[\![\tilde{d}_1]\!], [\![\tilde{d}_2]\!], \cdots , [\![\tilde{d}_n]\!]\}$ individually with $\bar{\mathbf{t}}$, σ and ξ as follows:

$$[\![\xi d_1]\!] = [\![\tilde{d}_1]\!]^{\frac{\xi}{\sigma^2}} \cdot [\![(-1)\frac{\xi \bar{t}_1}{\sigma^2}]\!] = [\![\tilde{d}_1]\!]^{\frac{\xi}{\sigma^2}} \cdot [\![\bar{t}_1]\!]^{(-1)\frac{\xi}{\sigma^2}}$$

$$= ([\![\tilde{d}_1]\!] \cdot [\![\bar{t}_1]\!]^{-1})^{\frac{\xi}{\sigma^2}} = [\![\frac{\tilde{d}_1}{\sigma^2} - \frac{\bar{t}_1}{\sigma^2}]\!]^{\xi}$$

$$= [\![\frac{\tilde{d}_1 - \bar{t}_1}{\sigma^2}]\!]^{\xi}$$

$$[\![\xi d_2]\!] = [\![\tilde{d}_2]\!]^{\frac{\xi}{\sigma^2}} \cdot [\![(-1)\frac{\xi \bar{t}_2}{\sigma^2}]\!] = [\![\tilde{d}_2]\!]^{\frac{\xi}{\sigma^2}} \cdot [\![\bar{t}_2]\!]^{(-1)\frac{\xi}{\sigma^2}}$$

$$= ([\![\tilde{d}_2]\!] \cdot [\![\bar{t}_2]\!]^{-1})^{\frac{\xi}{\sigma^2}} = [\![\frac{\tilde{d}_2}{\sigma^2} - \frac{\bar{t}_2}{\sigma^2}]\!]^{\xi} \tag{8}$$

$$= [\![\frac{\tilde{d}_2 - \bar{t}_2}{\sigma^2}]\!]^{\xi}$$

$$\cdots \cdots$$

$$[\![\xi d_n]\!] = [\![\tilde{d}_n]\!]^{\frac{\xi}{\sigma^2}} \cdot [\![(-1)\frac{\xi \bar{t}_n}{\sigma^2}]\!] = [\![\tilde{d}_n]\!]^{\frac{\xi}{\sigma^2}} \cdot [\![\bar{t}_n]\!]^{(-1)\frac{\xi}{\sigma^2}}$$

$$= ([\![\tilde{d}_n]\!] \cdot [\![\bar{t}_n]\!]^{-1})^{\frac{\xi}{\sigma^2}} = [\![\frac{\tilde{d}_n}{\sigma^2} - \frac{\bar{t}_n}{\sigma^2}]\!]^{\xi}$$

$$= [\![\frac{\tilde{d}_n - \bar{t}_n}{\sigma^2}]\!]^{\xi},$$

where $[\![\xi d_i]\!]$ denotes the element of scaled physiological case. Since ξ is large enough, $\frac{\xi}{\sigma^2}$ is guaranteed to be an integer. So $[\![\xi d_i]\!]$ is also an integer. Moreover, the scaled physiological case $[\![\xi \mathbf{d}]\!]$ can be written as

$$[\![\xi \mathbf{d}]\!] = ([\![\xi d_1]\!], [\![\xi d_2]\!], \cdots, [\![\xi d_n]\!]). \tag{9}$$

(5) After normalizing the PU's physiological case, HC will classify it.

4.4   Privacy-Preserving Physiological Case Classification

In the classification phase, PU only considers whether the physical condition is normal, so we presented the classification function of two-class problem involved SVM in this scheme. In addition, according to the description of Sect. 4.1, the kernel function in this scheme should be $(\mathbf{t}_i^T \cdot \mathbf{d})$. The specific classification process is as follows:

(1) By using the scaled physiological case $[\![\xi \mathbf{d}]\!] = ([\![\xi d_1]\!], [\![\xi d_2]\!], \cdots, [\![\xi d_n]\!])$ and normalized training case samples $\mathbf{t}_i = (t_{i1}, t_{i2}, \cdots, t_{in})$, HC computes the linear kernel in the form of ciphertext as

$$\begin{aligned}
[\![\mathbf{k}_i]\!] &= [\![\xi d_1]\!]^{\xi t_{i,1}} \cdots [\![\xi d_n]\!]^{\xi t_{i,n}} \\
&= [\![\xi t_{i,1} \cdot \xi d_1]\!] \cdots [\![\xi t_{i,n} \cdot \xi d_n]\!] \\
&= [\![\xi t_{i,1} \cdot \xi d_1 + \cdots + \xi t_{i,n} \cdot \xi d_n]\!] \\
&= [\![\xi \mathbf{t}_i^T \cdot \xi \mathbf{d}]\!].
\end{aligned} \tag{10}$$

It is clearly shown that HC calculates the kernel function value without any interaction with PU.

(2) HC calculates the ciphertext value of the decision function as follows:

$$\begin{aligned}
[\![d(\mathbf{d})]\!] &= [\![\xi^3 (\sum_{i \in S} \alpha_i y_i \mathbf{t}_i^T \mathbf{d} + b)]\!] \\
&= [\![\sum_{i \in S} \xi(\alpha_i y_i) \xi^2 (\mathbf{t}_i^T \cdot \mathbf{d}) + \xi^3 b]\!] \\
&= [\![\sum_{i \in S} \xi(\alpha_i y_i)(\xi \mathbf{t}_i^T \cdot \xi \mathbf{d}) + \xi^3 b]\!] \\
&= [\![\sum_{i \in S} \xi(\alpha_i y_i)(\xi \mathbf{t}_i^T \cdot \xi \mathbf{d})]\!] \cdot [\![\xi^3 b]\!] \\
&= [\![\xi^3 b]\!] \cdot \prod_{i \in S} [\![\mathbf{k}_i]\!]^{\xi(\alpha_i y_i)}.
\end{aligned} \tag{11}$$

Note that, the $[\![d(\mathbf{d})]\!]$ cannot be sent directly to PU, because he/she can decrypt it using his/her private key to obtain the secret information of SVM (e.g., $\alpha_i, y_i, \mathbf{t}_i, b$, etc.). Therefore, we describe how to transmit the physiological case classification result to PU without disclosing the secret information of SVM in the next steps.

(3) Suppose $|d(\mathbf{d})| < 2^l$ and $l \in \mathbb{Z}_N$, HC can calculate the ciphertext of the classification label using integers A, B and $d(\mathbf{d})$ as

$$
\begin{aligned}
[\![cl]\!] &= [\![d(\mathbf{d})]\!]^A \cdot [\![B]\!] \\
&= [\![A \cdot d(\mathbf{d}) + B]\!],
\end{aligned} \tag{12}
$$

where $|A \cdot d(\mathbf{d}) + B| < 2^{\kappa'-2}$, integers A and B are used only once, and new A and B are generated in each initialization phase.

(4) Next HC sends the classification label $[\![cl]\!]$ to PU.

(5) Upon receiving $[\![cl]\!]$, PU can recover the classification label by using his/her private key and obtain the diagnosis result. The correctness of the diagnostic query is described as follows:

5 Security Analysis

In this section, we conduct a security analysis of the proposed EPPC scheme. In this regard, the analysis will focus on how the proposed EPPC scheme can realize the privacy preservation and the source authentication of the PU's physiological data and the confidentiality of the HC's SVM classification parameters.

Theorem 1. *(Privacy): The privacy of PU's physiological data is protected in the proposed EPPC scheme.*

Proof. In EPPC, the PU's physiological data $\tilde{\mathbf{d}} = (\tilde{d}_1, \tilde{d}_2, \cdots, \tilde{d}_n)$ is encrypted before it is sent to HC. Specifically, each element of $\tilde{\mathbf{d}}$ individually using OU homomorphic cryptosystem as $[\![\tilde{\mathbf{d}}]\!] = (g^{\tilde{d}_1} \cdot h^{\tilde{r}_1} mod N, g^{\tilde{d}_2} \cdot h^{\tilde{r}_2} mod N, \cdots, g^{\tilde{d}_n} \cdot h^{\tilde{r}_n} mod N) = ([\![\tilde{d}_1]\!], [\![\tilde{d}_2]\!], \cdots, [\![\tilde{d}_n]\!])$. Since OU cryptosystem is secure of indistinguishability under the condition of chosen-plaintext attack (IND-CPA [7], the physiological data $\tilde{\mathbf{d}} = (\tilde{d}_1, \tilde{d}_2, \cdots, \tilde{d}_n)$ is semantic secure and privacy-preserving. Therefore, even if the adversary breaks into the database of HC, he/she can only obtain the ciphertext of PU's physiological data instead of identifying its specific content. After receiving $[\![\tilde{\mathbf{d}}]\!] = ([\![\tilde{d}_1]\!], [\![\tilde{d}_2]\!], \cdots, [\![\tilde{d}_n]\!])$ from PU, HC normalizes $[\![\tilde{\mathbf{d}}]\!]$ as

$$
\begin{aligned}
[\![\xi \mathbf{d}]\!] &= ([\![\frac{\tilde{d}_1 - \bar{t}_1}{\sigma^2}]\!]^\xi, [\![\frac{\tilde{d}_1 - \bar{t}_2}{\sigma^2}]\!]^\xi, \cdots, [\![\frac{\tilde{d}_n - \bar{t}_n}{\sigma^2}]\!]^\xi) \\
&= ([\![\xi d_1]\!], [\![\xi d_2]\!], \cdots, [\![\xi d_n]\!]).
\end{aligned} \tag{13}
$$

However, since HC does not have the private key (p', q') and the calculation of each element in Eq. 13 is performed by HC without interacting PU, even if HC obtain the data, he still cannot identify the PU's physiological data $\tilde{\mathbf{d}} = (\tilde{d}_1, \tilde{d}_2, \cdots, \tilde{d}_n)$. Therefore, the privacy of PU's physiological data is protected in the proposed EPPC scheme.

Theorem 2. *(Authentication): The authentication of the PU's physiological case is fulfilled in the proposed EPPC scheme.*

Proof. The description of the proposed EPPC scheme shows that the PU's physiological case is signed by BLS short signature [1] as $\delta = x_u \cdot H(\llbracket \tilde{\mathbf{d}} \rrbracket \| Y \| T)$. Because the BLS signature scheme is provably secure under the CDH problem in the random oracle model, the physiological case received by HC coming from a valid PU is assured. Meanwhile, the adversary $\mathcal{A}$'s malicious behaviors (such as falsifying signatures, modifying physiological case, etc.) can also be easily detected. Therefore, from the above analysis, the authentication of the PU's physiological case is achieved in the proposed EPPC scheme.

Theorem 3. *(Confidentiality): The training set and the SVM classification parameters of HC are confidential in the proposed EPPC scheme.*

Proof. In the privacy-preserving physiological case classification phase of the EPPC scheme, HC uses OU homomorphic cryptosystem to encrypt the linear kernel and the decision function, respectively. Specifically, the ciphertext of the linear kernel is $\llbracket \mathbf{k}_i \rrbracket = \llbracket \xi \mathbf{t}_i^T \cdot \xi \mathbf{d} \rrbracket$ and the ciphertext of the decision function is $\llbracket d(\mathbf{d}) \rrbracket = \llbracket \xi^3 b \rrbracket \cdot \prod_{i \in S} \llbracket \mathbf{k}_i \rrbracket^{\xi(\alpha_i y_i)}$. Because it is very rare to obtain the private key (p', q'), even if the adversary $\mathcal{A}$ steals the decision function $\llbracket d(\mathbf{d}) \rrbracket$ or linear kernel $\llbracket \mathbf{k}_i \rrbracket$, he is impossible to obtain the training set and the HC' SVM classification parameters.

Security analysis shows that our EPPC scheme can meet the security requirements.

6   Performance Evaluation

In this section, we will compare schemes EPPC and [10] from the following two aspects: computational cost and communication overhead.

6.1   Security Comparison

Firstly, we compare the security of scheme EPPC with that of related scheme [10]. The results of the comparison of performance characteristics are shown in Table 1. The two schemes all use homomorphic encryption scheme to encrypt data, and use SVM to obtain classification label. Therefore, patient privacy and SVM classification parameters can be protected in both schemes. However, the lack of message authentication in scheme [10] makes it possible for the patient's physiological data to be forged. In addition, scheme EPPC has obvious advantages in terms of the efficiency of obtaining diagnostic results. The specific analysis is as follows.

Table 1. Security comparison

Metrics/parameters	[10]	EPPC
Privacy preservation of patient	Yes	Yes
Privacy preservation of SVM classification parameters	Yes	Yes
Homomorphic encryption	Yes	Yes
SVM	Yes	Yes
Message authentication	No	Yes
Diagnostic efficiency	Low	High

6.2 Experimental Setup

All our evaluations were performed on Intel Core i7-6700 @3.10 GHz with 8 GB RAM. Our system ran Java with Win 10 64-bit. We performed our experiments using the code we wrote. The runtime of hash operation can be ignored because it is much smaller than the runtime of other operations. In addition, multiplication can be ignored compared to exponentiation in the same group operation. Here, we chose the same security parameters $|p'| = |q'| = 512$ bits to realize the same level of security.

6.3 Computational Cost

In this subsection, we analyze the computational cost of PU and HC in each phase.

We will compare the computational costs with the scheme [10]. Table 2 presents the comparison results.

Table 2. Comparison of computational costs

Schemes	PU/Patient	HC/Server	Total
[10]	$2nT_{ep}$	$(3n + |S| + 6)T_{ep}$	$(5n + |S| + 6)T_{ep}$
EPPC	$2nT_{eou} + T_m$	$(2n + |S| + 1)T_{eou} + 2T_b$	$(4n + |S| + 1)T_{eou} + T_m + 2T_b$

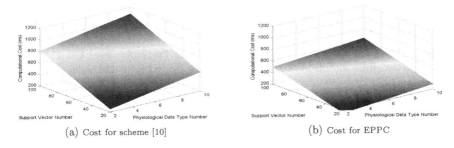

(a) Cost for scheme [10]   (b) Cost for EPPC

Fig. 2. Comparison of computational costs.

A comparison diagram of the total computational costs can be obtained from Table 2, as shown in Fig. 2. Figure 2 (a) and (b) represent the total computational cost of scheme [10] and EPPC, respectively. It is clearly shown that the EPPC scheme significantly decreases the computational cost for both patients and HC.

6.4   Communication Overhead

In this subsection, we analyse the communication overhead of the proposed EPPC scheme. It mainly focuses on the communication between PU and HC. First, we consider the PU-to-HC communication, in which PU generates his/her physiological case and sends this case to HC. Next, we consider the HC-to-PU communication, where HC generates the ciphertext of classification label and send it to PU.

Table 3 shows the communication overhead results of the two schemes.

Table 3. Comparison of communication overheads

Schemes	PU/Patient to HC/Server	HC/Server to PU/Patient	Total
[10]	$2048n$	2048	$2048(n+1)$
EPPC	$1536n + 260$	1536	$1536n + 1796$

7   Conclusion

In this paper, we have proposed an efficient and privacy-preserving physiological case classification scheme for e-healthcare system. This system employs the OU homomorphic cryptosystem, signature authentication technology and SVM algorithm to efficiently classify the physiological cases without leaking the privacy of PU and HC. The scheme can not only protect the physiological data of PU, but also prevent the leakage of HC classification parameters. What's important, the scheme has lower computational cost and communication overhead, so PU can quickly obtain diagnostic results in this scheme. Experimental results illustrate that the proposed EPPC scheme characterizes more efficient and practical.

Acknowledgment. We are grateful to the anonymous reviewers for their invaluable comments. Mingwu Zhang is the corresponding author. This research was supported by the National Natural Science Foundation of China (NSFC) under Grant No.61672010, the Ph.D research startup foundation of Hubei University of Technology under Grant No. BSQD2019023, and the open research project of The Hubei Key Laboratory of Intelligent Geo-Information Processing under Grant No. KLIGIP-2017A11.

References

1. Boneh, D., Lynn, B., Shacham, H.: Short signatures from the weil pairing. J. Cryptol. **17**(4), 297–319 (2004)

2. Byvatov, E., Schneider, G.: Support vector machine applications in bioinformatics. Appl. Bioinform. **2**(2), 67–77 (2003)

3. Kumar, P., Lee, H.J.: Security issues in healthcare applications using wireless medical sensor networks: a survey. Sensors **12**(1), 55–91 (2012)

4. Li, M., Lou, W., Ren, K.: Data security and privacy in wireless body area networks. IEEE Wirel. Commun. **17**(1), 51–58 (2010)

5. Liu, X., Deng, R., Choo, K, R., Yang, Y.: Privacy-preserving outsourced support vector machine design for secure drug discovery. IEEE Trans. Cloud Comput. **8**, 610–622 (2018). https://doi.org/10.1109/TCC.2018.2799219

6. Liu, X., Deng, R., Choo, K, R., Yang, Y.: Privacy-preserving reinforcement learning design for patient-centric dynamic treatment regimes. IEEE Trans. Emerg. Top. Comput. (2019). https://doi.org/10.1109/TETC.2019.2896325

7. Okamoto, T., Uchiyama, S.: A new public-key cryptosystem as secure as factoring. In: Proceedings of International Conference on the Theory and Applications of Cryptographic Techniques, pp. 308–318 (1998)

8. Punithavathi, P., Geetha, S., Karuppiah, M., Islam, S.H., Hassan, M.M., Choo, K.K.R.: A lightweight machine learning-based authentication framework for smart IoT devices. Inf. Sci. **484**, 255–268 (2019)

9. Rahulamathavan, Y., Phan, R.C.W., Veluru, S., Cumanan, K., Rajarajan, M.: Privacy-preserving multi-class support vector machine for outsourcing the data classification in cloud. IEEE Trans. Dependable Secure Comput. **11**(5), 467–479 (2014)

10. Rahulamathavan, Y., Veluru, S., Phan, R.C., Chambers, J.A., Rajarajan, M.: Privacy-preserving clinical decision support system using gaussian kernel based classification. IEEE J. Biomed. Health Inform. **18**(1), 56–66 (2014)

11. Tong, S., Koller, D.: Support vector machine active learning with applications to text classification. J. Mach. Learn. Res. **2**(1), 999–1006 (2002)

12. Vapnik, V.N.: An overview of statistical learning theory. IEEE Trans. Neural Netw. **10**(5), 988–999 (1999)

13. Wang, D., Cheng, H., He, D., Wang, P.: On the challenges in designing identity-based privacy-preserving authentication schemes for mobile devices. IEEE Syst. J. **12**(1), 916–925 (2018)

14. Wang, D., Li, W., Wang, P.: Measuring two-factor authentication schemes for real-time data access in industrial wireless sensor networks. IEEE Trans. Ind. Inform. **14**(9), 4081–4092 (2018)

15. Wang, G., Lu, R., Shao, J., Guan, Y.: Achieve privacy-preserving priority classification on patient health data in remote eHealthcare system. IEEE Access **7**(1), 33565–33576 (2019)

16. Xu, C., Wang, J., Zhu, L., Zhang, C., Sharif, K.: PPMR: a privacy-preserving online medical service recommendation scheme in ehealthcare system. IEEE Internet Things J. **6**(3), 5665–5673 (2019)

17. Yi, X., Bouguettaya, A., Georgakopoulos, D., Song, A., Willemson, J.: Privacy protection for wireless medical sensor data. IEEE Trans. Dependable Secure Comput. **13**(3), 369–380 (2016)

18. Zhang, L., Zhang, Y., Tang, S., Luo, H.: Privacy protection for e-health systems by means of dynamic authentication and three-factor key agreement. IEEE Trans. Ind. Electron. **65**(3), 2795–2805 (2018)

19. Zhang, Y., Lang, P., Zheng, D., Yang, M., Guo, R.: A secure and privacy-aware smart health system with secret key leakage resilience. Secur. Commun. Netw. **2018**(4), 1–13 (2018)

A Multi-class Detection System for Android Malicious Apps Based on Color Image Features

Hua Zhang[1], Jiawei Qin[1(✉)], Boan Zhang[1], Hanbing Yan[2], Jing Guo[2], and Fei Gao[1]

[1] State Key Laboratory of Networking and Switching Technology, Beijing University of Posts and Telecommunications, Beijing 100876, China
qinjiawei@bupt.edu.cn
[2] The National Computer Network Emergency Response Technical Team/Coordination Center of China, Beijing, China

Abstract. The visual recognition of Android malicious applications(Apps) is mainly focused on the binary classification using gray-scale images, while the multi-classification of malicious App families is rarely studied. If we can visualize the Android malicious Apps as color images, we will get more features than using grayscale images. In this paper, a method of color visualization for Android Apps is proposed and implemented. Based on this, combined with deep learning models, a multi-classifier for the Android malicious App families is implemented, which can classify 131 common malicious App families. Compared with the App classifier based on the gray-scale visualization method, it is verified that the classifier using the color visualization method can achieve better classification results. This paper uses three classes of Android App APK features: classes.dex file, class name collection and API call sequence as input for App visualization, and analyzes the classifier detection accuracy and detection time under each input characteristics. According to the experimental results, we found that using the API call sequence as the color visualization input feature can achieve the highest detection accuracy rate, which is 96.01% in the ten malicious family classification and 100% in the binary classification.

Keywords: Android malicious Apps · Deep learning · Visualization · Multi-class detection

1 Introduction

The openness of the Android system, while helping it win the market, has also brought it huge risks. According to the CVE [2](Common Vulnerabilities Exposures) 2018 annual report, the Android system ranks second in the vulnerability list with 611 vulnerabilities. They have brought more opportunities to malicious App developers. As a large amount of user data is connected to the

D. Wang et al. (Eds.): SPNCE 2020, LNICST 344, pp. 186–206, 2021.
https://doi.org/10.1007/978-3-030-66922-5_13

Internet via mobile phones and spread on the network, the target of hacking is gradually shifting from traditional PCs to mobile devices. As a result, more and more researchs [6, 20, 21, 24, 26, 30] focused on analyzing Android malicious Apps.

A difficult but important issue in the Android malicious App family classification is how to classify malicious Apps in the presence of a large number of families and achieve high accuracy. With the proliferation of Android malicious Apps, there are more and more Android malicious App families. How to distinguish the endless Android malicious App families has become a greater challenge. Existing research shows that malicious behaviors between malicious App families overlap more and more. The detection standards manually formulated after feature extraction cannot distinguish between families with high similarity, and the accuracy of fingerprint-based methods is getting lower and lower [30].

Using machine learning methods to classify Android malicious Apps has achieved high accuracy [6, 8, 10, 13, 15, 25]. However, due to its feature generation engineering that relies on expert knowledge, it is difficult for the above-mentioned classifiers to maintain a high accuracy rate after the changes of malware behavior trigger method. Joshua et al. [11] used a machine learning method of classification regression tree to study a family classifier that can classify 33 malicious App families manually labeled in the AMG [29] data set, achieving 95% accuracy. Wang [24] and others proposed the use of deep learning detection methods to implement Android malware detection systems, nonetheless, it did not study the implementation of multi-classification of malware. Andronio [6] analyzed the behavioral characteristics of Android ransomware and implemented a detection model for ransomware.

In exploring the visualization of malicious software, Nataraj et al. [19] drew the gray-scale image of Windows malicious software in a linear way, and Applied GIS T (Gabor filter) to image to obtain features. Then using K-Nearest Neighbor (KNN) algorithm as an automatic classification technology to classify 25 malicious software families and reached a accuracy rate of 98%. Jung et al. [16]used gray-scale image and convolution neural network model to conduct binary classification experiments on Android malware and benign. Nonetheless, no research has been done on color visualization or Android malicious App family classification. It focused on the benefits of visualizing the "data" section of the classes.dex file. And at the end of the article, it was pointed out that the direction of future research is the method of color image visualization.

In the common deep learning model, three-channel color images are used as training samples. For deep learning classifiers, compared to grayscale images, color image visualization theoretically has a higher dimension and more processing flows, so more features are learned and classification accuracy is higher. However, in the existing research, there is no method to classify the Android malicious family using only color image visualization.

In this paper, we classify Android malicious into multiple families by color visualization combined with deep learning. We propose a method of color visualizing Android App, study the features suitable for color visualization, and verify the effect of this method on the classification of a large number of malicious App families with overlapping malicious behaviors. The specific contributions of this paper as follows:

- A method of color visualization Android App is proposed and applied to malicious App family classification. In view of the better performance of the deep learning classification model on color picture classification tasks, this paper studies the effect of using gray image features and color image features in the Android malicious App family classification, which validates the feasibility of the App of color image visualization to the Android malware families, and proposes a color image visualization method for Android malicious family classification.
- The influence of different features on Android malicious App family classification is studied, and the features that are most suitable for color visualization are obtained. We analyzed the possible collections of Android malicious App features, and selected three more common collections as experimental objects: *classes.dex* file, App class name collection, and App interface call(API) collection. We performed color visualization on each feature, and conducted classification experiments. Using the deep learning method to study the performance of the three features in classification time and accuracy. Finally, according to the experiment results, it is judged that using API call sequence characteristics is the best choice.
- A classifier is implemented for a large number of malicious App families with overlapping malicious behaviors. After analyzing the characteristics of malicious App families, it is found that the increasing number of malicious App families brings difficulties to family classification: the similarity between families increases, and similar malicious behaviors overlap. We used color visualization combined with convolutional neural networks and deep residual networks to classify 131 malicious App families and reached a classification accuracy of 96.36%.

2   Related Work

Android malicious App visualization is a new trend in recent years. One of the common methods for the visualization of binary files comes from the paper by Conti et al. [9]. They used four different ways to visualize binary files. The first method is to draw each byte linearly to generate gray-scale images, where empty bytes are described as black pixels and the 0xff bytes are described as white pixels. The second method is to color a portion of the bytecode to indicate the presence or absence of a particular byte value. This method is especially useful for finding compressed portions or ascii code portions. Third, the traditional hex editor is implemented, which converts the binary to hexadecimal and then colors it. Fourth, using dot plots to show the cross entropy of a file, a dot plot is a way to visualize the similarity or self-similarity of data.

In the exploration of malware visualization, Gennisse et al. [12] used a partial color visualization method to study Android malicious family classification.

Zhang et al. [27] decompiled the executable file to get the opcode sequence, and then converted these sequences into the form of an image, and finally performed further feature extraction and recognition through CNN. They did not characterize the executable file and directly used all the data, which may lead to false positives in model identification. Kancherla et al. [17] converted the executable files to grayscale images, and then selected the model based on the intensity and texture-based feature selection for malware recognition. Grayscale images retain fewer features than color images, which can reduce the accuracy of malware identification. Nataraj et al. [19] linearly mApped grayscale images of Windows malware in the same way as Conti. The GIST (Gabor filter) was Applied to the image to obtain features. The K-Nearest Neighbor (KNN) algorithm was used as the automatic classification technology, and the classification accuracy rate of the 25 malware families reached 98%. In theory, the characteristics of color images are more abundant than grayscale images, and the accuracy of classification for deep learning should be higher. However, there is a lack of research on the use of color images for the classification of Android malicious families.

In recent years, there are more and more studies focusing on Android malware Apps. However, many studies only focus on the two categories of "malicious" and "benign". DroidDolphin [25] used dynamic analysis techniques such as Droid-Box [3] to extract thirteen features from the collected Apps, and constructed a detection system using support vector machine (SVM) model. Crowdroid [8] used dynamic analysis to extract API (App Programming Interface) calls as features and K-means clustering to detect malware. RiskRanker [13] classified Apps into high-risk, medium-risk and low-risk to judge malicious Apps. We find that there are few papers focusing on family classification.

In researchs of multi-family classification, DroidLegacy [10] focused the part of malicious families using piggybacking technology to embed malicious code in benign Apps during repackaging, however this type of malware is not representative of all malware. Dendroid [23] used text mining technology and data flow characteristics to construct a malicious family detection system based on App code structure analysis. It classified 33 families and achieved good results. However, no further research has been done on more families.

In the face of the endless stream of Android malicious App families, how to implement a family classification for most common malicious Apps becomes a problem: as the number of malicious families increases, the malicious behaviors of different families overlap [15]. Different malware families with higher malicious similarity are more difficult to distinguish, and the accuracy of the classifier will also decrease. Due to the lack of reliable manual annotation data sets, some papers use labeled data for a large number of family classification experiments, nevertheless, the results obtained are often questionable. RevealDroid [11] used the classification regression tree algorithm, combined with packet-level and method-level API calls, reflections, and Native code at package and method levels as features, and it successfully classified 33 families on AMC datasets. However, they did not further choose a reliable database by manual classification for more research. Instead, they used the AV [22] classifier to classify and label the collected unlabeled data, so the accuracy of this machine classifier has been questioned, RevelDroid also pointed this out in the paper.

3   Prerequirement

3.1   Malicious App Behavior of Android

Android malicious Apps refer to Android Apps with malicious intentions, which do great harm to mobile phones and users.Malicious App activities can be divided into four stages, the first is the "infection" stage. Malicious Apps often disguise as normal Apps, the common form is the free version of paid Apps, users often misinstall such malicious Apps. After the "infection" is the "destruction" stage, Apps may cause damage to the system, such as enhancing the permissions of malicious Apps, deleting mobile files, locking mobile phones and modifying passwords, which can prevent the normal use of users. "leak" stage can occur simultaneously with destruction, malicious Apps may collect user information and send it to the designated server. Finally, in the "last propagation" stage, malicious Apps may use infected mobile phones to send links or e-mails, alluring unaware friends to download them to click on or download Apps, so as to achieve the purpose of dissemination of malicious Apps. Generally speaking, an App can be judged as malicious if it has the following behaviors [29]:

- Consume the user's mobile phone fee and occupying the mobile phone system resources, causing other Apps to not work properly and preventing users from using the system normally other Apps unable to work properly, hindering the user's use of the system.
- Record the user's screen (such as screen capture or screen recording) without his or her permissions, and obtain private information such as user account and password.
- Allow others to remotely control the user's mobile phone without the user's permission.
- Intimidate the user, such as setting the lock screen to "You will be jailed" and modify the power-on password.

3.2   Android Malicious App Family

Android malicious App family refers to a kind of malicious Apps with the same behavior, which is the product of the detailed division of malicious Apps according to their behavior. The ten common malicious App families are as follows:

1. *Geinimi*: accept remote instructions, control mobile phones, can read and delete short messages, mute phone ringtone, automatically download files and collect information from mobile phone then pass it back to the server.
2. *FakeInstaller*: send paid short messages to certain numbers and cause user fees to be consumed, which is abundant in repackaged versions of popular Apps.

3. *DroidKungFu*: allow attackers' remote access to the infected phones, and can use the root vulnerability to disguise themselves. Common functions include deleting an executable file, opening a web page, downloading and installing an App, opening a URL, launching other programs and so on.

4. *Plankton*: transmit the user's private information, such as the mobile phone IMEI and the user's browsing history data to the remote server, and modify browser home page, add bookmarked.

5. *Opfake*: forge the interface, let the user think the software is a normal App, and steal user information.

6. *GinMaster*: gain access by rooting the devices, thereby steal sensitive user information and send it to the server, and install other software without the user's permission.

7. *Kmin*: send the IMEI information of the device to the remote server. At the same time, they will further threaten the security of the mobile phone by calling according to the remote command and blocking the short message from the operator, which will consume a lot of money.

8. *BaseBridge*: are similar to the *Kmin* Family, but they can kill anti-virus software processes running in the background.

9. *Adrd*: are similar to the *Geinimi*, but they can change the settings of mobile phones.

10. *DroidDream*: get information through rooting mobile devices, download malicious Apps silently in the background, usually run at night while the device is charging in order to avoid the monitoring of power consumption by the detection software.

In addition, there are many other malicious App families, such as the Nickyspy family: record dial-in and dial-out information for infected mobile phones, record users' GPS information, and send text messages to other numbers. Zsone family: automatically send text messages to subscribe for paid content, thus achieving the purpose of consuming telephone charges. Obad family: elevate system privilege to prevent being uninstalled and send text messages to value-added service numbers for profit. Zitmo family: steal verification code sent from the bank. The differences between these families vary, and the large overlap of malicious behavior makes it difficult to distinguish some of them.

4   Our Approach

4.1   Select Features

The size of different Android apps varies widely. If the entire App file is visualized, the visualized image sizes may differ by hundreds of times, which will bring a huge burden to the classification task of the images. Therefore, we need to select the features that can represent the behavior of the App and then perform color visualization.

In the internal structure of an Android App, in addition to the *dex* file that stores the code and the *AndroidManifest.xml* file that stores the configuration information, there is *res* directory that stores resource files such as image files and audio files. This part of the file has nothing to do with the code logic of the App, it is only stored as a resource of the App, and does not affect the behaviers of the program. A small number of malicious Apps may hide malicious code in image files. Such Apps are beyond the scope of this paper, so the resource file is not included in the selection of visualization features.

The basis for our classification of malicious App families is that each Android App has different performance in *classes.dex*, and *AndroidManifest.xml*, which reflect different characteristics and behaviors to distinguish malicious programs from different families [7,10,18].

classes.dex is a bytecode file that compiles Java files into classes and saves them, it contains the package name, classes, methods, variables and Application interfaces (APIs) of the Android App. Most of App's functional behaviors are implemented based on APIs, so we choose it as one of its features.

Every activity component, service component, content provider component, and broadcast receiver component in the Android App needs to be registered in the *AndroidManifest.xml* file. In addition, it also contains some permissions and *SDK* information. So it is part of the features.

4.2   Android App Color Visualization

The purpose of Android App color visualization is to convert the extracted features consist of sequence of APIs of the App and the information of the *AndroidManifest.xml* file into representations of color images. Figure 1 shows the detailed process of color visualization.

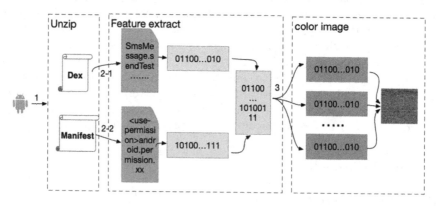

Fig. 1. Android App for color visualization process.

Unzip. As described above, in order to get API call sequence and *Androidmanifest.xml* file of the App, we need to decompress the Apk file. After decompression we get the above two files.

Feature Extraction. We use the tool androguard [1] to reverse the *classes.dex* file, and build the control flow graph(CFG) of the App from the bytecode obtained by the reverse. We extract the API call sequence from each node of the CFG, that is, the API call sequence for the i-th node is

$$NAPI_i = [API_1, API_2, \ldots API_n] \tag{1}$$

Then we map the connection relationship of nodes in CFG to the API call sequences from all nodes. In this way, API call sequence of the entire App is

$$AApp = [NAPI_1, NAPI_2, \ldots NAPI_n] \tag{2}$$

For the *Androidmanifest.xml* file, we use the method of bytecode restoration to restore it to readable text content. In order to convert the above two features into a visualized image, we need to convert the features of the string into numeric values. We use the linear rendering visualization method [9] to visualize them.

Color Visualization. The common binary file visualization method is to convert each byte to a value between 0 and 255, each value corresponds to a pixel in the image (0 is black, 255 is white). For image classification, more image channels mean that more pixels and more features that can be learned. The color visualization conversion method used in this paper is to represent a byte value with three channels of pixels. We use a "blue-green-yellow" color image instead of "black-white" in a grayscale image to represent a range of pixels. The generated image is no longer a single-channel grayscale image, but a three-channel color image, and the value of each channel is not simply repeated.

(a) Grayscale image.   (b) Color image.

Fig. 2. Grayscale image and color image of the same App. (gray image (Figure a on the left) and color image (Figure b on the right)).

As shown in Fig. 2, gray image (Fig. 2a) and color image (Fig. 2b) are generated from the same Android malicious App. the color image successfully maps the original "black-white" of the gray image to the "blue-green-yellow" color range. By analyzing the image file, the original single-layer channel gray image is transformed into three-layer channel color image, which contains more abundant information.

4.3   Malware Detection

Figure 3 shows the classification process of the Android malware multi-classifier. We roughly divide this process into two parts, which are the color visualization of the application and classification process using machine learning models. The details are described as follows.

Color Visualization. For an App to be detected, we need to decompress it and to get the *classes.dex* file and *Androidmanifest.xml* file. Then through the feature conversion process described in the previous section, the App file is color visualized to a color image, which is the input to the image classifier in the next process.

Classification. Algorithms that are commonly used in image classification include support vector machines(SVM), K nearest neighbors, neural networks, random forests. However, based on previous experiments results, deep residual network (ResNet) [14] have get better performance in image classification than the above algorithms. Therefore, we choose ResNet to process color image features. We use a network structure with fewer hidden layers, it contains two convolutional layers, two residual modules and two fully connected layers.

Result. The purpose of this paper is to achieve multi-classification of Android malware, therefore, the output of our system is the malicious family name of the app to be detected.

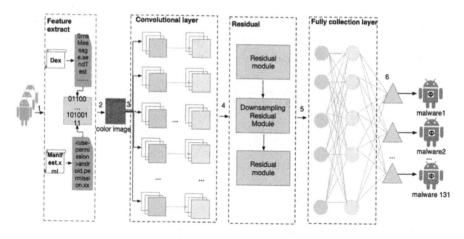

Fig. 3. Overview of malicious app multi-classification system based on color visualization.

5   Experiments

5.1   Characterization of Gray and Color Images

We select the *FakeInstaller* [4] and *Plankton* [5] families in the Drebin [7] data set as experimental data, a total of 1120 Apps, and form a training set and a test set according to the ratio of about 8:2. The size of training sample set of *FakeInstaller* is 403, test sample set size is 100. The size of training sample set of *Plankton* is 497, test sample set size is 120. A single-channel grayscale image and color image are generated for each App in the training set and the test set according to the steps of extracting features, and visualization. After obtaining image features, they are trained and classified by using ResNet. Also the number of training rounds is 100.

Binary Classification of Single Channel Grayscale Image and Three-Channel Grayscale Image. In order to make a comprehensive comparison with the grayscale visualized images, we manually add three-channel grayscale images. We copy the single-channel grayscale image twice and superimpose them with the original image to form a three-channel grayscale image. As shown in Fig. 4, 4(a) is a single-channel grayscale image, and 4(b) is a three-channel grayscale image.

(a) Single channel grayscale im- (b) Three-channel grayscale image.   age.

Fig. 4. Grayscale images of different channels of the same App.

The single-channel grayscale image classification result is shown in Fig. 5(a), and the three-channel grayscale image classification result is shown in Fig. 5(b). The abscissa is the number of training rounds, and the ordinate is the accuracy and loss. The accuracy of single-channel grayscale images is 81.36%, and the classification accuracy of three-channel grayscale images is 85.00%.

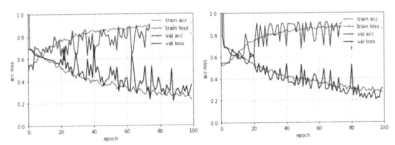

(a) The experiment results of sin- (b) The experiment results of three-
gle channel grayscale image classifi- channel grayscale image classifica-
cation.   tion.

Fig. 5. The experiment results of different channel gray image classification.

The accuracy of three-channel grayscale image is higher than single-channel grayscale image. It proves that multi-channel image is more effective for identifying Android malware Apps. Although the classification accuracy has been improved, the improvement of accuracy is only increased by 3.64%. It is proved that the simple repetition of single-channel images does not contribute much to the classification effect.

Binary Classification of Three-Channel Color Images. The results of the three-channel color image classification are shown in Fig. 6. The classification accuracy rate is 90.91%. The classification accuracy is improved by 9.55% compared with the single-channel grayscale image, also the accuracy compared with the three-channel grayscale image is increased by 5.91%. In the case of the same number of channels, the color image can help the CNN model to learn and classify better than the purely superimposed gray image. It has more features than the grayscale visualization image, and is more suitable for classification.

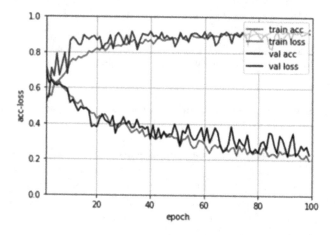

Fig. 6. The results of three-channel color image classification.

5.2   Multi-classification of Color Images Using Different DL Models

The Drebin [7] dataset collected a total of 5,560 samples from 179 different malicious App families. Some Apps cannot be opened with androguard [1], and further inspection revealed these Apps have been corrupted, possibly due to improper storage or file corruption during decompression. In the end, we get 131 malicious App families, each of which has a sample size greater than 2. We choose Convolutional Neural Network (CNN) [28] and Deep Residual Network (ResNet) [14] for multi-classification respectively. Among them, the CNN uses a 5-layer network structure and uses ReLU as the activation function for training. ResNet is trained by using a 20-layer network structure.

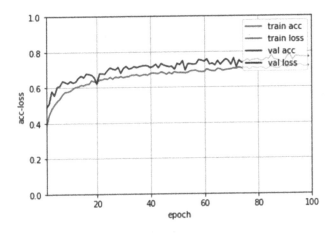

Fig. 7. Experimental results of 131 family classification by using CNN.

CNN. Figure 7 shows the results of a CNN classification experiment on 131 Android malicious families. The accuracy of classification can reach 76.68%. Since the loss function values are all greater than 1, so they are not shown in the figure.

ResNet. It can be seen from Fig. 8 that when the number of training iterations is small, the ResNet is not able to classify the image well, and the phenomenon of overfitting appears. After 80 complete trainings, the model can well distinguish most families, the accuracy and loss curve have stabilized, and the model classification accuracy has reached 96.36%.

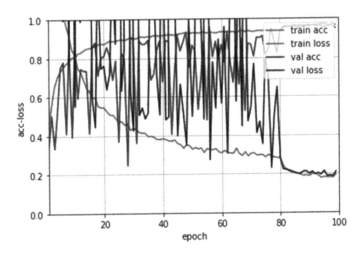

Fig. 8. Classification experimental process for 131 malicious sample families by DRNN(Deep residual neural network).

The details of classification accuracy of each family are shown in Table 1. It can be seen from the table that the ResNet can maintain a high accuracy rate, and identify more families.

Table 1. Accuracy of classification of 131 malicious families.

No	Family	DRNN	CNN	No	Family	DRNN	CNN	No	Family	DRNN	CNN
0	Fjcon	100.00%	0.00%	50	CellSpy	100.00%	0.00%	100	SpyBubble	100.00%	0.00%
1	Fakengry	100.00%	0.00%	51	Tesbo	100.00%	0.00%	101	Geinimi	98.91%	23.91%
2	Sakezon	87.50%	0.00%	52	Ceshark	71.43%	0.00%	102	Koomer	100.00%	0.00%
3	Placms	83.33%	0.00%	53	Fakelogo	100.00%	0.00%	103	Jifake	66.67%	0.00%
4	Generic	100.00%	0.00%	54	Nyleaker	88.89%	0.00%	104	CgFinder	0.00%	0.00%
5	Gonca	100.00%	0.00%	55	Gmuse	100.00%	0.00%	105	Nisev	100.00%	0.00%
6	GGtrack	100.00%	0.00%	56	Coogos	87.50%	0.00%	106	Stiniter	62.50%	0.00%
7	Anti	100.00%	0.00%	57	SeaWeth	83.33%	0.00%	107	Imlog	100.00%	23.26%
8	Mobilespy	85.71%	0.00%	58	Dialer	100.00%	0.00%	108	Replicator	100.00%	0.00%
9	FakeTimer	100.00%	0.00%	59	YcChar	100.00%	0.00%	109	Kiser	100.00%	0.00%
10	Antares	100.00%	0.00%	60	Yzhc	100.00%	89.19%	110	Spitmo	100.00%	100.00%
11	SheriDroid	100.00%	0.00%	61	TrojanSMS.H	100.00%	0.00%	111	EICAR-Test	33.33%	0.00%
12	FinSpy	100.00%	0.00%	62	Dabom	100.00%	0.00%	112	Iconosys	98.03%	91.45%
13	RootSmart	100.00%	0.00%	63	MobileTX	100.00%	100.00%	113	GinMaster	97.05%	92.63%
14	DroidDream	98.77%	100.00%	64	Plankton	99.04%	89.92%	114	Luckycat	100.00%	0.00%
15	KsApp_	100.00%	0.00%	65	ISmsHider	50.00%	0.00%	115	Kidlogger	100.00%	0.00%
16	DroidKungFy	100.00%	95.50%	66	Adsms	100.00%	0.00%	116	SMSZombie	90.00%	0.00%
17	SendPay	100.00%	94.92%	67	DroidRooter	100.00%	0.00%	117	FakePlayer	93.75%	0.00%
18	Gamex	100.00%	0.00%	68	FakeDoc	96.21%	75.00%	118	Moghava	100.00%	0.00%
19	LifeMon	100.00%	0.00%	69	NickyRCP	50.00%	0.00%	119	TheftAware	100.00%	0.00%
20	Saiva	100.00%	0.00%	70	Boxer	100.00%	100.00%	120	Vidro	100.00%	0.00%
21	Copycat	100.00%	0.00%	71	Dogowar	100.00%	0.00%	121	Fidall	100.00%	0.00%
22	Penetho	100.00%	0.00%	72	FakeRun	93.44%	55.74%	122	Steek	71.43%	0.00%
23	SmsWatcher	100.00%	0.00%	73	TrojanSMS.C	100.00%	0.00%	123	Typstu	100.00%	0.00%
24	RediAssi	100.00%	0.00%	74	TigerBot	100.00%	0.00%	124	SpyHasb	92.31%	0.00%
25	Aks	100.00%	0.00%	75	Hispo	100.00%	0.00%	125	AccuTrack	50.00%	0.00%
26	Raden	100.00%	0.00%	76	Fsm	0.00%	0.00%	126	Stealthcell	100.00%	0.00%
27	Mobinauten	25.00%	0.00%	77	Loozfon	50.00%	0.00%	127—	Hamob	92.86%	21.43%
28	BaseBridge	99.39%	87.84%	78	Lemon	100.00%	0.00%	128	Pirates	100.00%	0.00%
29	Dougalek	100.00%	100.00%	79	BeanBot	50.00%	0.00%	129	DroidSheep	100.00%	0.00%
30	Nickspy	81.82%	0.00%	80	Xsider	100.00%	0.00%	130	Gapev	100.00%	0.00%
31	Kmin	97.96%	60.54%	81	GAppusin	98.28%	55.17%				
32	ExploitLinux	95.59%	80.88%	82	Rooter	100.00%	0.00%				
33	Proreso	100.00%	0.00%	83	SpyPhone	100.00%	0.00%				
34	Vdloader	93.75%	0.00%	84	Glodream	98.53%	64.71%				
35	SmForw	100.00%	0.00%	85	Spyset	25.00%	0.00%				
36	Stealer	100.00%	0.00%	86	Spyoo	100.00%	0.00%				
37	FaceNiff	100.00%	0.00%	87	Mania	100.00%	0.00%				
38	SMSreg	100.00%	24.39%	88	FakeInstaller	99.78%	97.19%				
39	Tapsnake	100.00%	0.00%	89	Opfake	100.00%	97.83%				
40	Bige	100.00%	0.00%	90	Zsone	75.00%	0.00%				
41	CrWind	0.00%	0.00%	91	Nandrobox	100.00%	30.77%				
42	GPSpy	100.00%	0.00%	92	QPlus	100.00%	0.00%				
43	Cosha	100.00%	0.00%	93	Fauxcopy	0.00%	0.00%				
44	FarMap	80.00%	0.00%	94	Trackplus	100.00%	0.00%				
45	FoCobers	100.00%	80.00%	95	Ackposts	50.00%	0.00%				
46	SerBG	100.00%	14.29%	96	Fatakr	100.00%	100.00%				
47	PdaSpy	100.00%	0.00%	97	Zitmo	92.86%	21.43%				
48	SpyMob	100.00%	0.00%	98	Flexispy	50.00%	0.00%				
49	FakeFlash	100.00%	0.00%	99	Adrd	100.00%	79.12%				

5.3   Color Visualization Experiments with Different Features

The features selected in this paper are (1) *classes.dex* file obtained by decompilation of the APK file; (2) the set of class name extracted from the class.dex file;

(3) API call sequence extracted from the App. We will measure the impact of different visualization features on malicious application family classification by detecting accuracy and efficiency.

Binary Classification Experiment. We selected *FakeInstaller* and *Plankton* families with a total of 1,550 Apps. 80% of the samples are classified into the training set and 20% into the test set, so the number of *FakeInstaller* training samples is 740, the number of test samples is 185. the number of *Plankton* training samples is 500, and the number of test samples is 125.

Table 2. Android malware classification experiment results with different features for color visualization features.

Feature	Time(s)			Accuracy
	Feature extraction	Color visualization	Total	
classes.dex	1.16	1	2.16	90.91%
Set of class name	1.19	1	2.19	100%
Sequence of API call	1.20	1	2.20	100%

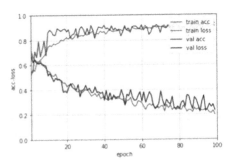

(a) Visualized image of the classes.dex file.   (b) Experimental results of class.dex visual classification.

Fig. 9. Android malware classification experiment results by using the classes.dex file with color visualization features.

Color Visualization of Classes.dex File. Figure 9(a) shows the color visualization image of the *classes.dex* file. Since the it contains all the code of the Android App, the visualization image has more details. There are obvious texture in the figure and different colors that represent different binary numbers. The pictures generated by malicious Apps of the same family have certain similarities in texture and color, which is the basis for image characterization as a method of classification of Android malicious Apps.

Accuracy. As shown in Fig. 9(b), the classification accuracy rate can reach 90.91%. The loss value is relatively high and can reach 20%. This is because the **classes.dex** file includes the code from the third-party libraries, and same third-party libraries code will cause the same characterization results in different apps, which is one of the factors affecting the accuracy of classification.

Efficiency. As shown in Table 2, the average time required to decompile an App file into a *classes.dex* file is 1.16 s. Depending on the size of App files, the maximum decompilation time is 12.93 s and the minimum is 0.10 s. Due to the limitation of image size, instead of visualizing all bytes in large files, we chose to partially visualize up to 89401 bytes. The visualization process takes an average of 1 s per App.

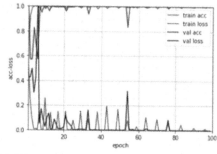

(a) Color visualization of class name collection

(b) Experimental results of visual classification of class name sets

Fig. 10. The experiment results of Android malware classification by using the set of class name with color visualization features.

Color Visualization of Set of Class Names in Apps. The set of class names in the App is a type of code extracted from the App file, which explain the class invocation of the App. A class name can be used as a description of the App's single behavior, and a collection of invocations can represent behaviors of the entire App in macro. Therefore, it can be used as a feature of the App for color visualization. The image after the feature visualization is shown in Fig. 10(a).

Accuracy. The classification results for the two families *FakeInstaller* and *Plankton* are shown in Fig. 10(b). The results show that the classification accuracy rate reaches 100%. It can be seen that the visualization result using the class name set as input is more conducive to learn from the image which is useful information and which is useless information.

Efficiency. As shown in Table 2, the average time to extract all the App class names from the App is 1.19 s, and the time taken depends on the size of the App. The minimum is less than 1 s and the highest is 5.54 s. The time-consuming aspect of color visualization of features is that the extracted set of class name are relatively small, and the time-consuming is 1 s.

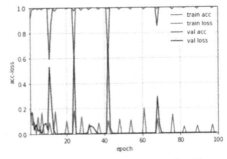

(a) Color visualization image of API call sequence.

(b) Experimental results of the classification of color visualization features by using the sequence of APIs.

Fig. 11. The experiment results of the sequence of APIs.

Color Visualization of API Call Sequence. We use the API call sequence as a visual feature input. APIs can better reflect the internal logical structure of the Android App, which has a positive impact on the improvement of classification accuracy. Due to the need to analyze the internal code structure of the Android App, it takes slightly more time than simply extracting the App class name. The color visualization image of the API call sequence is shown in Fig. 11(a).

Accuracy. The classification results for the two families are shown in Fig. 11(b). The classification result using the API sequence as the input of visualization can reach 100%. The occasional accuracy fluctuations in the figure may be due to the increase in similarity of different Apps to a certain extent due to the third-party libraries.

Efficiency. As shown in Table 2, the average time taken by the App analysis to extract the features of the API call sequence is 1.20 s. The time taken by the API call is determined by the size of the App. The maximum time is 13.16 s and the shortest time is 0.1 s. It also takes 1 s in color visualization.

An Experiment on the Classification of ten Malicious Families. In order to further verify the differences between the two types of features of the class name set and the API call sequence, we conducted multiple classification experiments using these two features. The malicious App families used include *FakeInstaller, DroidKungFu, Plankton, Opfake, GinMaster, BaseBridge, Iconosys, Kmin, FakeDoc, Geinime*, a total of 4005 malicious Apps. The number of samples in the training data set is 80%, and the rest are used in the test data set.

Color Visualization of Set of Class Names in Apps. This experiment mainly verifies the accuracy of classification. The results in the experiment is shown in Fig. 12. After 100 complete iterations, the classification accuracy reaches 91.40%.

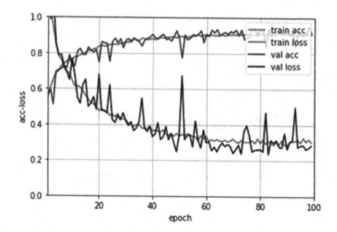

Fig. 12. Experimental results of the classification of ten malicious families visualized by the collection of class name.

There may be more features that overlap in different families, which leads to a decrease in accuracy.

Color Visualization of API Call Sequences. As shown in Fig. 13, we use API call sequence as visualization. The accuracy of family classification is 96.01%. Compared with the results of binary classification, the accuracy is reduced.

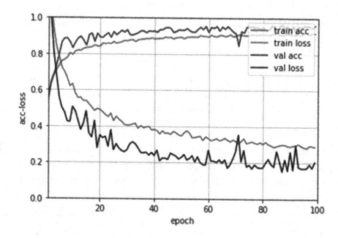

Fig. 13. Experimental results of the classification of ten malicious families visualized by the Apps' API call sequence)

In the above two experiments, the accuracy of each family is shown in Table 3. API call sequence has higher accuracy in most families, except in *FakeDoc* and

FakeInstallers. We compared the malicious behavior of the two families and found that Apps in *FakeDoc* download the malware without the user's consent, Also Apps in *FakeInstallers* send the SMS without the user's permission. The behavior of the two malicious families is similar. This phenomenon leads to low accuracy.

Table 3. Accuracy of classification of ten malicious families.

Family	Class name	API call
DroidKungFu	99.2%	99.2%
PlanKton	91.1%	92.2%
FakeDoc	89.8%	84.5%
Geinimi	72.1%	87.5%
Iconosys	88.1%	95.2%
GinMaster	81.1%	87.2%
BaseBridge	81.1%	98.3%
Kmin	89.5%	92.2%
FakeInstaller	98.2%	94.2%
Opfake	99.1%	99.1%

From the two aspects of classification accuracy and time consumption, the feature of API call sequence has the best effect. It achieves the highest accuracy in the two classifications and ten classes of experiments, and the time spent is relatively short, so that it has become the current optimal choice.

6   Discussions and Limitations

Obfuscation. More and more Apps are used obfuscation. For malicious Apps, they are used obfuscation to hide malicious behaviors. Here are: (1) encoding classes and methods into meaningless strings; (2) adding some useless APIs to Apps; (3) storing malicious APIs in the form of ascii code. We extract APIs belong to the Android system, these APIs cannot be obfuscated, so for the first two obfuscation techniques, our method can get the APIs. For the third obfuscation technique, we cannot get the APIs.

Packer. Some malicious Apps use packing technology to hide malicious code. In our feature extraction, there is no unpacking process for these Apps, so we can not analysis these Apps. But for our current research on the unpacking method, we can already use the method of memory insertion to realize the automatic unpacking process. So in the future researchs, we will implement detection for packed malicious Apps.

7 Conclusion

We present a method for multi-classification of Android malicious App families using color visualization. Experiments in this paper proves that compared to single-channel images, deep learning models can more easily learn features from three-channel images, thereby achieving higher classification accuracy. We use ResNet to implement a multi-classification of 131 malicious families, and find that the best classification results can be achieved when using API calls as features of malicious App color visualization. All in all, in terms of using the deep learning model to classify Android malicious App families, compared to the traditional single-channel gray-scale image for App visualization, the color image visualization method has obvious advantages.

Acknowledgement. This work was supported in part by the National Key R&D Program of China under Grant No. 2018YFB0804703.

References

1. Androguard. https://github.com/androguard/androguard. Accessed 8 Jan 2019
2. Common vulnerabilities and exposures. https://cve.mitre.org/. Accessed 4 June 2018
3. Droidbox. https://github.com/pjlantz/droidbox. Accessed 10 July 2019
4. Fakeinstaller. https://www.mcafee.com/blogs/other-blogs/mcafee-labs/fakeinstaller-leads-the-attack-on-android-phones/. Accessed 18 Aug 2019
5. Plankton. https://news.ncsu.edu/2011/06/wms-android-plankton/. Accessed 18 Aug 2019
6. Andronio, N., Zanero, S., Maggi, F.: Heldroid: dissecting and detecting mobile ransomware. In: Bos, H., Monrose, F., Blanc, G. (eds.) RAID 2015. LNCS, pp. 382–404. Springer, Heidelberg (2015). https://doi.org/10.1007/978-3-319-26362-5_18
7. Arp, D., Spreitzenbarth, M., Hubner, M., Gascon, H., Rieck, K., Siemens, C.: Drebin: effective and explainable detection of Android malware in your pocket. Ndss **14**, 23–26 (2014)
8. Burguera, I., Zurutuza, U., Nadjm-Tehrani, S.: Crowdroid: behavior-based malware detection system for Android. In: Proceedings of the 1st ACM Workshop on Security and Privacy in Smartphones and Mobile Devices, pp. 15–26 (2011)
9. Conti, G., Dean, E., Sinda, M., Sangster, B.: Visual reverse engineering of binary and data files. In: Goodall, J.R., Conti, G., Ma, K.-L. (eds.) VizSec 2008. LNCS, vol. 5210, pp. 1–17. Springer, Heidelberg (2008). https://doi.org/10.1007/978-3-540-85933-8_1
10. Deshotels, L., Notani, V., Lakhotia, A.: Droidlegacy: automated familial classification of Android malware. In: Proceedings of ACM SIGPLAN on Program Protection and Reverse Engineering Workshop, pp. 1–12 (2014)
11. Garcia, J., Hammad, M., Malek, S.: Lightweight, obfuscation-resilient detection and family identification of Android malware. ACM Trans. Softw. Eng. Methodol. (TOSEM) **26**(3), 1–29 (2018)

12. Gennissen, J., Cavallaro, L., Moonsamy, V., Batina, L.: Gamut: sifting through images to detect Android malware (2017)
13. Grace, M., Zhou, Y., Zhang, Q., Zou, S., Jiang, X.: Riskranker: scalable and accurate zero-day Android malware detection. In: Proceedings of the 10th International Conference on Mobile Systems, Applications, and Services, pp. 281–294 (2012)
14. He, K., Zhang, X., Ren, S., Sun, J.: Deep residual learning for image recognition. In: Proceedings of the IEEE Conference on Computer Vision and Pattern Recognition, pp. 770–778 (2016)
15. Hsiao, S.W., Sun, Y.S., Chen, M.C.: Behavior grouping of android malware family. In: 2016 IEEE International Conference on Communications (ICC), pp. 1–6. IEEE (2016)
16. Jung, J., Choi, J., Cho, S.J., Han, S., Park, M., Hwang, Y.: Android malware detection using convolutional neural networks and data section images. In: Proceedings of the 2018 Conference on Research in Adaptive and Convergent Systems, pp. 149–153 (2018)
17. Kancherla, K., Mukkamala, S.: Image visualization based malware detection. In: 2013 IEEE Symposium on Computational Intelligence in Cyber Security (CICS), pp. 40–44 (2013)
18. Lin, C.M., Lin, J.H., Dow, C.R., Wen, C.M.: Benchmark dalvik and native code for Android system. In: 2011 Second International Conference on Innovations in Bio-inspired Computing and Applications, pp. 320–323. IEEE (2011)
19. Nataraj, L., Karthikeyan, S., Jacob, G., Manjunath, B.: Malware images: visualization and automatic classification. In: Proceedings of the 8th International Symposium on Visualization for Cyber Security, pp. 1–7 (2011)
20. Pektaş, A., Acarman, T.: Learning to detect android malware via opcodesequences. Neurocomputing (2019)
21. Qiu, J., et al.: Data-driven Android malware intelligence: a survey. In: Chen, X., Huang, X., Zhang, J. (eds.) ML4CS 2019. LNCS, vol. 11806, pp. 183–202. Springer, Cham (2019). https://doi.org/10.1007/978-3-030-30619-9_14
22. Sebastián, M., Rivera, R., Kotzias, P., Caballero, J.: AVCLASS: a tool for massive malware labeling. In: Monrose, F., Dacier, M., Blanc, G., Garcia-Alfaro, J. (eds.) RAID 2016. LNCS, vol. 9854, pp. 230–253. Springer, Cham (2016). https://doi.org/10.1007/978-3-319-45719-2_11
23. Suarez-Tangil, G., Tapiador, J.E., Peris-Lopez, P., Blasco, J.: Dendroid: a text mining approach to analyzing and classifying code structures in Android malware families. Expert Syst. Appl. 41(4), 1104–1117 (2014)
24. Wang, W., Zhao, M., Wang, J.: Effective Android malware detection with a hybrid model based on deep autoencoder and convolutional neural network. J. Ambient Intell. Hum. Comput. 10(8), 3035–3043 (2019)
25. Wu, W.C., Hung, S.H.: Droiddolphin: a dynamic android malware detection framework using big data and machine learning. In: Proceedings of the 2014 Conference on Research in Adaptive and Convergent Systems, pp. 247–252 (2014)
26. Xiao, X., Zhang, S., Mercaldo, F., Hu, G., Sangaiah, A.K.: Android malware detection based on system call sequences and LSTM. Multimedia Tools Appl. 78(4), 3979–3999 (2019)
27. Zhang, J., Qin, Z., Yin, H., Ou, L., Hu, Y.: IRMD: malware variant detection using opcode image recognition. In: 2016 IEEE 22nd International Conference on Parallel and Distributed Systems (ICPADS), pp. 1175–1180 (2016)
28. Zhang, K., Zuo, W., Chen, Y., Meng, D., Zhang, L.: Beyond a Gaussian denoiser: residual learning of deep CNN for image denoising. IEEE Trans. Image Process. 26(7), 3142–3155 (2017)

29. Zhou, Y., Jiang, X.: Android malware genome project. http://www.malgenomeproject.org/. Accessed 4 June 2018
30. Zhou, Y., Jiang, X.: Dissecting android malware: characterization and evolution. In: 2012 IEEE Symposium on Security and Privacy, pp. 95–109. IEEE (2012)

Authentication and Access Control

PUF-Based Two-Factor Group Authentication in Smart Home

Sai Ji[1,3], Rongxin Qi[1,3(✉)], and Jian Shen[1,2,3]

[1] Nanjing University of Information Science and Technology, Nanjing, China
q_qirongxin@126.com
[2] Cyberspace Security Research Center, Peng Cheng Laboratory, Shenzhen, China
[3] Jiangsu Engineering Center of Network Monitoring, Nanjing, China

Abstract. Various IoT-based applications such as smart home, intelligent medical and VANETs, have been put into practical utilization. Smart home is one of the most concerned environments, which allows users to remotely access and control smart devices via a public network. With development of the mobile network and smart devices, more services can be provided to users by smart devices. To securely access devices and obtain collected data over the public network, multi-factor authentication schemes for smart home have obtained wide attention. However, most of these schemes cannot withstand impersonation attack, physical device lost attack, privileged-insider attack, smart card lost attack and so on. Besides, high communication and computational costs weaken the system performance, which causes that most authentication schemes are not suitable for resource-constrained smart devices. To mitigate the aforementioned drawbacks, we proposed a two-factor anonymous group authentication scheme to implement secure access to multiple devices simultaneously using chinese remainder theorem and secret sharing technology. Our scheme also utilizes fuzzy extractor to extract personal biometric information, which helps uniquely validate authorized users in smart home. Our scheme can support various security features and withstand the most well-known attacks in smart home. Performance analysis indicates that the proposed scheme can efficiently reduce communication/computational costs when the user accesses multiple devices simultaneously.

Keywords: Smart home · Secret sharing · Authentication · Fuzzy extractor

1 Introduction

With the rapid development of IoT technology, various IoT-based applications such as smart home, intelligent medical and VANETs, have emerged. In these applications, smart home has obtained wide attention in recent years due to its convenience, efficiency and other properties, which provides basic and practical

© ICST Institute for Computer Sciences, Social Informatics and Telecommunications Engineering 2021
Published by Springer Nature Switzerland AG 2021. All Rights Reserved
D. Wang et al. (Eds.): SPNCE 2020, LNICST 344, pp. 209–222, 2021.
https://doi.org/10.1007/978-3-030-66922-5_14

home control services for the users. The smart home is a dwelling that connects major appliances and service, and permits them to be accessed via the public network [14]. In most existing schemes, smart home is usually composed of user equipment (e.g., smartphone), home gateway (HG) and lots of smart devices (e.g., surveillance camera, lighting controller, temperature sensors) [12]. The smart devices are interconnected to collect the data in smart home and exchange the collected data with the user via the public network. HG acts as the communication medium between the user and smart devices.

Smart devices are generally easy to suffer from various attacks such as impersonation attack, physical device lost attack and privileged-insider attack during the execution of the protocol. Once these devices are broken, user privacy will be compromised. For example, the unauthorized users may access the surveillance cameras and control them to monitor the resident in smart home. In addition, most of these IoT devices such as sensors, have the limited resources to execute complex computational operation [5]. In recent years, many Elliptic Curve Cryptography (ECC)-based schemes [10,13] have been proposed to enhance the authentication security. However, these schemes generally require to perform complex computational operations, which are not suitable for the resource-constrained devices. In addition, some schemes cannot provide most security features and functionalities such as user anonymity, perfect forward secrecy and dynamic device addition. To solve the security and privacy issues in IoT environments, a large number of authentication schemes have been proposed [11,19,20]. In most of the existing schemes, the computational cost and communication cost too high to be suitable for resource-constrained [19]. If the user wants to access multiple smart devices simultaneously, it is necessary to frequently verify the authenticity of the user and send access request to corresponding smart devices in a short time, which may lead to network delay and even congestion. Therefore, it is crucial to design an efficient and lightweight authentication scheme to establish the secure session key between the user and smart devices in smart home. Group authentication schemes are put forward to solve aforementioned issues. Group authentication scheme based on secret sharing can authenticate multiple the smart devices belonging to the same group simultaneously. Considering the security of the parameters stored in the smart devices, physical unclonable function (PUF) is utilized to prevent stolen devices attack. PUF can be utilized to assist smart devices to generate biometric key, which efficiently protect the security smart devices [1]. Therefore, we propose a PUF-based two-factor group authentication scheme for smart home. Our scheme supports many well-known features such as un-traceability, user anonymity, forward secrecy. The smart devices are allowed to dynamically join or leave the group.

Our Contributions

- A PUF-based anonymous group authentication scheme is presented in our paper. Our scheme is suitable for the resource-constrained smart devices only using lightweight operation and symmetric cryptography. Furthermore, the proposed scheme meets many security requirements such as user anonymity, un-traceability and withstand many known attacks.

- The dynamic joining and leaving of smart devices from deployed network are both supported by our proposed scheme. The illegitimate smart devices fail to attain the group key without the secret share. The new smart device just register itself before joining the deployed network.
- The physical security of smart devices is guaranteed by physical unclonable function technology. The output of PUF depends on the physical micro-structure of the physical device. PUF has the characteristics of tamper-resistant, unclonability, unpredictability.
- The issue of repeated user authentication is solved by utilizing secret sharing technology. The user can authenticate the multiple smart devices simultaneously and establish secure group session key, which effectively reduces communication and computational costs.

1.1   Related Work

Smart home allows the authorized users to remotely access devices and obtain information collected by these devices. To address security and privacy issues in IoT, a large number of researchers [6,9,21] have studied many authentication schemes for smart home.

In 2011, Vaidya et al. proposed a novel authentication and key establishment mechanism which is based on ECC. Although their scheme satisfies more security requirements compared to schemes, their scheme is not suitable for resource-constrained home area networks. Therefore, many schemes focus on providing more security features while reducing resource cost of schemes. To solve communication security issue in WSNs, Xue et al. [21] utilized temporary credentials to implement authentication between the user and sensing nodes for WSNs in 2013. Their scheme is lightweight to be suitable for the sensing nodes using hash function and bit-wise XOR operation. However, He et al. [6] thought their scheme fails to resist offline password guessing attack, impersonation attack and tampering attack. In 2013, He et al. [6] proposed an improved authentication scheme which overcomes the security threats in Xue's scheme and only increases little computational cost. In 2014, Turkanovic et al. [17] focused on a scenario where the user accessing a single targeted sensor in WSNs does not need to interact with HG. Meanwhile, Kalra et al. [8] found that Xue's scheme is vulnerable to smartcard lost attack. Kalra et al. [8] proposed a novel authentication scheme based on password and smartcard, which can resist most known attacks and has lower cost than other schemes. However, their scheme do not consider resisting sensing node capturing attack and privileged-insider attack. In 2018, Shen et al. [15] adopted the cloud to enhance the capabilities of devices and established a lightweight authentication scheme without certificates for WBANs.

The entity in IoT environment has similar features to the sensing nodes in traditional WSNs. Due to the heterogeneity and dynamics, the higher security and privacy requirements need to be satisfied in IoT environment. Kuma et al. [9] proposed an anonymous authentication framework for smart home using only hash function and symmetric cryptography. Kumar et al. firstly considered the features of anonymity and unlinkability for smart home and their scheme can resist many known attacks. Challa *et al.* [4] proposed a novel signature-based authenticated key establishment scheme for generic IoT environment. The user can not only communicate with smart devices but also with other users through HG. In 2018, Srinivas *et al.* [16] proposed an anonymous three-factor authentication and key agreement scheme which supports credentials update, user revocation and new devices addition.

2   Preliminaries

2.1   Chinese Reminder Theorem [22]

It is assumed that there are n prime positive integers $p_1, p_2, \cdots, p_n$. Let P be the product of n prime positive integers as $P = \prod_{i=1}^{n} p_i$ and $P_i = P/p_i$, where $i = 1, 2, \cdots n$. Let P_i^{-1} be the modular multiplicative inverse of $P_i \bmod p_i$ and satisfy $P_i P_i^{-1} \equiv 1 \,(\bmod\ p_i)$. Then, let $a_i, i = 1, 2, \cdots, n$. be any n positive integers. The Eq. (1) has an unique general solution mod P.

$$
\begin{aligned}
X &\equiv a_1 \bmod p_1 \\
X &\equiv a_2 \bmod p_2 \\
&\vdots \\
X &\equiv a_n \bmod p_n
\end{aligned}
\tag{1}
$$

The general solution of Eq. (1) is calculated in the Eq. (2).

$$
\begin{aligned}
X &= a_1 P_1^{-1} P_1 + a_2 P_2^{-1} P_2 + \cdots + \\
&\quad a_n P_n^{-1} P_n (mod P) \\
&= \sum_{i=1}^{n} a_i P_i^{-1} P_i \,(\bmod P) \\
&= a_1 + a_2 + \cdots + a_n (mod P)
\end{aligned}
\tag{2}
$$

2.2   Physical Unclonable Function [18]

PUF which is based on complex physical system is a function $F : C \to R$ $(C : \{0,1\}^{\lambda_1}, R : \{0,1\}^{\lambda_2})$. The challenges and their corresponding response are called challenge-response pairs. PUF has the following properties:

1. **Unclonable:** For all $c \in C$, there is no function F' satisfing $F'(c) = F(c)$. The probability of duplicating function F with a cloned function F' in probabilistic polynomial time (PPT) is negligible.
2. **Computable:** It is feasible to compute the $r_i = F(c_i)$ in probabilistic polynomial time for all $c_i \in C$.
3. **Unpredictable:** For all $c \in C$, the probability of adversary $\mathcal{A}$ correctly guessing response r of function F corresponding to challenge c in PPT is negligible. The output of function F is a random string uniformly chosen from $\{0,1\}^{\lambda_1}$.

4. **Tamper-proofing:** For all $c, c' \in C$, even the Hamming distance between c and c' is equal to t (t is sufficiently small) or less, the probability of outputting the similar results is negligible. Therefore, PUF is able to resist tampering attacks.

2.3   Fuzzy Extractor [13]

Fuzzy extractor takes a low-entropy value containing noise as input and outputs the same uniform random value as long as the inputs values are close. Fuzzy extractor is utilized to extract the user's biometric information and the smart device's information. It is assumed that fuzzy extractor is composed of two algorithms defined in a tuple $\langle M, l, t \rangle$.

Gen(): It is a probabilistic algorithm. The user takes his/her biometrics BIO_i from the metric space M as the input of algorithm *Gen*, and it outputs the biometric key $\sigma_i \in \{0,1\}^l$ and the parameter τ_i.

Rep(): It is a deterministic algorithm. *Rep* takes the biometrics $BIO_i' \in M$, reproduction parameter τ_i and t as the input (t is the fault tolerance value and sufficiently small). The algorithm *Rep* can reproduce the biometric key σ_i as $Rep\left(BIO_i', \tau_i\right) = \sigma_i$, where the Hamming distance between twice inputs is t or less.

3   Authentication Scheme Construction

Network Model. The authentication scheme in smart home consists of the user U_i, home gateway (HG), lots of smart devices SD_j and key generation center (KGC). All the entities are defined as follows.

- KGC : KGC is a trusted key generation center and is utilized to distribute sensitive parameters for the user, HG and lots of smart devices securely.
- HG : It is a trusted entity and cannot be compromised by the adversary $\mathcal{A}$.
- U_i : The user U_i is owner of the smartphone UE_i which has capabilities to extract U_i's biometrics and verify U_i's identity. U_i can access the smart devices after registering with the KGC. It is assumed that $\mathcal{A}$ may attain authentication credentials in the UE_i.
- SD_j : Smart devices can execute the commands and collect the information in smart home. Every smart device has a unique identity and cannot be forged physically by $\mathcal{A}$. All the smart devices have the PUF module which protects them from physically capturing attack.

Threat Model. Under the network model mentioned above, It is assumed that $\mathcal{A}$ in our scheme has same capabilities as the adversary in Dolev-Yao (DY) threat model [7]. The capabilities of $\mathcal{A}$ in our scheme are enumerated as follows:

- $\mathcal{A}$ can eavesdrop, intercept, modify, inject and delete all the messages transmitted via the public network.

Table 1. Notations and Descriptions

Notations	Descriptions
U_i, SD_j and HG	i^{th} user, j^{th} smart device and home gateway
UE_i	i^{th} user equipment
ID_i, ISD_j and ID_{HG}	U_i's, SD_j and HG's identity
PW_i	U_i's password
BIO_i	U_i's biometrics
$Gen(\cdot), Rep(\cdot)$	Generation and reproduction algorithm of fuzzy extractor
σ_i, R_j	U_i's biometrics key, SD_j's physical key
τ_i, x_i, h_j	Public parameters
T_i	Current timestamp
ΔT	Maximum communication delay
K_{HG}	HG's secret key
K_i	Symmetric key between U_i and HG
GSK	Group session key between the user and smart devices
S	Secret value utilized for secret sharing
s_j	SD_j's secret share
PUF	Physical unclonable function
$H(\cdot)$	One-way hash function
$\oplus, \|$	Concatenation and bit-wise XOR operation, respectively

- $\mathcal{A}$ can store or resend the messages which are intercepted or forged.
- $\mathcal{A}$ can impersonate as the legitimate user or the smart device to participant during the running of the scheme.
- $\mathcal{A}$ can obtain the credentials stored in users' smartphones and smart devices, and launch various types of attacks on the protocol. However, it cannot compromise the group session key during the running of the scheme.

In addition, the adversary $\mathcal{A}$ also has some abilities in CK-adversary model proposed by Canetti et al. [2,3]. Under the CK-adversary model, the reveal of the ephemeral state information or other sensitive information have no influence on the session security and long-term secrets. Therefore, it is necessary to guarantee that the security of other sessions cannot be affected even through ephemeral secret is compromised (Table 1).

3.1 Smart Device Registration Phase

SDRP1. The smart device SD_j, $j = 1, 2, \cdots, n$. utilizes the PUF and fuzzy extractor to extract the information to register itself. The smart device SD_j firstly select random nonce c_j and compute $r_j = F(c_j)$. SD_j computes $(R_j, h_j) = Gen(r_j)$ to generate secret key R_j and sends R_j to KGC securely.

SDRP2. When receiving the registration from the smart device SD_j, $j = 1, 2, \cdots, n$. KGC chooses the identity ISD_j for each smart device and randomly selects a polynomial $f(x)$ of degree $t - 1$: $f(x) = a_0 + a_1 x + \cdots + a_{t-1} x^{t-1} \mod p$, such that all the coefficients $a_j, j = 0, 1, 2, \cdots, t - 1.$, and $s = f(0)$ are in finite field $GF(p)$. KGC computes $H(s)$ and $s_j = f(x_j)$ (x_j is public system information related to the smart device SD_j). KGC randomly selects a prime positive integer p_j, $j = 1, 2, \cdots, n$. corresponding to smart device SD_j. Then, KGC computes $P = \prod_{j=1}^n p_j$, $P_j = P/p_j$, $j = 1, 2, \cdots, n$. and $\chi = \sum_{j=1}^n P_j P_j^{-1}$ ($P_j P_j^{-1} \equiv 1 \mod p_j$, $\chi \mod p_j \equiv 1$). Finally, KGC calculates $RP_j = R_j \oplus p_j$, $share_j = R_j \oplus s_j$ and sends ISD_j, RP_j, $share_j$ to corresponding smart device SD_j securely.

3.2   User Registration Phase

URP1. U_i firstly chooses a ID_i and high entropy password PW_i, and imprints personal biometric information BIO_i using the fuzzy extractor in user equipment UE_i. UE_i adopts key generation algorithm $Gen(\cdot)$ to generates corresponding biometric key σ_i and public parameter τ_i as $Gen(BIO_i) = (\sigma_i, \tau_i)$. To protect the PW_i and σ_i, UE_i randomly generates a nonce a and take personal credentials ID_i, PW_i, σ_i and a as input to compute $RPW_i = h(ID_i \parallel PW_i \parallel \sigma_i) \oplus a$. Finally, UE_i securely sends request $\langle ID_i, RPW_i \rangle$ to KGC.

URP2. When getting the request $\langle ID_i, RPW_i \rangle$ from U_i, KGC firstly generates a 1024-bit long-term secret value K_{HG} and calculates $K_i = H(ID_i \parallel K_{HG})$, $TPW_i = K_i \oplus RPW_i$. Then, KGC generates the anonymous identity TID_i corresponding to ID_i and securely sends the information $\langle TID_i, TPW_i \rangle$ to UE_i. Finally, KGC deletes the information RPW_i and TPW_i from its database.

URP3. Upon receiving the response $\langle TID_i, TPW_i \rangle$ from KGC, UE_i computes $A_i = H(ID_i \parallel PW_i \parallel \sigma_i)$, $rPW_i = TPW_i \oplus a$, $B_i = H(ID_i \parallel A_i \parallel \sigma_i)$. Then, UE_i stores $\langle TID_i, rPW_i, B_i, \tau_i, Gen(\cdot), Rep(\cdot), H(\cdot), t \rangle$ in its memory. Finally, UE_i deletes TPW_i, RPW_i, A_i from UE_i so as to prevent user equipment from compromising sensitive information.

3.3   Home Gateway Registration Phase

HG choose a identity ID_{HG} and sends the registration request to KGC. Upon receiving the request from HG, KGC issues a long-term secret key K_{HG}, the user identity ID_i, temporal identity TID_i, public parameters $h_j, x_j, j = 1, 2, \cdots, n$ and $H(s)$ to HG securely.

3.4   Login and Authentication Phase

LAP1. U_i firstly inputs ID_i and high entropy password PW_i^* and imprints personal biometrics BIO_i^* into UE_i. UE_i computes $\sigma_i^* = Rep(BIO_i^*, \tau_i)$ by the reproduction algorithm if the Hamming distance between two biometrics is t or less. Then, UE_i calculates $A_i^* = H(ID_i \parallel PW_i^* \parallel \sigma_i^*)$, $B_i^* = H(ID_i \parallel A_i^* \parallel \sigma_i^*)$. UE_i utilized B_i^* to validate the user locally. After verifying the user's identity successfully, UE_i calculates symmetric key $K_i = A_i \oplus rPW_i^*$. UE_i randomly generates a nonce r_i and T_1. UE_i then calculates $M_1 = K_i \oplus r_i$, $M_2 = H(M_1 \parallel ID_i \parallel TID_i \parallel r_i \parallel T_1)$. UE_i sends $\langle TID_i, M_1, M_2, T_1 \rangle$ to HG via an open channel.

LAP2. Upon receiving the login request, HG firstly checks the freshness of the timestamp T_1. If it is true, HG retrieves ID_i, K_{HG} and computes $K_i^* = H(ID_i \parallel K_{HG}) = K_i$, $r_i^* = K_i^* \oplus M_1$, $M_3 = H(M_1 \parallel ID_i \parallel TID_i \parallel r_i^* \parallel T_1)$, and checks if $M_2 = M_3$. If valid, continue the session. Otherwise, HG terminates session immediately. Then, HG randomly generates a nonce r_{HG} and a timestamp T_2, and computes $m_{HG} = r_{HG} \times \chi$. HG calculates $M_4 = E_{r_{HG}}(ID_i, r_i^*, H(K_i))$, $M_5 = H(ID_i \parallel r_{HG} \parallel r_i^* \parallel H(K_i) \parallel M_4 \parallel T_2)$. Finally, HG broadcasts the message $\langle M_4, M_5, m_{HG}, T_2 \rangle$ to all the smart devices via an open channel.

LAP3. Upon receiving message, SD_j firstly checks the freshness of the message by timestamp T_2. If it is valid, SD_j calculates $F(c_j^*) = r_j^*$, $R_j^* = Rep(r_j^*, h_j)$, $p_j = RP_j \oplus R_j^*$, $s_j^* = share_j \oplus R_j^*$, $r_{HG}^* = m_{HG} \bmod p_j$ ($\chi \bmod p_j \equiv 1$, r_{HG} is a shared group key of all the legitimate smart devices). Then, SD_j decrypts M_4 as $ID_i, ID_{HG}, r_i^*, H(K_i)$ using shared group key r_{HG}^*, and compute $M_6 = H(ID_i \parallel r_{HG}^* \parallel r_i^* \parallel H(K_i) \parallel M_4 \parallel T_2)$ and check whether $M_5 = M_6$. If it is valid, SD_j terminates the request. Otherwise, SD_j generate a timestamp T_3 and calculates $M_{7_j} = E_{r_{HG}^*}(s_j, ISD_j)$, $M_{8_j} = H(s_j \parallel M_{7_j} \parallel ISD_j \parallel r_{HG}^* \parallel T_3)$. Finally, SD_j sends message $\langle M_{7_j}, M_{8_j}, T_3 \rangle$ to HG.

LAP4. After receiving $\langle M_{7_j}, M_{8_j}, T_3 \rangle$ from smart devices $SD_j, j = 1, 2, \cdots, m$. HG checks the freshness of timestamp T_3. If it is valid, HG can obtains s_j, ISD_j by using r_{HG} to decrypt M_{7_j}, and computes $s' = \sum_{j=1}^{m} s_j \prod_{r=1, r \neq j}^{m} \frac{-x_r}{x_j - x_r}$, HG also checks whether $H(s') = H(s)$. If it is true, continues the session. Otherwise, HG computes M_{9_j} and checks whether $M_{8_j} = M_{9_j}$ to verify SD_j. If it matches, the message is from valid SD_j. Otherwise, HG marks the invalid smart devices and terminates the session. Then, HG computes $M_{10} = H(H(s) \parallel r_{HG})$, $M_{11} = E_{r_{HG}}(M_{10})$, $M_{12} = H(M_{10} \parallel M_{11})$. Finally, HG sends $\langle M_{11}, M_{12} \rangle$ to smart devices.

LAP5. Each SD_j extracts M_{10} using shared group key r_{HG}^*, computes $M_{13} = H(M_{10} \parallel M_{11})$ and checks whether $M_{12} = M_{13}$. If it is valid, each SD_j computes $GSK = H(r_{HG}^* \parallel H(K_i) \parallel r_i^* \parallel ID_i \parallel M_{10})$, $M_{14} = H(r_{HG}^* \parallel ID_{HG} \parallel GSK)$. Finally, each SD_j sends message $\langle M_{14} \rangle$ to HG.

LAP6. HG encrypts parameters as $M_{15} = E_{K_i^*}(M_{10}, r_{HG}, r_i^*, ID_{HG})$, and generates a timestamp T_4, a new anonymous identity TID_i^{new}. HG calculates $M_{16} = H(K_i^* \parallel TID_i \parallel T_4) \oplus TID_i^{new}$, $M_{17} = H(M_{15} \parallel M_{16} \parallel r_i^* \parallel T_4)$, Finally, HG sends the message $\langle M_{15}, M_{16}, M_{17}, T_4 \rangle$ to UE_i.

LAP7. UE_i firstly checks the freshness of timestamp T_4 when receiving the message $\langle M_{15}, M_{16}, M_{17}, T_4 \rangle$. UE_i then utilizes long-term secret key K_i to decrypt M_{15} and obtains $(M_{10}, r^*_{HG}, r^*_i, ID_{HG})$. Then, UE_i checks whether $r_i = r^*_i$. If it matches, U_i calculates $GSK^* = H(r^*_{HG} \parallel H(K_i) \parallel r_i \parallel ID_i \parallel ID_{HG} \parallel M_{10})$, $M_{18} = H(r^*_{HG} \parallel ID_{HG} \parallel GSK^*)$, $M_{19} = H(M_{18} \parallel M_{15} \parallel r_i \parallel T_4)$. UE_i checks if $M_{17} = M_{19}$. If it matches, the group session key is established successfully. Finally, UE_i replaces $TID^{new}_i = H(K^*_i \parallel TID_i \parallel T_4) \oplus M_{16}$ with new anonymous identity TID^{new}_i.

3.5 Biometrics and Password Update Phase

U_i provides personal credentials ID_i, PW^{old}_i and BIO^{old}_i to UE_i. UE_i utilizes these credentials validate the authenticity of U_i. If the credentials are valid, the credentials will be updated. When passing the validation, U_i enters the new credentials PW^{new}_i and biometrics BIO^{new}_i. UE_i utilizes these new credentials to compute new parameters and updates these parameters without the help of KGC.

4 Security Analysis

The widespread Real-or-Random (ROR) model proposed by Abdalla *et al.* is adopted to establish our security model in this section.

1. **Participants:** Let $\prod^u_{U_i}$, $\prod^v_{SD_j}$, $\prod^t_{HG}$ represent instances u, v and t of participant U_i, SD_j and HG, respectively.
2. **Partnering:** If the next conditions are satisfied, The instances $\prod^u_{U_i}$ and $\prod^v_{SD_j}$ are said to be partners [7].
   (i) both instance $\prod^u_{U_i}$ and $\prod^v_{SD_j}$ are accepted,
   (ii) both instances $\prod^u_{U_i}$ and $\prod^v_{SD_j}$ authenticate each other,
   (iii) the instance $\prod^u_{U_i}$ and the instance $\prod^v_{SD_j}$ are only partners each other.
3. **Freshness:** The instance $\prod^u_{U_i}$ or $\prod^v_{SD_j}$ is *fresh* if the session key SK is not compromised to $\mathcal{A}$.
4. **Adversary:** $\mathcal{A}$ has all the capabilities as adversary in Dolev-Yao (DY) threat model [7] and also has some capabilities defined in CK-adversary model [2, 3]. Furthermore, $\mathcal{A}$ can make queries as $Execute(\prod_u, \prod_v)$, $Reveal(\prod^t)$, $Send(\prod^t, m)$, $CorruptUserEquipment(\prod^t_{U_i})$, $CorruptSmartDevice(\prod^t_{SD_j})$ and $Test(\prod^t)$ to challenger to obtain the sensitive information. These queries are utilized to construct a series of game. After games, $\mathcal{A}$ guesses a bit b' and wins the game only if $b' = b$. $Succ$ represents that $\mathcal{A}$ wins the game. The advantage of $\mathcal{A}$ in breaking the IND-CCA of our scheme $\mathcal{P}$ in PPT time is $Adv^{IND-CCA}_{\mathcal{P},\mathcal{A}}(\mathcal{K}) = |2 \cdot Pr[Succ] - 1|$. The proposed scheme $\mathcal{P}$ is secure under the ROR model when $Adv^{IND-CCA}_{\mathcal{P},\mathcal{A}}(\mathcal{K})$ is negligible.

Theorem 1. *Let $\mathcal{A}$ be the adversary running in the polynomial time t against our authentication scheme $\mathcal{P}$ in the random oracle. Let Dic, q_h, q_{send}, q_e, $|Hash|$, $|Dic|$, m and l^r represent the a uniformly distributed password dictionary, the number of $Hash$ oracles, the number of $Send$ oracle, the number of $Execute$ oracles, the space of hash function, the size of Dic, the bit length of biometrics key σ_i and the bit length of the random nonce, respectively. The advantage of $\mathcal{A}$ in breaking scheme $\mathcal{P}$ in PPT is defined as follows*

$$Adv_{\mathcal{P},\mathcal{A}}^{AKA}(\mathcal{K}) \leq \frac{q_h^2}{|Hash|} + \frac{(q_{send} + q_e)^2}{2^{l^r}} + \frac{q_{send}}{2^{m-1} \cdot |Dic|} + \frac{2}{q} \cdot Adv_{\mathcal{P},\mathcal{A}}^{IND-CPA}(\mathcal{K}).$$

Un-traceability and User Anonymity. It is assumed that $\mathcal{A}$ has capabilities of intercepting all the message during the execution of the authentication phase over the public channel. The user's identity ID_i is protected by hash function $H(\cdot)$ and symmetric cryptography. It is computationally infeasible for $\mathcal{A}$ to attain identity without secret parameters r_{HG}, r_i, B_i, σ. Therefore, our scheme guarantees the feature of user anonymity. Moreover, the transmitted message generally involves the current timestamp and random nonce and U_i temporary identity TID_i is updated when session is completed successfully. Therefore, it is also computationally infeasible for $\mathcal{A}$ to track the user's activity in each session. In conclusion, the un-traceability and user anonymity are both guaranteed in our scheme.

Session Key Security. The session key GSK is calculated by both all the authenticated smart devices and the user U_i. The message M_{14} contains the session key. Suppose that $\mathcal{A}$ intercepts the message and tries to forge GSK' by random nonces r_i', r_{HG}. However, $\mathcal{A}$ does not know the parameters $ID_i, H(K_i), M_{10}$, it is impossible for $\mathcal{A}$ to compute GSK due to the collision resistance property of $H(\cdot)$. Thus, our scheme guarantees session key security successfully.

Replay Attack. It is assumed $\mathcal{A}$ is capable to intercept all the message between the user, HG and smart devices. The transmitted messages usually involve the random nonces and timestamps. Even if $\mathcal{A}$ intercepts the messages and replays these messages shortly after, they can not pass the verification of timestamps due to maximum communication delay ΔT. Thus, our scheme can resist replay attack.

Smart Device Impersonation Attack. It is supposed $\mathcal{A}$ intercepts the transmitted message during the execution of the scheme. $\mathcal{A}$ needs to generate the valid information. However, $\mathcal{A}$ does not know the sensitive parameters to obtain the authentication parameters. Furthermore, the smart device is protected by physical unclonable function, which cannot be forged on hardware. It is computationally infeasible to impersonate the smart device in PPT. Therefore, our scheme can withstand smart device impersonation attack.

Ephemeral Secret Leakage Attack. In our scheme, a secure group session key $GSK^* = H(r_{HG}^* \parallel H(K_i) \parallel r_i \parallel ID_i \parallel ID_{HG} \parallel M_{10})$ is established between a user and smart devices during the login and authentication phase. M_{10} is composed of long-term secret $H(S)$ and short-term secret r_{HG}. In particular, the secret S is computed by secret reconstruction algorithm of secret sharing technology. In addition, $ID_{HG}, ID_i, H(K_i)$ are the long-term secrets and r_i is a short-term secret. On the one hand, it is assumed that the short-term secrets r_{HG}, r_i are revealed to $\mathcal{A}$. However, it is computationally infeasible to compute the GSK due to lack of the long-term secrets. On the other hand, it is assumed that $\mathcal{A}$ can obtain the long-term secrets. Even through $\mathcal{A}$ obtain some secret shares s_j from the smart devices, it is computationally infeasible to construct the secret S and then calculate the M_{10}. The short-term secrets r_{HG}, r_i are randomly generated by the HG and U_i. It is also hard for $\mathcal{A}$ to compute GSK without the short-term secrets r_{HG}, r_i. Therefore, $\mathcal{A}$ cannot compute the current session key unless both all the long-term secrets and short-term secrets are compromised simultaneously. Our scheme can thwart ephemeral secret leakage attack.

5   Performance Analysis

We evaluate the communication and computational cost in our authentication scheme compared to other schemes [4,11,19]. The proposed scheme is simulated using Pair-Based Cryptography (PBC) library and GNU Multiple Precision Arithmetic (GMP) library. C language is utilized on Ubuntu 16.04 with 2.50 GHz Intel(R) Core(TM) i5-4200M CPU, and 8 GB of RAM. We suppose that the bit length of identities, random nonces, timestamps, hash function operation are 128bits, 128 bits, 32bits, 160 bits, respectively. It is also assumed that $|\lambda_1| = 128$, $|\lambda_2| = 160$ and AES-128 is adopted for symmetric cryptography, where λ_1, λ_2 denote the length of input and output of physical unclonable function, respectively. Table 2 show the total communication cost of our scheme and associated three schemes [4,11,19].

Table 2. Communication cost comparison.

Scheme	One device accessing cost (bits)	n devices accessing cost (ms)
[4]	2016	$2016n$
[11]	2048	$2048n$
[19]	2592	$2592n$
Our scheme	3296	$1376+1920n$

We compare the total execution time with other schemes [4,11,19] during the login and authentication phase. It is assumed that T_h, $T_{E/D}$, T_{fe}, T_{xor}, T_{ecm}, T_{mm}, T_{mac} and T_{hmac} denote the computational cost required for a hash function, a symmetric cryptography using AES-128, a fuzzy extraction operation, a XOR operation, a point multiplication operation using ECC, a modular multiplication operation, a message authentication code (MAC) operation and a hashed MAC operation, respectively. The bit-wise XOR operation is not considered in the evaluation as the its computational cost is less than other operations. Besides, it is assumed that $T_h \approx T_{mac} \approx T_{hmac}$, $T_{fe} \approx T_{ecm}$ in our experiment according to [19]. The computational cost of T_h, $T_{E/D}$, T_{fe}, T_{mm} and T_{ecm} is 0.0026 ms, 0.00325 ms, 1.989, 0.171 ms and 1.989 ms (ms is the abbreviation of milliseconds), respectively. The computational cost of accessing a single and multiple devices for the related scheme and our scheme is described in the Table 3.

Table 3. Computational cost Comparison.

Scheme	One device accessing cost (ms)	n devices accessing cost (ms)
[4]	$T_{fe} + 16T_h + 13T_{ecm}$	$(T_{fe} + 16T_h + 13T_{ecm})n$
[11]	$T_{fe} + 19T_h + 8T_{E/D} + 3T_{ecm}$	$(T_{fe} + 19T_h + 8T_{E/D} + 3T_{ecm})n$
[19]	$T_{fe} + 21T_h + 8T_{E/D}$	$(T_{fe} + 21T_h + 8T_{E/D})n$
Our scheme	$T_{fe} + 20T_h + 8T_{E/D}$	$T_{fe} + 8T_h + 4T_{E/D} + (4T_{E/D} + 12T_h)n$

6   Conclusion

In this paper, we proposed a PUF-based two-factor anonymous group authentication scheme for smart home based on secret sharing technique and Chinese Remainder Theorem. The proposed scheme can withstand most of several known attacks, which is proved under ROR model and security discussion. Compared with other related schemes, our scheme can effectively reduce the resource cost during the login and authentication phase. In addition, our smart devices protected by physical unclonable function are secure against device capturing attack.

References

1. Banerjee, S., Odelu, V., Das, A.K., Chattopadhyay, S., Rodrigues, J.J.P.C., Park, Y.: Physically secure lightweight anonymous user authentication protocol for Internet of Things using physically unclonable functions. IEEE Access **7**, 85627–85644 (2019)
2. Canetti, R., Krawczyk, H.: Analysis of key-exchange protocols and their use for building secure channels. In: Pfitzmann, B. (ed.) EUROCRYPT 2001. LNCS, vol. 2045, pp. 453–474. Springer, Heidelberg (2001). https://doi.org/10.1007/3-540-44987-6_28. http://dl.acm.org/citation.cfm?id=647086.715688
3. Canetti, R., Krawczyk, H.: Universally composable notions of key exchange and secure channels. In: Knudsen, L.R. (ed.) EUROCRYPT 2002. LNCS, vol. 2332, pp. 337–351. Springer, Heidelberg (2002). https://doi.org/10.1007/3-540-46035-7_22
4. Challa, S., et al.: Secure signature-based authenticated key establishment scheme for future IoT applications. IEEE Access **5**, 3028–3043 (2017)

5. Chiang, M., Zhang, T.: Fog and IoT: an overview of research opportunities. IEEE Internet of Things J. **3**(6), 854–864 (2016)
6. He, D., Kumar, N., Chilamkurti, N.: A secure temporal-credential-based mutual authentication and key agreement scheme for wireless sensor networks. In: International Symposium on Wireless and pervasive Computing (ISWPC), pp. 1–6, November 2013
7. Dolev, D., Yao, A.: On the security of public key protocols. IEEE Trans. Inf. Theory **29**(2), 198–208 (1983)
8. Kalra, S., Sood, S.K.: Advanced password based authentication scheme for wireless sensor networks. J. Inf. Secur. Appl. **20**, 37–46 (2015). Security, Privacy and Trust in Future Networks and Mobile Computing
9. Kumar, P., Braeken, A., Gurtov, A., Iinatti, J., Ha, P.H.: Anonymous secure framework in connected smart home environments. IEEE Trans. Inf. Forensics Secur. **12**(4), 968–979 (2017)
10. Li, X., Niu, J., Bhuiyan, M.Z.A., Wu, F., Karuppiah, M., Kumari, S.: A robust ECC-based provable secure authentication protocol with privacy preserving for industrial Internet of Things. IEEE Trans. Ind. Inf. **14**(8), 3599–3609 (2018)
11. Li, X., Peng, J., Niu, J., Wu, F., Liao, J., Choo, K.R.: A robust and energy efficient authentication protocol for industrial Internet of Things. IEEE Internet of Things J. **5**(3), 1606–1615 (2018)
12. Jiang, L., Liu, D.-Y., Yang, B.: Smart home research. In: Proceedings of 2004 International Conference on Machine Learning and Cybernetics (IEEE Cat. No. 04EX826), vol. 2, pp. 659–663, August 2004
13. Odelu, V., Das, A.K., Goswami, A.: A secure biometrics-based multi-server authentication protocol using smart cards. IEEE Trans. Inf. Forensics Secur. **10**(9), 1953–1966 (2015)
14. Ricquebourg, V., Menga, D., Durand, D., Marhic, B., Delahoche, L., Loge, C.: The smart home concept: our immediate future. In: 2006 1ST IEEE International Conference on E-Learning in Industrial Electronics, pp. 23–28, December 2006
15. Shen, J., Gui, Z., Ji, S., Shen, J., Tan, H., Tang, Y.: Cloud-aided lightweight certificateless authentication protocol with anonymity for wireless body area networks. J. Netw. Comput. Appl. **106**, 117–123 (2018)
16. Srinivas, J., Das, A.K., Wazid, M., Kumar, N.: Anonymous lightweight chaotic map-based authenticated key agreement protocol for industrial Internet of Things. IEEE Trans. Dependable Secure Comput. 1 (2018)
17. Turkanović, M., Brumen, B., Hölbl, M.: A novel user authentication and key agreement scheme for heterogeneous ad hoc wireless sensor networks, based on the Internet of Things notion. Ad Hoc Netw. **20**, 96–112 (2014)
18. Wallrabenstein, J.R.: Practical and secure IoT device authentication using physical unclonable functions. In: 2016 IEEE 4th International Conference on Future Internet of Things and Cloud (FiCloud), pp. 99–106, August 2016
19. Wazid, M., Das, A.K., Odelu, V., Kumar, N., Jo, M.: Design of secure user authenticated key management protocol for generic IoT networks. IEEE Internet of Things J. **PP**(99), 1 (2017)
20. Ye, X., Huang, J.: A framework for cloud-based smart home. In: Proceedings of 2011 International Conference on Computer Science and Network Technology, vol. 2, pp. 894–897, December 2011

21. Xue, K., Ma, C., Hong, P., Ding, R.: A temporal-credential-based mutual authentication and key agreement scheme for wireless sensor networks. J. Netw. Comput. Appl. **36**(1), 316–323 (2013). http://www.sciencedirect.com/science/ARTICLE/pii/S1084804512001403

22. Zhang, J., Cui, J., Zhong, H., Chen, Z., Liu, L.: PA-CRT: Chinese remainder theorem based conditional privacy-preserving authentication scheme in vehicular ad-hoc networks. IEEE Trans. Dependable Secure Comput. 1 (2019). https://doi.org/10.1109/TDSC.2019.2904274

An Authentication Framework in ICN-Enabled Industrial Cyber-Physical Systems

Yanrong Lu[1,2(✉)], Mengshi Zhang[3], and Xi Zheng[4]

[1] School of Computer Science and Technology, Civil Aviation University of China, Tianjin, China
luyanrong1985@163.com
[2] Tianjin Key Laboratory of Intelligence Computing and Novel Software Technology, Tianjin University of Technology, Tianjin, China
[3] Facebook Inc., Menlo Park, CA, USA
[4] Autonomous Systems Research Center Australia, Brisbane, Australia

Abstract. Industrial Cyber-Physical Systems (ICPS), as a new industrial revolution, are to provide advanced intellectual foundation for next generation industrial systems. While such systems present substantial security challenges for the host-centric communication with the growing trend of sensor data streams. Information Centric Networking (ICN) architecture suggests features exploitable in ICPS applications, reducing delivery latency and promoting quality of services that applies broadly across Industrial Internet. Emerging available solutions for secure communication, however, few of them have thoroughly addressed concerns related to securing access due to the dependence on an online provider server. In this work, we propose a concrete authentication framework for ICN ICPS based on proxy signature, which guarantees authentic sensor data access only to legitimate users and does not require interaction between users. This framework would help lower the level of the complexity of the entire system and reduce the cost of authentication by leveraging edge cache. We prove the security of the proposed authentication scheme and present performance analysis to show its efficiency.

Keywords: Information Centric Networking (ICN) · Industrial Cyber-Physical System (ICPS) · Authentication

1 Introduction

The emergence Industrial 4.0 enables existing factory world to be more flexible, efficient and smart. Currently, wireless technology speeds up the deployment of Industrial 4.0 at a reasonable economic cost and with reduced energy

Supported in part by the National Natural Science Foundation of China under Grants 61802276 and in part by the Opening Foundation of Tianjin Key Laboratory of Intelligence Computing and Novel Software Technology, Tianjin University of Technology, China.

D. Wang et al. (Eds.): SPNCE 2020, LNICST 344, pp. 223–243, 2021.
https://doi.org/10.1007/978-3-030-66922-5_15

and resources consumption [17]. With such direction, industries have shifted their concern to intelligent realizing and controlling processes with an acceptable quality of experience for their users. Based on these benefits, Industrial Cyber-Physical Systems (ICPS) as a new paradigm provides a variety of advantages for cooperative communication to significantly improve service quality and reduce expenditure for the infrastructure. Despite its many attractive features including adaptability, reliability, data integration, automation and optimization, ICPS is faced with high complexity, inefficient distribution control as well as security and privacy issues [2,3].

Moreover, according to Cisco's global mobile data traffic forecast, by 2021 [1], the amount of device will generate 77 billion tb of traffic per month. ICPS communication networks will be an important part of communication between devices. This implies that huge amount of data generated by device and advanced sensor technology deployed in ICPS, which has posed significant challenges on the current host centric-based communication model. The quest for the ever-increasing demand of bandwidth has not been addressed adequately by the existing Internet infrastructure. On the other hand, such tremendous data delivery leads to complicate network management and ICPS application development (e.g. smart grid, smart manufacturing, smart health), which mismatches quality of user experience. Along these lines, one of the most challenging tasks is to offer reliable communication from the source towards destination enabling real-time monitor of supply and demand balances with lower latency.

To cope with these forementioned challenges, a scalable networking architecture that satisfies ICPS communication requirements is imperative. Information Centric Networking (ICN), as one of promising approaches, shifts the current Internet infrastructure from a host-centric paradigm to content-based model in which information source is separated from where information is needed. This decoupling of information from the exact location enables a higher degree of flexibility, especially in supporting device-to-device content sharing [5,6]. Thus, ICN is considered as one of an ideal underlying communication stack for ICPS since it has a great advantage in eliminating transmission latency.

Taking smart grid, a typical ICPS applications, as an example. There could be a power server which is in charge of monitoring, managing, and coordinating the energy consumption, and allows its subscribed users to benefit from real-time electricity consumption data. With the integration of ICN into ICPS, a terminal user only needs to express its interest with content name from multiple devices, to the specific sensor or actuator that provides the information instead of the hosts. Since a copy of the data is cached in along router once an energy data is fetched from the power server, thus allowing further requests with the same name to be fulfilled quickly.

Yet, in spite of those advantages, the deployment of ICN on ICPS poses some new challenges, among which authentication is a high-ranking one. In the IP-based solution, before a user retrieves some sensor data from devices, the session connection is provided such that end-to-end trust is easily established [4, 10,12,18]. However, due to the in-network cache in ICN, the unpredictability with which repositories provide data leads to difficulty in preventing fake data injection attacks. Similarly, any requests can be satisfied by the routers results in attackers pretend to be a trusted individual to gain access and manipulate the system. To combat these threats, users and routers need to verify data before retrieving or caching them while the identities of users also need to be verified. Therefore, a concrete and efficient authentication instantiation is needed for the implement of ICN in ICPS model.

In this paper, we present an authentication scheme under ICN-enabled architecture to provide user-to-user communication service for ICPS. We describe an ICPS framework that addresses authentication of users and supports user-to-user authentication based on ICN paradigm. To authenticate and save computational consumption on ICPS devices, proxy signature as a building block is enforced so that content integrity is preserved. To separate authentication from ICPS devices, we leverage the edge router close to the users, to authenticate the users' requests. It is believed that by taking advantage of these operations, content transmission can be more robust, with lower latency and less complexity as compared with existing work. To summarize, we make four-fold contributions:

- We provide a two-layer security framework for ICN-based ICPS. The upper layer provides registration service, while the lower layer employs sensor data transmission. It obviates the need for a direct connection with ICPS devices and lower the level of complexity of the system;
- We develop a proxy signature that enables computational savings of ICPS devices. It preserves the privacy privileges of original signer and proves unforgeable in the random oracle;
- We propose a session-based scheme that relies on the proxy signature to ensure user authentication. It ensures that authentic sensor data are only available by legitimate users;
- We present security and performance of the proposed approach to show its robustness and effectiveness compared to existing work.

The roadmap of this article is organized as follows. We state our system model, threat model, and objectives in Sect. 2. We then present our proxy signature in Sect. 3 and the construction of our whole scheme in Sect. 4. Security analysis is shown in Sect. 5 and performance analysis is presented in Sect. 6. Finally, we make a conclusion in Sect. 7.

1.1 Related Work

Current solutions for access security in ICN framework are content authenticity in the sense that a user signs on the requested content, such as [21] or content confidentiality in the sense that a user encrypts the requested content, such as [9,13]. Content authenticity, for example, Zheng et al. [21] proposed a certificateless-based signature with revocation to prevent false data injection. The scheme realizes provider authentication so that any receiver could identify the source of the data. Moreover, it largely decreases content verification overhead compared to public key signature algorithms. Current content confidentiality, for example, Fotiou et al. [9] suggested an identity-based cryptography (IBC) based proxy re-encryption scheme. Since there is no need to manage (e.g., issue and revoke) public-key certificates in IBC schemes, this scheme significantly reduces the computational overhead compared to public key infrastructure-based schemes. Still, it needs to pre-store encryption keys and thus is impractical for large-scale deployments. Li et al. [13] presented an attribute-based encryption for access control enforcement in ICN, and the symmetric key used to encrypt the data generated by attributes of an authorized user. Despite its low overhead, revocation remains a challenge in such approach, for the private keys corresponding to each attribute have to be regenerated and redistributed during revocation. Other schemes that are considered under proxy broadcast encryption [15] and re-encryption [7], indeed these schemes contribute for computational cost, but has vulnerabilities such as denial-of-service attack scenario.

Coming up with different solutions, many researchers have contributed to the hybrid approaches. Xue et al. [19] designed a mechanism built upon group signature and symmetric cryptographic operations, where all of keys are generated by a trusted provider. With such design, user privacy and data confidentiality are both achieved. However, each user needs to get the necessary secret information through a secure channel from the provider, which requires the provider and users to be directly connected thus offsetting the benefits of in-network caching. Nunes-Tsudik [16] proposed a scheme in which the authentication request is sent to the network through Kerberos enforcement aiming to reduce the vulnerability of the user. A hybrid authentication and authorization based on trusted platform that could provide considerable robustness for content security. Practically, due to the introduction of fully trusted model, there is a problem of single point failure. Conventional end-to-end communication also makes it inefficient in practice.

Various existing proposals have been adopted ICN in ICPS application to provide low latency communication. For example, De Silva et al. integrated ICN within Internet of things (IoT) to decouple the control plane and the data plane to satisfy the communication requirements of the lighting control. Mick et al. [14] suggested an IoT-based ICN on-boarding authentication procedure to handle the data communication. In their approach, the ICN controller was used to key management and centralized server. While this scheme provides a flexible mechanism, it pre-stores a shared key in a physical device, which may significantly undermine its scalability and make data access unfriendly.

Although existing research can ensure some level of access security, those approaches do not fully address in-network caching strategy. This is because most of existing work either happen on a single endpoint or require user to interact with each other. Another concern is that a heavy burden for ICPS devices with limited computational and processing capacities to generate the required signatures with asymmetric cryptography. Additionally, user privacy immediately becomes a paramount task once ICPS devices are connected to the

Internet since the envisioned ICN routers could deduce sensitivity of the message and infer the identities of those who are involved in the message transmission.

Motivated by such observation, the availability of ICN routers allows us to support as an edge cooperative way at the first-hop and last-hop routers in the ICN architecture. As such, exploring a delegation model to the ICPS is a natural solution to reduce computation overhead on ICPS users. To enforce sensor data authenticity and integrity, proxy signature presents a promise to the minimal use of computation overhead on ICPS devices. It allows a proxy to generate a digest of cryptographic on behalf of the original signer. Anyone accessible to the public keys of the original signer and proxy signer can verify the authenticity of the purpoted signature afterwards. Proxy signature can be classified into three delegation types: full delegation, partial delegation and delegation by warrant.

The existing research on proxy signature addresses applications areas different from this article, such as grid computing [8]. There is still few concerns on authentication in ICN ICPS with proxy signature. In this paper, we use proxy signature as a building block to create an authentication framework for ICN ICPS. Our framework ensures that legitimate users to access the authentic sensor data.

2   Problem Statement

2.1   System Model

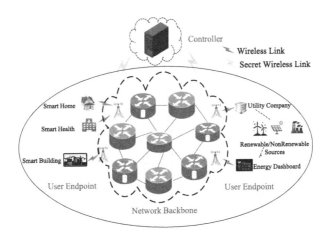

Fig. 1. Two-layer ICN-ICPS architecture.

Figure 1 depicts a basic network architecture of ICN ICPS. The architecture is composed of two levels: the upper level is a controller that provides system infrastructure; the lower level consists of the physical devices, where ICN is leveraged as communication pillar to handle per packet. In the lower level, users use the two packets types, Interest and Data packets, to communicate and exchange information, respectively. ICN routers as intermediate forwarders are used to do a match between the Interest and Data source, and cache some information to be available for later requests.

There are four types of participants involved in the system, controller, users, edge routers and intermediate routers. In particular, users refer to consumers or ICPS devices who use terminal or application to subscribe or provide the energy service. One of two edge routers, one adjacent to the consumers and another closer to the ICPS devices, to be implemented as edge servers, respectively. We refer to the first edge router as consumer gateway and the second as device proxy. The gateway is enforced access policies to decide whether an interest or data packet should be forwarded or dropped based on the verification results. The proxy is allowed to perform data analysis and then produce certain operations based on the request on behalf of the ICPS devices. Intermediate routers are distributed in the system to forward the packets between users.

2.2   Threat Model

Under the framework shown in Fig. 1, we assume that all participants are registered with a controller, before they are approved to get access the system resource. The controller is not necessarily to be a strong trusted third party. This property makes our model realistic and fine-grind, since a centralized entity necessitates centralized trust and represents a single point of failure. We also assume that there are two types of adversaries: passive adversary and active adversary. Passive attacks may be mounted from the routers who have collected plenty of interests information to learn "who is requesting" and "who is replaying". In contrast the passive adversary, the active adversary has more powerful capabilities so that they could launch some stronger attacks for any packet transmissions at the communication level, such as, capture/analyze interests, modify requests and responds, and also masquerade as legitimate users to send interests.

2.3   Objectives

Our design goal is to propose a secure authentication scheme based on the defined threat model for ICN-ICPS at reasonable cost. For this target, the proposed scheme must hold security attributes. It is desired that the proposed scheme to cover the following requirements:

Integrity: Making sure that the received data is completed, unmodified during transmissions.

Authenticity: Ensuring that the data comes from the ones they claim to be. Any users should be held non-repudiation for their actions.

Authentication: Identifying both users to be adequately trusted and preventing adversaries to mount masquerade attack by performing traffic analysis and capturing packets.

Anonymity: Protecting real identities of the users so that other related privacy information is exposed as little as possible to the intermediate routers.

Key Establishment: Providing a random session key to be contributed by only the user and its responder. This indicates that the negotiated key should be inaccessible to the third party.

Notation: $h_i(\cdot)$ be hash functions. Q_0, s_0 be system's public key and private key. Q_A, s_A be an entity A's public key and private key. id_A be an entity A's identity. $MAC(\cdot)$ be a message authentication code. $r \leftarrow S$ denotes an element is randomly chosen from the set S. m and σ be a content and its signature. $\mathbb{G}_1$ be a cyclic additive group over prime finite field $\mathbb{F}_p$; $\mathbb{G}_2$ be a cyclic multiplicative group; P be the generator of $\mathbb{G}_1$; q be the prime order of $\mathbb{G}_1$ and $\mathbb{G}_2$; e be a pairing from $\mathbb{G}_1$ to $\mathbb{G}_2$; The Computational Diffie-Hellman (CDH) problem: Given $\langle P, aP, bP \rangle$ with uniformly random choices of $a, b \leftarrow \mathbb{Z}_q^*$, compute $abP \leftarrow \mathbb{G}_1$.

3 Building Block: Proxy Signature

In this section, we propose a proxy signature that relies on non-trusted-aided third party and prove its security is equivalent to the CDH problem in the random oracle model. Then, we analyze the performance in terms of signature size and computation overhead.

3.1 Construction

The proxy signature consists of a `Setup` algorithm, two sub-algorithms `PartialKeyGen` and `KeyGen`, and a proxy designation issuing algorithm which involves three algorithms: `ProxyKeyGen`, `Sign`, and `Verify`.

`Setup`: Algorithm 1 takes security parameter κ and returns the system parameters params and master-key. Generally, this algorithm is run by a carrier called Private Key Generator (PKG). We assume throughout that params are publicly known, while the master-key will be known only to the PKG.

Algorithm 1: Setup

Input: 1^κ

Output: s, params$= \langle \mathbb{G}_1, \mathbb{G}_2, e, P, Q_0, h_1, h_2, h_3, h_4, MAC \rangle$

1: Generate $\langle \mathbb{G}_1, \mathbb{G}_2, e \rangle$.

2: Choose
$$P \leftarrow \mathbb{G}_1, \quad s \leftarrow \mathbb{Z}_q^*, \quad h_1, h_3 : \{0,1\}^* \rightarrow \mathbb{G}_1, \quad h_2, h_4 : \{0,1\}^* \rightarrow \mathbb{Z}_q^*.$$

3: Compute $Q_0 \leftarrow sP$.

PartialKeyGen: The PKG runs Algorithm 2 to generate a partial key for each entity A. The algorithm inputs params, master-key, a secret value $x_A \leftarrow \mathbb{Z}_q^*$ selected by entity A and an identifier for entity A with a string $id_A \leftarrow \{0,1\}^*$, as input. Each partial private key is calculated by hashing an identity id_A and $x_A P$ to append them with the master-key. Usually, the output is transported to entity A over a confidential channel.

Algorithm 2: PartialKeyGen

Input: params, id_A, $x_A P$

Output: p_A corresponding to A

1: Transmit A $\rightarrow$ PKG: $\langle id_A, x_A P \rangle$.

2: Compute
$$D_A \leftarrow h_1(id_A, x_A P), \quad p_A \leftarrow s D_A.$$

KeyGen: Algorithm 3 takes as input params, an entity A's partial private key p_A and A's secret value x_A and constructs A's private key s_A and public key Q_A, respectively. Each full private key is a combination of each partial private key p_A and a secret random value x_A. And a public key is computed by $x_A P$. This algorithm is run by an entity A itself who is the only entity in possession of x_A.

Algorithm 3: KeyGen

Input: params, p_A, x_A

Output: s_A / Q_A corresponding to A

1: Compute
$$s_A \leftarrow \langle p_A, x_A \rangle, \quad Q_A \leftarrow x_A P.$$

ProxyKeyGen: Algorithm 4 requires only two short communication flows between an original signer and a proxy signer. It begins to run at the original signer side and gets as input params, the original signer's secret value x_A and a warrant m_ω that includes delegation period, message types, the identity information of a proxy signer id_B, the public key of an original signer Q_A and a proxy signer Q_B. The original signer commits itself to a proxy transcript proxy-trans: $\langle m_\omega, \varpi \rangle$ and then transmits them to the proxy signer who is designated to check it using the original signer's public key. Consequently, the proxy signer will get a proxy signing key pair $SK_B \leftarrow \langle x_B, p_B + x_A H_2 P \rangle$, $PK_B \leftarrow \langle id_B, Q_B \rangle$ if the check is correct.

Algorithm 4: `ProxyKeyGen`

Input: params, x_A, m_ω, Q_A, Q_B

Output: true: $\langle SK_B, PK_B \rangle$; **false:** failure.

1: Compute

   $H_2 \leftarrow h_2(id_B, m_\omega, Q_A, Q_B), \ \varpi \leftarrow \frac{x_A D_B}{x_A + H_2}$.

2: proxy-trans: $\langle m_\omega, \varpi \rangle$.

3: Compute

   $H_2 \leftarrow h_2(id_B, m_\omega, Q_A, Q_B)$.

4: Check if $e(\varpi, Q_A + H_2 P) = e(Q_A, D_B)$ **then**

5:     return **true**

6: **else**

7:     return **false**

8: **end if**

`Sign`: Algorithm 5 runs at the proxy side who signs a message m on behalf of the original signer. It requires params, a proxy secret key SK_B, proxy-trans, the public-key of the proxy signer. The resulting signature on the message m consists of the triple $\langle R, V \rangle$.

Algorithm 5: `Sign`

Input: params, SK_B, PK_B, proxy-trans

Output: σ

1: Choose $r \leftarrow \mathbb{Z}_q^*$.

2: Compute

   $R \leftarrow rP, \ H_3 \leftarrow h_3(m, id_B, R, Q_B)$,

   $V \leftarrow p_B + x_A H_2 P + x_B H_2 P + r H_3$.

3: $\sigma \leftarrow \langle R, V \rangle$.

`Verify`: Algorithm 6 runs at the verifier side and checks the validity of a signature of a given message m with respect to params and a set of public key $\langle PK_B, Q_A \rangle$, respectively. Notice that if σ is a valid signature on a message m, it outputs **true** and **false** otherwise.

3.2   Comparison

We compare signature length and computational complexity with the most recent signature schemes [11,18,20] in Table 1. In Table 1, $\mathbb{G}_1$ and Exp. in signature length represent the size of the group element and modulus, respectively,

---

Algorithm 6: Verify

Input: params, σ, m, Q_A, PK_B

Output: true: success; **false:** failure.

1: Check **if**

$$e(V, P) = e(Q_0, D_B) \cdot e(Q_A, H_2 P) \cdot e(Q_B, H_2 P) \cdot e(R, H_3)$$

then

2:     return **true**;

3: **else**

4:     return **false**;

5: **end if**

---

and pair. and $\mathbb{G}_1$ in computation denote the computation complexities of a pairing and a group operation. Here we assume that the cost of an exponential computation is equal to that of elliptic curve multiplication.

Table 1. Comparison of signature length and computational complexity of signature protocols

Protocol	Signature Length	Computational Complexity		
		KeyGen	Sign	Verify
Xiong-Qin [18]	$3(\mathbb{G}_1)$	$10(\mathbb{G}_1)$	$8(\mathbb{G}_1)$	$5(\text{Pair}) + 3(\mathbb{G}_1)$
Hwang et al. [11]	$3(\mathbb{G}_1) + 5(\text{Exp.})$	$3(\mathbb{G}_1)$	$4(\text{Pair}) + 7(\mathbb{G}_1)$	$6(\text{Pair}) + 5(\mathbb{G}_1)$
Zhang et al. [20]	$1(\mathbb{G}_1) + 1(\text{Exp.})$	$4(\mathbb{G}_1)$	$1(\mathbb{G}_1)$	$2(\text{Pair}) + 3(\mathbb{G}_1)$
Ours	$2(\mathbb{G}_1)$	$3(\mathbb{G}_1)$	$5(\mathbb{G}_1)$	$4(\text{Pair}) + 1(\mathbb{G}_1)$

Signature Length: To achieve equivalent secure level, the size of prime p is considered to be at least 512 bits. And for achieving a fair comparison, using a supersingular elliptic curve $y^2 = x^3 + ax$ group $\mathbb{G}_1$ of order q with embedding degree two, a modulus size that satisfying $q|p + 1$ for any odd $q = 3(mod)4$. Table 1 summarizes the comparison results. Observe that our signature length is sightly longer than [20] but shorter than [11,18] (i.e., we use the fact that bit length of q with 160 bits and the group element in $\mathbb{G}_1$ with 512 bits). Therefore, the bandwidth consumption on the proposed signature is approximately the same as the state-of-the-art.

Computational Complexity: We compare KeyGen, Sign and Verify algorithms in Table 1 which are the main stages in these schemes. In general, the cost of a pairing operation is several times than a scalar multiplication in $\mathbb{G}_1$. Thus, the number of pairing operations is a key performance metric. Table 1 summarizes the results and omits other operations due to their trivial complexity.

Accordingly, the key generation needs three multiplication operations, the signing operation requires three multiplication operations and no pairing computations. The verifier can collapse the $e(Q_A, H_2P)$ and $e(Q_B, H_2P)$ pairings into a single $e(Q_A + Q_B, H_2P)$ term. Thus verifying a signature requires one multiplication operation and four pairing computations. Observe that our signature only incurs a minor cost than [20] but obviously outperforms signature systems [11,18]. However, it is worth noting that our signature algorithm is secure against forgery attack which lays down a concrete design foundation for the whole scheme.

4   Our Solutions

To satisfy the security demands of ICN-ICPS, the proposed proxy signature as a building block to design authentication scheme, preserving the anonymity of user. We first give a basic overview of our design and then describe it.

4.1   Overview

Our framework presents two techniques: proxy signature and session-based variant, that provides authentication for ICPS traffic. The proxy signature is used to authenticate the claimed providers to their consumers without revealing their actual identities, while the session-based variant aims to verify the legitimacy of the requesting consumer.

A full specification of our scheme consists of five steps: The first step is the registration procedure between entities (users, gateway, and proxy) and the controller which served as the PKG (See `PartialKeyGen` in Sect. 3.1). The second step is performed at the consumer side who initiates an Interest packet, its neighbor gateway forwards the packet to the routers until reaching the ICPS device. The third step is performed at the ICPS device side who delegates its signing rights to a designated proxy (See `ProxyKeyGen` in Sect. 3.1). The fourth step is performed at the proxy side who generates a cryptographic digest of the data on behalf of the ICPS device and returns it to the gateway along the original path (See `Sign` in Sect. 3.1). This step happens at the time that there are no cache hit on router nodes. Upon successful receiving the data packet, the last step is the communication interaction between the consumer and its gateway. The consumer is granted to access the content once its identity is valid (See Sect. 4.2). Figure 2 illustrates the interactions between the users, gateway, routers, and proxy and their corresponding procedure execution.

To summarize, all routers may infer content exchanged using cache, but cannot link cache to a specific user. Our design not only does not require interactions with the provider in existence, but also achieve authentication with no identity information leakage on both users.

4.2 The Complete Authentication Scheme

Prior to beginning the communication, the controller initializes the system and publicizes the system parameters params, both participants need to identify themselves to the controller and acquire a pair of key $\langle Q_C, s_C \rangle$, $\langle Q_{GW}, s_{GW} \rangle$, $\langle Q_{EP}, s_{EP} \rangle$, $\langle Q_P, s_P \rangle$ for consumer, gateway, edge proxy and provider according to KeyGen algorithm described in Sect. 3.1.

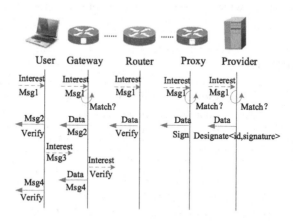

Fig. 2. An overview of our solutions. Here match denotes ICN router checks cache.

Consider a scenario, where a consumer attempts to obtain the power data for different purposes (e.g., real-time energy charging and distribution) from the resource providers (which can be the public administration or a private company). The consumer only intends to receive the valid content without any forging, altering or making a false content. On the other hand, the content is required to be legally consumed to prevent malicious consumers from launching security attacks against legitimate consumers. Table 2 shows the interactions between consumer and its gateway, in which the gateway has privilege of immediately responding to the consumer due to it has the matched content that satisfies the consumer requirements.

A consumer initiates a topic to broadcast its interest Msg1 in its vicinity. Starting from the name root tree [?], / delimits the boundaries of the components. The *domain* indicates that a specific area coverage that a provider entity can offer service. The *energy* specifies the main service that belongs to the provider. The *access* specifies the macro category service that offers further details useful related with the provider to create a more specific answer to the included into the Data. The *query* specifics the consumer demand.

We describe the case when the gateway is trying to achieve a connection with the consumer for the first time. Note that we assume here that the gateway has retrieved a Data packet from the proxy using the Interest paradigm from the network. Generally, this step is carried out when the requested content cannot be satisfied in any routers.

Table 2. Message exchange outline

Msg1: **Interest:**
    /domain/energy/Access/Query
Msg2: **Data:**
    Name: */domain/energy/Access/Query*
    Signature Info: $\langle Signature \rangle$
    Content: Session Invite Packet
Msg3: **Interest:**
    /domain/authenticate/id/Communicate/Req
Msg4: **Data:**
    Name: */domain/authenticate/id/Communicate/Req*
    Signature Info: $\langle Signature \rangle$
    Content: Response Packet

As mentioned-above, the gateway invites a session by sending an invitation connection Data packet Msg 2 which contains three components of content packets, i.e., name, signature and content itself together with a randomly generated identity *id*, used later by the gateway to identify itself to the consumer. Specifically, the gateway follows Algorithm 7 which outputs a $\langle signature \rangle$ that includes a set of public values with the help input parameters due to the fact that ICN requires every content to be authenticated.

Algorithm 7: Msg2 performed by gateway

Input: params
Output: $\langle Signature \rangle$
  1: Choose $r_{GW} \leftarrow \mathbb{Z}_q^*$.
  2: Compute
     $U \leftarrow r_{GW}P,\ W \leftarrow p_{GW} + r_{GW}Q_0.$
  3: $\langle Signature \rangle : \langle U, W \rangle.$

After obtaining the Data packet within a tolerable time interval, the consumer has to communicate with the nearby gateway to reply a request message to be able to pass the authentication step. The consumer responds with Msg3 by running algorithm 8 that takes the gateway's set of attributes such as public key Q_{GW}, identity *id* and private key s_C of the consumer associated with other essential public parameters params as inputs and outputs Req as the components of the reply messages.

The consumer constructs an Interest packet Msg3, where **authenticate** field explicits the sub-categories-of-services that accepts of the invite contains the identity *id* of who it corresponds to the gateway which allows a direct, bidirectional `communicate` path between them. It then returns the Data packet to the gateway inviting the session.

Algorithm 8: Msg3 performed by user

Input: params, s_C, Q_{GW}, $\langle Signature \rangle$

Output: true: Req: $\langle E, S \rangle$; **false:** failure.

1: Check **if** $e(W, P) = e(D_{GW}, Q_0) \cdot e(U, Q_0)$ **then**

2:　　Pick $r_C \leftarrow \mathbb{Z}_q^*$.

3:　　Compute
$$E \leftarrow r_C Q_{GW}, \ S \leftarrow r_U P - x_U D_{GW}.$$

4:　　return **true**

5: **else**

6:　　return **false**

7: **end if**

Algorithm 9 is performed by the gateway when receiving the Interest packet from the invited consumer. After check the equality, the gateway constructs a Data packet Msg4 that includes the content m, and τ encoded in the $\langle signature \rangle$ along with the signature σ together. It returns the Data packet to the consumer requesting the content.

Algorithm 9: Msg4 performed by gateway

Input: params, Req, p_{GW}

Output: true: output $\langle Signature \rangle$; **false:** failure

1: Derive: $r_C P \leftarrow x_{GW}^{-1} E$,

2: **if** $e(p_{GW}, Q_C) \cdot e(S, Q_0) = e(r_C P, Q_0)$ **then**

3: Compute
$$K \leftarrow h_4(r_{GW} P, r_C P,$$
$$r_{GW} r_C P), \ \tau \leftarrow MAC(r_{GW} P, K).$$

4:　　return **true**

5: **else**

6:　　return **false**

7: **end if**

Finally, the consumer executes algorithm 10 with the purpose of verifying the validness of the gateway. If the validation information does not match, the consumer discards the session. Otherwise, the consumer starts to verify the authenticity of the signature originated from the proxy. The consumer rejects the packet if the algorithm outputs false, and accept it otherwise.

Algorithm 10: Msg 4 checked by user

Input: params, σ

Output: true: success; **false:** failure

1: $K' \leftarrow h_4(r_{GW}P, r_C P, r_C r_{GW} P)$

2: **if** $(r_{GW}P, \tau, K') \rightarrow 1$ **then**

3: verify σ

4: **if** (verification==successful) **then**

5:   return **true**

6: **else**

7:   return **false**

8: **else**

9: return **false**

10: **end if**

---

5 Security Analysis

5.1 Security Properties Analysis

Our scheme guarantees that the controller only can know partial private keys, avoiding it impersonating as legitimate entity. This policy makes the controller more applicable in real ICN-ICPS scenarios. In what follows, we discuss the authentication scheme with respect to the objectives listed in Sect. 2.3.

Authenticity: An adversary $\mathcal{A}$ cannot forge a signature that is attributed to a legitimate party such that the party cannot repudiate.

Theorem 1. *Suppose hash functions* h_i *(i = 1, 2, 3) are random oracles. If a polynomial time adversary* $\mathcal{A}$ *has an advantage* $\epsilon(\kappa)$ *in forging a signature* σ, *then there exists an algorithm* $\mathcal{C}$ *that can break CDH problem with an advantage at least* $(\epsilon(\kappa)/2)(1 - q_s(q_{h_3} + q_s)/2^\kappa)(e(q_r + 1))^{-1}$ *by making* q_{h_3}, q_s, q_r *queries to the* h_3, *signing and revealpartialkey oracles.*

The security proof can be found in the appendix due to space limitation.

As such, proxy signature ensures that the content contained in the Data packet is valid with respect to the public keys of the proxy and the provider, where different content names can be leveraged to generate different signatures.

Authentication: The issue is mainly twofold. One is to verify the legitimacy of a requesting consumer in case a malicious adversary can launch an impersonation to hurt other innocent consumers' benefits. The other is to check the validity of the provider in case a malicious adversary injects poison content into the system. For the legitimacy of a consumer, only its gateway can authenticate its accessible by comparing $e(p_{GW}, Q_C) \cdot e(S, Q_0)$ to $e(r_C P, Q_0)$ due to the private key p_{GW} is only known by itself. If it is not valid, any malicious Interest packets from the consumer cannot be verified and will be dropped. For the validity of the provider, we take proxy signature to provide authenticity such that the sign in content indeed originates from the claimed provider. Before that, the gateway is required to be authenticated since frequent content access provided makes it effortless to become the goal of adversaries. Receiving a right τ signals that the gateway is the correct one so that the consumer authenticates the proxy together provider by verifying σ under the public keys of both. The unforgeability guarantee that the system can be convinced that the proxy belongs to its allocated delegation rights.

Anonymity: This property focuses on preventing adversaries from discovering the real identities of both users. In the session-based approach, the real identity of the consumer is only related to its private key which is the result of encryption with the private key of the controller and a random value as inputs. Thus, it is impossible to obtain the real identity due to adversaries have no information about these long-term private keys. Moreover, considering the problem of gateway's compromission, in the communication process of session connection, each consumer can get services without revealing its identity to its gateway. Therefore, if there exists gateway compromised, our scheme can still preserve the privacy of consumers' identities. Preventing adversaries from disclosing the real identity of the provider can be achieved via the proxy signature. Since the provider is anonymous to generate a warrant but tell the proxy nothing on the identity of it. With such approach, adversaries neither know who produces the content nor identify who consumes it. Therefore, identity protection to pair-wise communications is ensured.

Key Establishment: The security of key implies a polynomial time adversary $\mathcal{A}$ cannot distinguish between an instance's real session key and a random value.

Theorem 2. *The authentication scheme $\prod$ is secure, assuming the hardness of CDH problem and h_4 is a random oracle.*

Proof. Suppose that there exists $p(\kappa)$ participants and $s(\kappa)$ sessions. An oracle $\prod_{i,j}^{t}$ refers to the t-th instance of participant i involved with a partner with j in a session, which has a matching conversation to $\prod_{j,i}^{s}$ with a key as $h_4(K)$. Define the advantage of $\mathcal{A}$ to be $\epsilon(\kappa)$ and E be that $\mathcal{A}$ can query to the random oracle h_4, here $Pr[\mathcal{A}]$ is the success probability that $\mathcal{A}$ outputs a guess bit $\hat{b}$ such that $\hat{b} = b$ held by one of the oracles $\prod_{i,j}^{t}$. As we know $Pr[\mathcal{A}] = Pr[\mathcal{A}|E]Pr[E] + Pr[\mathcal{A}|\overline{E}]Pr[\overline{E}]$. Note that h_4 is a random oracle and the transcript is part of the input of h_4 to generate the session key, then

$\Pr[\mathcal{A}|E] = 1/2$. Then, we have $\Pr[\mathcal{A}] \leq (1 + \Pr[\overline{E}])/2$ and $\Pr[\mathcal{A}] \geq (1 - \Pr[\overline{E}])/2$. It follows $\Pr[\overline{E}] \geq 2\epsilon(\kappa)$. That is, $\mathcal{A}$ indeed chose oracle $\prod_{j,i}^{s}$ with a non-negligible advantage $(2\epsilon(\kappa))/(p(\kappa)^2 s(\kappa) q_{h_4})$, where q_{h_4} denotes the number of h_4 queries. This obviously contradicts to the CDH problem.

5.2 Comparison

The security features among the existing ICN schemes [16,19] are shown in Table 3. In terms of authentication, authenticity, anonymity, with trusted entity, security backbone, with random oracle, our scheme can achieve all of these properties, which cannot be reached by others.

Table 3. Security comparison of our authentication scheme with Xue *et al.* [19] and Nunes-Tsudik [16] schemes

Features	Xue *et al.* [19]	Nunes-Tsudik [16]	Ours
User authentication	$\checkmark$	$\checkmark$	$\checkmark$
Content authenticity			$\checkmark$
Anonymity	Partial		$\checkmark$
Without trusted entity			$\checkmark$
Security assumption	Strong DH		CDH
With random oracle	$\checkmark$		$\checkmark$

$\checkmark$ indicates that the property is satisfied

6   Performance Analysis

To gain insights on the benefits of the proposed scheme, we evaluate the performance of our scheme from the perspectives of efficiency at provider side and overall cost. We use a machine with a two-core 2.70 GHz processor and 8GB memory running Windows 10.0.18362.592 for experiment. For the efficiency at provider side, we compare our scheme with vanilla ICPS (Fig. 3). With vanilla ICPS, signatures are generated by the provider. In contrast, our scheme seamlessly integrates proxy signature that largely alleviates the burden of provider side. For the overall cost, we compare computation and communication overhead with existing ICN-based two-party type schemes [19] and [16] in Table 4. For simplicity, we used a minimal setup containing a single provider, proxy, gateway and consumer in each domain. In Table 4, Pair, G., Enc./Dec., and Mac in computation represent the computation complexities of a pairing, a group operation, a symmetric encryption/decryption and a message authentication code, respectively, and G., Exp., Mac and Enc. in communication denote the size of the group element, modulus, message authentication code and ciphertext.

Efficiency: Define a metric $\eta \leftarrow N_p/N_b$, where N_p and N_b are the number of the most time-consuming operation involved on proxy based model and basic scheme (without proxy), respectively. In the proxy way, signature generation can be further divided into two components: at provider side and at proxy side. Let H_p be the transition probability, then N_p during the time interval, Δt, can be represented by $RN_c\Delta t + H_pN_a\Delta t$, where R is the request arriving rate at provider side, N_c and N_a are the number of the most time-consuming operation needed at provider and proxy, respectively. Note that H_p is also equal to the hit rate because it equates the probability that the content is mismatched by other proxies. Thus, η is represented by $(RN_c+H_pN_a)/(RN_b)$. As shown in Fig. 3, the value H_p increases, the values of η becomes larger because more popular content can be satisfied in the proxy. It makes sense that the gap decreases as the amount of request increases since the increment of signature results in the generation of more content request. That is, more popular content is more favored for our scheme.

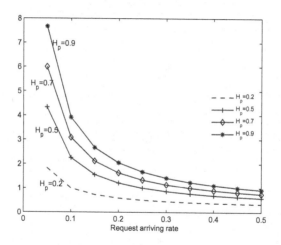

Fig. 3. The value of η with different request arriving rate.

Computation Overhead: Assume that the cost of a symmetric cryptographic computation is equal to that of a Mac computation, and ignore light computations, such as exponential in $\mathbb{Z}_q$, hash and map-to-point hash. Since both the consumer and gateway should check the state of a signature originated by the proxy, the signature verification consumption are considered for measuring the cost of both them. The proposed scheme takes seven pairings, five group operations in $\mathbb{G}_1$, and a Mac operation on average at consumer side, seven pairings, four group operations in $\mathbb{G}_1$ and a Mac operation on average at gateway side, two pairings and five group operations in $\mathbb{G}_1$ on average at proxy side, and one group operation in $\mathbb{G}_1$ on average at provider side. Observe that the computation overhead is more expensive than the competitive one. This degradation is forgivable since the proposed scheme can offer authentication and anonymity of users simultaneously, which, however, is not the case in the other work.

Communication Overhead: We compute the overall bit size of packets transmitted between users. It comprises seven group elements of $\mathbb{G}_1$ and a Mac. Using the type A elliptic curve described in Sect. 3.2, each element of $\mathbb{Z}_q$, $\mathbb{G}_1$ and $\mathbb{G}_1$ are 160, 512 and 1024 bits, respectively. The comparisons is shown in Table 4. Since our scheme will piggyback signature in Data packets, it will increase the size of communication overheads. However, from Table 4, we can see that the communication complexity of our scheme is higher than [16] but much lower than [19]. A trade-off for perfect authentication security is that both users need to be authenticated, which incurs more communication overhead than [16]. In terms of security features summarized in Table 4, the increased consumption is a substitution for increased security.

Table 4. Performance comparison of our authentication scheme with Xue *et al.* [19] and Nunes-Tsudik [16] schemes

	Xue *et al.* [19]	Nunes-Tsudik [16]	Ours
Computation	**User:** 3(Pair)+9(G.)+1(Dec.) **Gateway:** 5(pair)+8(G.)	**User:** 2(Dec.)	**User:** 7(Pair)+5(G.)+1(Mac) **Gateway:** 7(Pair)+4(G.)+1(Mac) **Proxy:** 2(Pair)+5(G.)
	Provider: 2(G.)+1(Enc.)	**Provider:** 4(Enc.)+1(Dec.)	**Provider:** 1(G.)
Communication	6(G.)+6(Exp.)+1(Enc.)	5(Enc.)	7(G.)+1(Mac)

7 Conclusion

We present a system framework to enforce security policies in ICN-ICPS such that it can provide end-to-end secure communication. Specially, we incorporate proxy signature to implement a session-based way featured with anonymous authentication. The security of the proposed scheme is proved under the random oracle model. Performance results show that the proposed scheme is an efficient solution that can provide authentication security service for ICN-ICPS user.

Although the proposed scheme tackles some of security problems in ICN-ICPS communication, it is not clear how to deal with the revocation key when users leave or join a topic. Another drawback is that the non-fulfillment of the proposed scheme in a testbed to analyze the performance for system metrics such as latency and energy consumption on user side. Notwithstanding its limitation, the work provides a comprehensive solution for user security in ICN-ICPS. However, these problems might be solved to apply subset difference mechanism and ndnsim and we leave it as future work due to the limitation of the length.

References

1. Cisco annual internet report 2018–2023 white paper. http://www.cisco.com
2. Cyber-physical systems: situation analysis of current trends, technologies, and challenges (2012). http://www.google.com. Accessed 2030
3. Ashibani, Y., Mahmoud, Q.H.: Cyber physical systems security: analysis, challenges and solutions. Comput. Secur. **68**, 81–97 (2017)
4. Azad, M.A., Bag, S., Perera, C., Barhamgi, M., Hao, F.: Authentic-caller: self-enforcing authentication in a next generation network. IEEE Trans. Industr. Inf. **16**(5), 3606–3615 (2020)
5. Chandrasekaran, G., Wang, N., Tafazolli, R.: Caching on the move: towards D2D-based information centric networking for mobile content distribution. In: IEEE 40th conference on Local Computer Networks (LCN), pp. 312–320 (2015)
6. Compagno, A., Conti, M., Droms, R.: OnboardiCNg: a secure protocol for on-boarding iot devices in ICN. In: Proceedings of the 3rd ACM Conference on Information-Centric Networking (ICN), pp. 166–175 (2016)
7. Fan, C.I., Chen, I.T., Cheng, C.K., Huang, J.J., Chen, W.T.: FTP-NDN: file transfer protocol based on re-encryption for named data network supporting nondesignated receivers. IEEE Syst. J. **12**(1), 473–484 (2018)
8. Foster, I., Kesselman, C., Tsudik, G., Tuecke, S.: A security architecture for computational grids. In: Proceedings of the 5th ACM Conference on Computer and Communications Security (CCS). pp. 83–92 (1998)
9. Fotiou, N., Polyzos, G.C.: Securing content sharing over ICN. In: Proceedings of the 3rd ACM conference on Information-Centric Networking (ICN), pp. 176–185 (2016)
10. Genge, B., Haller, P., Duka, A.V.: Engineering security-aware control applications for data authentication in smart industrial cyber-physical systems. Future Gener. Comput. Syst. **91**, 206–222 (2019)
11. Hwang, J.Y., Chen, L., Cho, H.S., Nyang, D.: Short dynamic group signature scheme supporting controllable linkability. IEEE Trans. Inf. Forensics Secur. **10**(6), 1109–1124 (2015)
12. Kim, Y., Kolesnikov, V., Thottan, M.: Resilient end-to-end message protection for cyber-physical system communications. IEEE Trans. Smart Grid **9**(4), 2478–2487 (2016)
13. Li, B., Huang, D., Wang, Z., Zhu, Y.: Attribute-based access control for ICN naming scheme. IEEE Trans. Dependable Secure Comput. **15**(2), 194–206 (2016)
14. Mick, T., Tourani, R., Misra, S.: LASeR: lightweight authentication and secured routing for ndn iot in smart cities. IEEE Internet Things J. **5**(2), 755–764 (2017)
15. Misra, S., Tourani, R., Natividad, F., Mick, T., Majd, N.E., Huang, H.: AccConF: an access control framework for leveraging in-network cached data in the ICN-enabled wireless edge. IEEE Trans. Dependable Secure Comput. **16**(1), 5–17 (2017)
16. Nunes, I.O., Tsudik, G.: KRB-CCN: lightweight authentication and access control for private content-centric networks. In: Preneel, B., Vercauteren, F. (eds.) ACNS 2018. LNCS, vol. 10892, pp. 598–615. Springer, Cham (2018). https://doi.org/10.1007/978-3-319-93387-0_31
17. Tramarin, F., Vitturi, S., Luvisotto, M.: A dynamic rate selection algorithm for IEEE 802.11 industrial wireless LAN. IEEE Trans. Ind. Inf. **13**(2), 846–855 (2016)
18. Xiong, H., Qin, Z.: Revocable and scalable certificateless remote authentication protocol with anonymity for wireless body area networks. IEEE Trans. Inf. Forensics Secur. **10**(7), 1442–1455 (2015)

19. Xue, K., Zhang, X., Xia, Q., Wei, D.S., Yue, H., Wu, F.: SEAF: a secure, efficient and accountable access control framework for information centric networking. In: 2018 International Conference on Computer Communications(INFOCOM), pp. 2213–2221. IEEE (2018)
20. Zhang, Y., Deng, R., Zheng, D., Li, J., Wu, P., Cao, J.: Efficient and robust certificateless signature for data crowdsensing in cloud-assisted industrial IoT. IEEE Trans. Industr. Inf. **15**(9), 5099–5108 (2019)
21. Zheng, Q., Li, Q., Azgin, A., Weng, J.: Data verification in information-centric networking with efficient revocable certificateless signature. In: 2017 IEEE Conference on Communications and Network Security (CNS), pp. 1–9 (2017)

Access Control for Wireless Body Area Networks

Gang Shen[1], Wenxiang Song[1], Yumin Gui[2(✉)], and Hanjun Gao[3]

[1] School of Computer Science, Hubei University of Technology, Wuhan 430068, China
shengang@hbut.edu.cn, 547521741@qq.com
[2] Department of Ophthalmology, Wuhan Puren Hospital, Wuhan 430080, China
5537468@qq.com
[3] China Nuclear Power Operation Technology Corporation, LTD.,
Wuhan 430070, China
gaohj@cnnp.com.cn

Abstract. Wireless body area network (WBAN) is a network providing healthcare, which is becoming more and more popular. However, the crucial issues of security and privacy in WBAN should still be considered. In this paper, we propose a secure access control scheme for WBAN, which is based on ciphertext policy attribute-based encryption (CP-ABE). Specifically, if the physician has attributes that satisfy the access structure set by the patient, he/she can decrypt the patient's physiological data. A secure two-party protocol is adopted to protect data from internal attacks. In addition, our scheme can implement the strategy that physicians at different levels can only access the corresponding information of patient, which is conducive to improving the efficiency of access. Security analysis indicates that proposed scheme can resist various security threats and achieve privacy preservation of patients' sensitive information. Compared with related schemes, our scheme is more secure and efficient.

Keywords: Wireless body area networks · Access control · Sensitive information · Privacy protection

1 Introduction

WBAN, a special wireless sensor network, is mainly formed by various low-energy, low-cost, heterogeneous, tiny sensors worn on the body. As a popular technology, WBAN has been widely used in e-health to monitor the patients physical health in real time [11]. The sensors in WBAN, such as electrocardiograph (ECG), electroencephalography (EEG), blood pressure, are used to collect and monitor the patients' various physiological signals. The physiological information processed by the sensor network will be transmitted to the remote health care servers or the personal server device via Internet [9].

WBAN plays a very important role in today's medical services because it can bring people the following benefits [1]: i) provide remote health monitoring

D. Wang et al. (Eds.): SPNCE 2020, LNICST 344, pp. 244–254, 2021.
https://doi.org/10.1007/978-3-030-66922-5_16

for patients; ii) achieve real-time diagnosis; iii) give much physical mobility for patients; iv) reduce expenses to medical server; v) store patients' medical data for real-time accessing. However, the disclosure of physiological information could bring a potential threat to patient's privacy [7], e.g., the patient's health status or personal information could be inferred by physiological information. In addition, if the patient's physiological information is maliciously modified, it may cause his/her condition to be misdiagnosed. Therefore, privacy-preserving technique must be employed in WBAN communications. To solve the privacy problem, many related schemes have been proposed for WBAN [8,12,14,15]. Zhang et al. [15] use the biometric signal as the key to encrypt the medical data, and the receiver can decrypt the data by using the same key. Li et al. [8] present an anonymous mutual authentication and key agreement scheme for WBAN. To protect the patient's identity privacy, location privacy and sensor deployment privacy, Zhou et al. [14] propose a secure and privacy-preserving key management scheme for cloud-based WBAN. They embed the symmetric structure of human body into the symmetric key mechanism of Blom by using blind technology and improved secret sharing technology. Tian et al. [12] introduce the scheme of access control of key policy attribute-based encryption (KP-ABE) [4] in WBANs. In their schemes, the ciphertext can be decrypted if the attribute set related to the ciphertext meet the access structures associated with user's key. However, comparing KP-ABE and CP-ABE [2], the latter is more suitable for data sharing systems. The reason is that data users have the right to decide access policies in CP-ABE [3,6,13]. Additionally, Hur [5] provides a CP-ABE scheme for a secure data sharing system, which solves the key escrow problem by using a secure two-party computation (2PC) protocol.

The above mentioned schemes cannot satisfy the requirements of resisting internal attacks and hierarchical access at the same time. Therefore, we propose a secure user access control scheme for WBAN in this paper. Specifically, the main **contributions** of this work are as threefold.

(1) Firstly, we propose a scheme based on CP-ABE, which employs the 2PC protocol mentioned in scheme [5] to ensure the confidentiality of patient's physiological information. Security analysis indicates that the proposed scheme can resist not only external attacks but also internal attacks.
(2) Secondly, in order to improve the diagnosis efficiency, the proposed scheme can realize hierarchical access, that is, physicians at different levels can only access the corresponding information of patient.
(3) Finally, the performance and security of the proposed scheme are analyzed. The results show that the proposed scheme is efficient and secure, and is suitable for WBAN.

The remaining part of the paper is organized as follows. We introduce some preliminary knowledge in Sect. 2. In Sect. 3, we propose the concrete scheme. The security analysis is described in Sect. 4. We also discuss the performance evaluation in Sect. 5. Finally, we draw our conclusions in Sect. 6.

2   Preliminaries

2.1   System Model

The entities in our scheme are as follows: a key generation center (KGC), a data server (DS), a physiological data owner (PDO) and a set of physicians, as shown in Fig. 1. We assume that KGC and DS could become the internal attackers under the influence of adversary.

(1) Key generation center (KGC): KGC is an honest-but-curious institution which generates parameters for CP-ABE. It is responsible for publishing, revoking, and updating physician's attribute key, and granting differential access permissions to physicians according to their attributes.
(2) Data server (DS): DS is an incompletely trusted institution for key generation, which is responsible for providing stored data access to external users.
(3) Physiological data owner (PDO): PDO has physiological data, and can upload the data into the DS for monitoring. In addition, PDO is responsible for dividing the information into different levels and defining access policy.
(4) Physicians: Physicians include resident physician, physician-in-charge, associate chief physician and chief physician. Physicians with different titles can access the corresponding level of physiological information on DS.

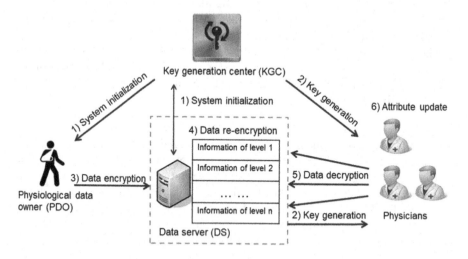

Fig. 1. System model.

2.2   Design Goals

Based on the aforementioned system model, we design a secure user access control scheme that has the following two desirable objectives.

(1) *Security.* The adversary cannot eavesdrop on the patient's information from the transmitted data, thus ensuring the patient's privacy.
(2) *Efficiency.* In the proposed scheme, physicians at different levels can only obtain corresponding information, which improves the efficiency of the system to a certain extent.

2.3   CP-ABE Scheme

CP-ABE scheme [2] consists of the following four algorithms:

(1) Setup: The public parameters PK and master key MK are generated by this algorithm.
(2) Encryption: Input a plaintext M, the public parameters PK and an access structure $\mathbb{T}$, output the ciphertext CT.
(3) Key generation: Input a set of attributes S, the master key MK, and the public parameters PK, output a decryption key SK.
(4) Decryption: Input the ciphertext CT, the parameters PK and the decryption key SK, output the plaintext M.

2.4   Access Tree

The access tree $\mathbb{T}$ is used to represent the structure of access control. The parameters in the tree are presented in Table 1. Ciphertext is associated with $\mathbb{T}$, and each non-leaf node of $\mathbb{T}$ denotes a threshold gate. Let $\mathbb{T}_x$ be an subtree of $\mathbb{T}$ and A be a set of attributes that satisfies $\mathbb{T}_x$. $\mathbb{T}_x(A)$ can be calculated using the following recursive process.

(1) $\mathbb{T}_x(A)$ will be calculated when x is a non-leaf node. Only if at least t_x children return 1, $\mathbb{T}_x(A)$ returns 1.
(2) When x is a leaf node, then $\mathbb{T}_x(A)$ returns 1 if and only if $\lambda_x \in A$.

Table 1. Notation description

Notation	Description
$\mathbb{T}$	Access structure
x	Non-leaf node
num_x	The number of children of node x
t_x	Threshold value
λ_x	The attribute associated with the leaf node x
$p(x)$	The parent of the node x in $\mathbb{T}$
$index(x)$	Return a number associated with node x
$\mathbb{T}_x$	The subtree of $\mathbb{T}$ rooted at the node x
A	A set of attributes, and $\mathbb{T}_x(\gamma) = 1$

3 The Proposed Concrete Scheme

The proposed concrete scheme includes six phases: system initialization, key generation, data encryption, data re-encryption, data decryption and attribute update.

3.1 System Initialization

In the proposed scheme, the PDO first divides his/her physiological data PF into n parts of different levels, that is $PF = \{PF_1, PF_2, \cdots PF_n\}$ [10]. Next, each physiological data PF_i is encrypted by using a symmetric key sk_i. That is, the information of level i is $CPF_i = Enc_{sk_i}(PF_i)$, where $Enc(\cdot)$ is a symmetric encryption algorithm. PDO encrypts a set of symmetric keys $\{sk_1, sk_2, \cdots, sk_n\}$ with the CP-ABE scheme and store them in DS with $\{CPF_1, CPF_2, \cdots, CPF_n\}$. Therefore, the physicians at different levels can obtain different levels of physiological information about PDO.

We assume the system is bootstrapped by KGC. Given a security parameter κ, KGC generates $(q, g, \mathbb{G}_1, \mathbb{G}_2, e)$ by running $\mathcal{G}(\kappa)$, where $g \in \mathbb{G}_1$. We define the Lagrange coefficients $\Delta_{i,\Phi}$ as $\Delta_{i,\Phi}(x) = \Pi_{j \in \Phi, j \neq i} \frac{x-j}{i-j}$, where $i \in \mathbb{Z}_q^*$ and a set Φ of elements in $\mathbb{Z}_q^*$. We also choose two hash functions $H_0 : \{0,1\}^* \to \mathbb{G}_1$ and $H_1 : \mathbb{G}_2 \to \mathbb{Z}_q^*$. Then, the pairs of master public and private key of KGC and DS are given, respectively. KGC picks a random number $\beta \in \mathbb{Z}_q^*$ as its master private key, and calculates the master public key $v = g^\beta$, so its pairs of master public and private key is $(PK_K = v, MK_K = \beta)$. DS selects a random number $\alpha \in \mathbb{Z}_q^*$, and the pairs of master public and private key is $(PK_D = e(g,g)^\alpha, MK_D = g^\alpha)$. DS also picks $\gamma \in \mathbb{Z}_q^*$, and publishes $PK_D' = g^\gamma$ as another master public key. Finally, the system parameters $\{\mathbb{G}_1, g, H_0, H_1\}$ are published, $\{\alpha, \beta, \gamma\}$ are kept.

3.2 Key Generation

Each physician runs following steps to obtain secret key set:

(1) To authenticate the user, the 2PC protocol will be used. KGC first chooses a random number $r_k \in_R \mathbb{Z}_q^*$ as the secret of the physician. Then, 2PC protocol returns the output $w = (\alpha + r_k)\beta$ to DS. DS randomly selects $\tau \in_R \mathbb{Z}_q^*$ and computes $X = g^{w/\tau} = g^{(\alpha+r_k)\beta/\tau}$, and then sends it to KGC. After receiving X, KGC computes $Y = X^{1/\beta^2} = g^{(\alpha+r_k)/\tau\beta}$, and sends it to the DS. DS outputs a personal key component $D = g^{(\alpha+r_k)/\beta}$ by computing Y^τ.

(2) We suppose that physician's attribute set is A as input by KGC. The output is that a group of attribute keys and secret value r_k identified by the group. For each attribute $j \in A$, KGC picks random $r_j \in \mathbb{Z}_q^*$. Then, it computes physician's attribute keys as follows:

$$SK_{K,p} = (\forall j \in A : D_j = g^{r_k} H_0(j)^{r_j}, D_j' = g^{r_j}). \tag{1}$$

(3) The physician's personal key $SK_{D,p} = D$ is outputted by DS. Whole secret key set of physician is as:

$$SK_p = (SK_{D,p}, SK_{K,p})$$
$$= (D = g^{(\alpha+r_k)/\beta}, \forall j \in A : D_j = g^{r_k} H_0(j)^{r_j}, D'_j = g^{r_j}).$$

(2)

Additionally, the DS also outputs another secret $SK'_p = H_0(ID_p)^\gamma$, which will be used to distribute the selective attribute group key, where ID_p denotes the identity of physician.

3.3 Data Encryption

We assume that the symmetric key is $sk_i \in \mathbb{G}_2$. Before sk_i is send to the DS for sharing, it will be encrypted under $\mathbb{T}$ of defining by PDO, who can obtain the ciphertext as follows:

(1) Choose a polynomial q_x for each node x in $\mathbb{T}$. Let the degree of q_x be d_x, where $d_x = t_x - 1$ and t_x is the threshold value.
(2) PDO picks a random $s \in \mathbb{Z}_q^*$ and sets $q_R(0) = s$ for the root node R. It sets $q_x(0) = q_{p(x)}(index(x))$ for any other x, and selects d_x to define q_x.
(3) The ciphertext CT is constructed as:

$$\text{CT} = (\mathbb{T}, C_1 = sk_i \cdot e(g,g)^\alpha), C_2 = v^s, \forall y \in B : C_y = g^{q_y(0)},$$
$$C'_y = H_0(\lambda_y)^{q_y(0)}).$$

(3)

where B is a group of leaf nodes in $\mathbb{T}$.
(4) Send encrypted physiological data CT to DS.

3.4 Data Re-encryption

The proxy re-encryption protocol is used by DS to achieve user revocation. The protocol is executed as follows:

(1) DS Chooses $K_{\lambda_y} \in \mathbb{Z}_q^*$, and re-encrypts CT as CT':

$$\text{CT}' = (\mathbb{T}, C_1 = sk_i \cdot e(g,g)^\alpha), C_2 = v^s, \forall y \in B : C_y = g^{q_y(0)},$$
$$C'_y = (H_0(\lambda_y)^{q_y(0)})^{K_{\lambda_y}}).$$

(4)

(2) DS selects random $\rho \in_R \mathbb{Z}_q^*$, and $\forall physician \in \mathbb{G}_1$, and computes $x_k = H_1(e(H_0(ID_p)^\rho, PK'_D))$.
(3) DS constructs the function of polynomial as:

$$f^y(x) = \prod_{i=1}^m (x - x_i) = \sum_{i=0}^m \alpha_i x^j (mod q).$$

(5)

(4) A random $r \in \mathbb{Z}_q^*$ is chosen by DS who constructs HM_y, and generates a message of head as:

$$HM = (g^\rho, \forall y \in B : HM_y).$$

(6)

where $HM_y = \{K_{\lambda_y} \cdot g^{r a_0}, g^{r a_1}, ..., g^{r a_m}\}$. The DS responds with (HM_y, CT') to physician after receiving data request query from the physician.

3.5 Data Decryption

If a physician is associated with an attribute λ_y, he/she can obtain attribute group key K_{λ_j} from HM_j as follows:

(1) Computes x_k, then computes K_{λ_j}.
(2) Then physician uses the attribute group keys to update its secret key as follows:

$$SK_p = (D = g^{(\alpha+r_k)/\beta}, \forall j \in \Lambda : D_j = g^{r_k}H_0(j)^{r_j}, D'_j = (g^{r_j})^{1/K_{\lambda_j}}). \quad (7)$$

where Λ denotes a set of attributes.
(3) Suppose x is a leaf node, and $\lambda_k \in \Lambda$, then physician computes

$$DecryptNode(CT', SK_p, x) = e(g,g)^{r_k q_x(0)}. \quad (8)$$

(4) Suppose x is a non-leaf node. The physician computes F_z as the output of $DecryptNode(CT', SK_p, z)$ for all x's children nodes z:

$$F_x = e(g,g)^{r_k q_x(0)}. \quad (9)$$

(5) Let $DecryptNode(CT', SK_p, R)$ is a function of the root node R of $\mathbb{T}$. Then, $DecryptNode(CT', SK_p, z) = e(g,g)^{r_k s}$. The sk_i can be recovered by calculating as $CT' = sk_i$.
(6) Finally, the physician can decrypt the information of level i ($PF_i = Dec_{sk_i}(CPF_i)$) by using the secret key sk_i.

3.6 Attribute Update

If physician can improve his/her access level by updating the attributes. Specifically, KGC update the physician's attribute set A to A'. Then, the physician's new key becomes SK'_p.

4 Security Analysis

In this section, we will conduct the security analysis of the proposed scheme.

4.1 Resist Internal Attacks

In the proposed scheme, KGC and DS cannot decrypt the physiological data alone even if they get the ciphertext. The reason is that KGC and DS do not know each other's master secrets due to the 2PC protocol, so they cannot generate physician's secret keys independently [5]. Therefore, the proposed scheme can resist internal attacks.

4.2   Resist External Attacks

In addition, data confidentiality against the outside adversary can be also guaranteed. The physician cannot obtain the desired value $e(g, g)^{r_k s}$ if the set of attributes do not meet $\mathbb{T}$ in the ciphertext. In order to decrypt a node x holding an attribute λ_x, the physician must obtain pair (C'_x, D'_x) from the ciphertext and its private key, respectively. C'_x is blinded, which means the physician cannot get the value $e(g, g)^{r_x q_x(0)}$, the updated attribute group key cannot be obtained by the revoked physician. Users whose attributes cannot meet the access policy will be not able to decrypt the ciphertext. Therefore, the proposed scheme can resist external attacks.

4.3   Collusion Resistance

In this scheme, since the value is randomized from the private key of particular user, value $e(g, g)^{\alpha s}$ cannot be recovered by the attackers of collusion.

5   Performance Evaluation

We first compare our scheme with scheme [12] in terms of the main characteristic. Then the performance of the computational cost of data encryption and decryption is analyzed.

5.1   Comparative Analysis

Tian et al. [12] propose a scheme of KP-ABE for access control in WBAN, in which the ciphertext can be decrypted if his/her attributes related to the ciphertext meet the access structures. However, our scheme is based on CP-ABE. Since data owners master the access policy in CP-ABE, it is more suitable for practical application than KP-ABE. For example, PDO can define the access policies based on the attributes owned by the physician's level. When the physician's attributes meet the access structure of the encrypted data, the physician can decrypt the patient's physiological data. Therefore, the access control based on CP-ABE is more efficient and flexible. Additionally, our proposed scheme also uses a 2PC protocol in phase of key generation to resist the internal attacks. We present the main performance comparison in Table 2. Compared with the schemes [12], the proposed scheme can better realize both the efficiency of user access control, and realize hierarchical access of patient's physiological data.

5.2   Performance Evaluation

In this paper, the evaluations are conducted on Intel Core i7-6700 @3.10 GHz with 8 GB RAM. The computation costs of our scheme are showed in Table 3, where k_1 denotes the number of attributes associated with physician's private key, k_2 denotes the number of physicians in an attribute group and k_3 denotes the

Table 2. Performance comparison

Characteristics	[12]	Proposed scheme
Security of system entities	No-mentioned	Honest-but-curious
Type of ABE	KP-ABE	CP-ABE
Hierarchical access	No	Yes
Data confidentiality	Yes	Yes
User revocation	Yes	Yes
Collusion attack resistance	Yes	Yes
Data re-encryption	No	Yes
Attribute updated	Yes	Yes
Resist internal attacks	No	Yes

Table 3. Comparison of computational costs

	Encryption	Decryption
Time	$(2k_3 + 1)T_{eG_1} + T_{eG_2}$	$k_1 k_2 T_{eG_1} + T_{eG_2} log k_3 + (2k_1 + 2)T_p$

number of attributes appeared in $\mathbb{T}$. In addition, T_{eG_1}, T_{eG_2} and T_P represent an exponentiation operation in $\mathbb{G}_1$, an exponentiation operation in $\mathbb{G}_2$ and a bilinear pairing operation, respectively.

Figure 2 shows the computational costs of data encryption and decryption, respectively. We can see that the proposed scheme is efficient for reducing the computational costs of data encryption and decryption.

In our scheme, the patient's physiological information is divided into n parts and encrypted as $\{CPF_1, CPF_2, \cdots, CPF_n\}$. We assume that the time taken for decrypting a physiological information is t. In the traditional ABE scheme, the time for a physician to obtain patient information is nt, while in our scheme, the time for a physician to obtain patient information is t. Therefore, our scheme can greatly improve the diagnostic efficiency.

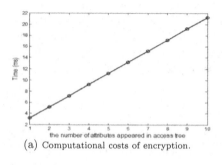

(a) Computational costs of encryption.

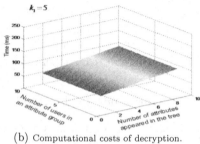

(b) Computational costs of decryption.

Fig. 2. Computational costs.

6   Conclusion

In this paper, we propose a secure access control scheme based on CP-ABE for WBAN. Specifically, the system employs the CP-ABE and a secure 2PC protocol to resist the internal attacks. Additionally, the physicians at different levels can only access the corresponding information of patient, which realize hierarchical access to improve the diagnosis efficiency. Security analysis and experimental results demonstrate that the proposed scheme can realize secure and efficient user access control in WBAN.

Acknowledgment. We are grateful to the anonymous reviewers for their invaluable comments. The Ph.D research startup foundation of Hubei University of Technology under Grant No. BSQD2019023.

References

1. Ali, M., Sadeghi, M., Liu, X.: Lightweight fine-grained access control for wireless body area networks. Sensors **20**(4), 1–22 (2020)
2. Bethencourt, J., Sahai, A., Waters, B.: Ciphertext-policy attribute-based encryption. IEEE Symposium on Security and Privacy, pp. 321–334 (2007)
3. Bobba, R.: Attribute-sets: a practically motivated enhancement to attribute-based encryption. Cryptology ePrint Archive, pp. 587–604 (2009)
4. Goyal, V., Pandey, O., Sahai, A., Waters, B.: Attribute-based encryption for fine-grained access control of encrypted data. In: Proceedings of ACM Conference on Computer and Communications Security, pp. 89–98 (2006)
5. Hur, J.: Improving security and efficiency in attribute-based data sharing. IEEE Trans. Knowl. Data Eng. **25**(10), 2271–2282 (2013)
6. Ibraimi, L., Petkovic, M., Nikova, S., Hartel, P., Jonker, W.: Mediated ciphertext-policy attribute-based encryption and its application. In: Youm, H.Y., Yung, M. (eds.) WISA 2009. LNCS, vol. 5932, pp. 309–323. Springer, Heidelberg (2009). https://doi.org/10.1007/978-3-642-10838-9_23
7. Kargl, F., Lawrence, E., Fischer, M., Lin, Y. Y.: Security, privacy and legal issues in pervasive eHealth monitoring systems. In: Proceedings of International Conference on Mobile Business, pp. 296–304. IEEE (2008)
8. Li, X., Ibrahim, M.H., Kumari, S., Sangaiah, A.K., Gupta, V., Choo, K.K.R.: Anonymous mutual authentication and key agreement scheme for wearable sensors in wireless body area networks. Comput. Netw. **129**(2), 429–443 (2017)
9. Meharouech, A., Elias, J., Mehaoua, A.: Future body-to-body networks for ubiquitous healthcare: a survey, taxonomy and challenges. In: Proceedings of International Symposium on Future Information and Communication Technologies for Ubiquitous Healthcare, pp. 1–6. IEEE (2015)
10. Shen, G., Su, Y., Zhang, M.: Secure and membership-based data sharing scheme in V2G networks. IEEE Access **6**(1), 58450–58460 (2018)
11. Salayma, M., AI-Dubai, A., Romdhani, I., Nasser, Y.: Wireless body area network (WBAN): a survey on reliability, fault tolerance, and technologies coexistence. ACM Comput. Surv. **50**(1), 31–338 (2017)
12. Tian, Y.: An attribute-based encryption scheme with revocation for fine-grained access control in wireless body area networks. Int. J. Distrib. Sens. Netw. **2014**(11), 1–9 (2014)

13. Yu, S., Wang, C., Ren, K, Lou, W.: Attribute based data sharing with attribute revocation. In: Proceedings of ACM Symposium on Information Computer and Communication Security, pp. 261–270 (2010)
14. Zhou, J., Cao, Z., Dong, X., Xiong, N., Vasilakos, A.V.: 4S: a secure and privacy-preserving key management scheme for cloud-assisted wireless body area network in m-healthcare social networks. Inf. Sci. **314**, 255–276 (2015)
15. Zhang, Z., Wang, H., Vasilakos, A.V., Fang, H.: ECG-cryptography and authentication in body area networks. IEEE Trans. Inf Technol. Biomed. **16**(6), 1070–1078 (2012)

FIDO – That Dog Won't Hunt

Michael Scott[(✉)]

MIRACL Labs, Dublin, Ireland
`mike.scott@miracl.com`

Abstract. FIDO is an authentication technology based on the mathematics of public key cryptography that emerged in the 1970s and the 1980s. It is promoted by a large industry backed consortium as the two-factor successor to the username/password mechanism, which is well understood as being no longer fit for purpose. But intrinsic to FIDO is the requirement for both client-side secure hardware and a vulnerable server-side credentials database. Here we propose a better solution which would ditch both of these requirements by separating the registration and authentication processes, and which provides true multi-factor authentication using more modern ideas that have emerged from cryptographic research.

Keywords: Authentication · FIDO · M-Pin · Public key substitution · PKI · Phishing attacks · Credential databases · MIRACL

1 Introduction

On joining any organisation (server) which limits access, an individual (client) is typically issued with certain credentials, possession of which they must establish in order to gain access. Alternatively they may generate their own credentials and have them endorsed by the organisation. Either way any method of authentication which merely involves the handing over of credentials for inspection, is vulnerable to a so-called phishing attack, where they are fooled into handing their credentials to a malicious party who may then masquerade as the legitimate individual.

The problem considered here is one of a client authenticating to a server (sometimes known as the Relying Party) over the internet. We assume that a method is already in place to allow the server to authenticate to the client. There will be a one-off registration phase where credentials are established, followed over time by multiple authentication attempts.

For extra security, two-factor authentication is often recommended. It is easy to underestimate just how much harder it is for an attacker to solve two completely unrelated problems in order to achieve their malign ends, and the extent to which such a prospect demoralises all but the most persistent attacker.

© ICST Institute for Computer Sciences, Social Informatics and Telecommunications Engineering 2021
Published by Springer Nature Switzerland AG 2021. All Rights Reserved
D. Wang et al. (Eds.): SPNCE 2020, LNICST 344, pp. 255–264, 2021.
https://doi.org/10.1007/978-3-030-66922-5_17

2   Starting Over

Consider the simplest of all authentication schemes. At registration, the client generates a memorised password (their credential), which is passed up to a server, which stores it next to the client's username in a database. To authenticate, the client sends his/her username and password to the server, which uses the username to look up the stored password and compares it with what has been received. It is assumed that the server is in a position to block multiple password guesses. Clearly this is not very secure, and wide open to both database hacking and phishing attacks. Basically client and server share a secret, the password. Such a protocol can come under attack at either the server or client end, or indeed during the transport of data between the two.

The first improvement is to insist on the server deploying the well known SSL/TLS protocol to ensure that an encrypted tunnel is created between client and server, which effectively solves the transport issue and authenticates the server to the client. Briefly the server has a private/public key pair, the client generates an ephemeral session key, encrypts it with the server's public key, and sends it up to the server. The server can recover the same session key by decrypting using their private key. To convince the client that they are using the right public key, they do not just take it on trust from the server, but rather extract it from an X.509 certificate which has been digitally signed by a recognised Certificate Authority (CA). This is well known tried-and-tested Public Key Infrastructure (PKI) solution which ensures server integrity and allows e-commerce to flourish on the Internet.

Unfortunately this is not enough to prevent sophisticated phishing attacks, as the wrong server can authenticate itself just as easily as the right one, if the client is not careful to establish exactly who they are talking to. We will refer to this method as the proto-authentication protocol.

3   Introducing Client-Side Secure Hardware

Now let us allow ourselves a secure hardware vault on the client side. This allows the client to store inside the vault a very large random 128-bit secret, that he does not need to memorise, or indeed ever be aware of. At registration it is transported to the server and stored next to their username. Now we can be a little cleverer. To authenticate, the server sends a random challenge to the client which gets its vault to encrypt it using the secret as a key, sends it back to the server which looks up the same secret from the database and uses it to decrypt. If the result is the same as the original challenge, authentication succeeds.

Note that we have come a long way thanks to secure hardware. The client experience is now passwordless, phishing attacks are impossible (as the challenge is different every time), and the client secret is no longer transmitted to the server with every authentication, in fact it never leaves the security of its vault. We have established possession of the secret without revealing anything useful about it. If a two factor authentication experience is desired – easy – arrange it such that the vault only performs its function on entry of a short PIN number. We note in passing that once a secure vault is assumed, it may also be used to store a biometric template of the client, so that presentation of that biometric would be an alternate mechanism for opening the vault. For the purposes of this paper we will assume the PIN entry option is used as a second factor.

Since client and server are intent on entering into a trusting relationship, it doesn't really matter which of them generates the large secret. So a simple elaboration would be for the client hardware to generate an ephemeral public/private key pair, and during registration send the public key to the server, which generates the secret, encrypts it with the public key, and sends it back to the hardware, which decrypts it with the private key that can then be immediately deleted. Now the secret is never visible at any endpoint on the client side.

Note that this simple idea gives us everything that FIDO gives us on the client side, at a fraction of the cost and complexity.

However it is not so clear that all is well on the server side. That credential database is still a major weakness. The mutual client/server secret is just sitting there, in the clear and inviting attack by a resourceful hacker.

3.1   Username/Password

The Username/Password protocol is another way of improving on our proto scheme, this time without requiring secure hardware. The basic idea is that a one-way hash of the password is stored on the server side, and when the client password is presented to the server, it is the hash of it that is compared with the value stored in the database. This certainly improves things at the server end, as passwords are no longer stored in their original form, and cannot easily be recovered from the hash because of its one-wayness.

In fact there is often a lot more than that to Username/Password as deployed by organisations like Facebook [11] and LinkedIn [8]. Commonly the stored credential is a combination of a salt and an iterated hash of the concatenation of username, salt, pepper, a secret key and the actual password. Such a method is recommended by NIST [2] section 5.1.1. Each component adds something to the overall security. By iterating the hash function maybe 10,000 times we introduce a tiny but acceptable delay into the authentication process, but we also increase the attackers workload by the same factor. The salt, typically 128 random bits, ensures that identical passwords do not result in identical credentials, and again slows down a brute-force attack which iterates through a large dictionary of common passwords, forcing the attacker to search the full dictionary as it attacks each individual account. Care is also taken to protect against active substitution attacks where an attacker can overwrite his victim's credential with one of his own choosing. By including the username in the hash, the attacker cannot simply substitute his own genuine credential for that of his victim.

Perhaps the most significant component is the "pepper". This is a global secret, typically stored in a HSM (Hardware Security Module), which must be contributed on-the-fly before the correct hash is calculated. It is there to mitigate against the most common failure mode of Username/Password, where the credential database is stolen, and an attacker can at their leisure go about attacking every individual's password. Without possession of the pepper component, the attacker cannot succeed.

Safely handling a single global secret that is so widely used is problematical but it is an option for larger, well equipped corporations. Even if the pepper is compromised (perhaps by an insider attack), the salting provides a second level of security. As a last line of defence, the choice of a strong password can still thwart the determined attacker.

We conclude that Username/Password, certainly if implemented carefully with the "pepper" component, is still capable of providing a secure authentication experience. Alas it is intrinsically one-factor, and strong passwords are, to put it mildly, inconvenient.

4   FIDO

FIDO was first proposed in 2013 by an alliance of industry leaders concerned by the shortcomings of Username/Password, and motivated by an acceptance that two factor authentication provides much more security.

Much progress has been reported on the standardisation and specification fronts. Adoption of the technology by large corporations and alliance members as an optional authentication mechanism has accelerated. However, from anecdotal evidence it would appear that adoption by edge users has been slower, and the hoped-for phasing out of username and password largely hasn't happened to date. One promising development has been the recent release of an open source implementation by Google to allow the creation of FIDO enabled hardware authentication tokens [3].

FIDO attempts to improve upon our original hardware based solution by using public key cryptography. The client vault generates a long term public/private key pair, and on registration supplies its username and public key to the server, which stores them in a credentials database. To authenticate, the client presents its username and the server responds with a random challenge. The client uses its private key to digitally sign this challenge, and sends this signature back to the server, who verifies it using the public key it extracts from the credential database. We acknowledge that there is a lot more to FIDO than this simple description. In particular there is an attestation mechanism used to attest to the properties of the secure hardware and its capabilities. However this is separate from the authentication process considered here. There is a lot of engineering involved in implementing FIDO. Browsers are modified to allow communication to/from a hardware authenticator via quite complex WebAuthn APIs, CTAP, and CBOR protocols.

But already we can see the big advantage: The credential database stores public keys rather than secrets or hashed passwords, which on the face of it sounds safer in the context of the hacker who steals it. Determining private keys from public keys is a much more daunting task. As before the client has authenticated without handing over its secret (the private key), and since the challenge is different every time, a phishing attack is again no longer possible.

The vault may exist as an authentication token that connects via USB to a laptop or desktop computer. More likely it exists as an extension to the processor architecture used by a modern smart mobile phone.

Since FIDO uses public key cryptography it might be assumed that therefore it also uses a PKI, as such is a normal requirement to ensure correct handling of public keys. The main problem solved by PKI is that of making the association between the public key and its true owner. In the absence of such a mechanism we have no way of knowing that the public key we are using is the right one, and therefore we are wide open to a public key substitution attack. However while FIDO could be integrated into an existing PKI scheme if one already exists for clients, this is not the recommended approach [7].

4.1   The Problem with PKI

PKI has a well deserved reputation for being difficult for clients to manage. Consider the email experience: most email clients offer a facility for clients to use PKI to encrypt and digitally sign their emails but as we know, hardly anyone uses this facility (unless mandated within large enterprises with large IT support departments). To the surprise of many, PKI encrypted/signed email never achieved mass adoption. For organisations which do support PKI down to the full enrolment of each individual client, the authentication problem is already largely solved, and there would be no need for FIDO – "Why use FIDO at all?" [7].

To avoid suffering the same fate adoption-wise as PKI encrypted emails, it is understandable that FIDO avoids the use of a full PKI [7]. So we have a problem here: A full PKI would have a negative impact on adoption, while anything less will introduce security weaknesses. Larger enterprises appear to be aware of the danger, and have gone to some lengths to integrate FIDO with existing PKI infrastructure [13]. But this will not be the norm as is made clear in [7], and certainly will not apply to the consumer market. There appears to some confusion around this issue. Some proponents of FIDO seem to be under the mistaken impression that FIDO always deploys a full PKI [16]. It doesn't (although confusingly PKI is used for the separate attestation process). As stated quite baldly in [7] "FIDO does not use digital certificates for users". Or as coyly stated in the website FAQ "FIDO takes a lightweight approach to asymmetric public-key cryptography".

Summarising [7] says "Broadly, FIDO is simpler (than PKI) for end-users to use, and for application developers to integrate into web and mobile applications. FIDO infrastructures can also be less expensive to operate and manage given that they do not require many artefacts generally associated with PKIs". See page 8 of this FIDO document to see which artefacts are left out.

4.2   So How Does FIDO Protect Public Keys? Spoiler – It Doesn't

Recall that a FIDO server is expected to maintain a database of user credentials, basically a credential ID and the associated public key that matches the private key stored back in the client's vault. Actually FIDO public keys are not really "public" in the usual sense. They reside in the server's database, which would not itself normally be visible to the curious onlooker. Indeed read access to the credential database is regarded as a "threat" in the most recent FIDO security reference draft [6].

The WebAuthn guide states that credential databases "are no longer as attractive to hackers, because the public keys aren't useful to them.", which is hardly an encouragement for the implementer to go to any great lengths to protect them. This leads to a classic false sense of security. "Databases are no longer targets because public keys are useless without the corresponding private keys" [17]. Not so fast!

The most devastating attack on the FIDO credential database is a public key substitution attack. The attacker inserts their own public key for that of their victim, and hence gains access to their account, which is just as devastating as a password file compromise in the Username/Password context. The credential database, no matter what its format, is always a tempting target for attack, as success compromises not just individual accounts, but all accounts. Such substitution attacks are already taken into account by the Username/Password community – see above. It is surely ironic that while they do take the issue of credential substitution seriously, the FIDO community apparently does not.

As explicitly admitted by FIDO's own Security Reference document [6] section 7.3.1/2.1.2, an attacker who gains write access to the server database can launch a public key substitution attack. The outcome results in a violation of FIDO's Strong User Authentication, which is of course the whole point of FIDO. The FIDO response is revealing. The attack is described as "outside of the scope of the FIDO specifications" [6]. So it is not an issue for FIDO, and if such an attack is successful it would not be regarded as a failure of FIDO. This does not, of course, make such an attack any less of a problem. Someone has to make sure this doesn't happen, but noone has any idea how to go about it other than to implement a fully fledged PKI, which FIDO itself deprecates. We note in passing that FIDO is explicitly ignoring the NIST recommendation [2] section 5.1.8.2 which states that public keys "SHALL be protected against modification".

In fact there is a growing awareness of the problem, and [7] can be read as a somewhat pained attempt to square this circle. "The implementation of FIDO can benefit organizations in many ways. It offers strong public-key based authentication that is equivalent to certificate-based authentication but without the overhead of maintaining complex and expensive public-key infrastructure." One of the authors of [7] himself has recognised and highlighted the threat [12], there described as the "substitution of keys" attack.

It is worth quoting his contribution to this discussion group thread [4] at length "This is why I shudder to think of the consequences for FIDO when I see companies who are not in the cryptographic key-management space (banks, e-commerce companies, CRM companies, etc.) who are under the mistaken impression that if they simply implement a protocol and put together a GUI on the implementation, they have themselves a FIDO Server. Who are they going to blame when their FIDO-based strong-authentication application is compromised? Themselves or FIDO?"

In his own supported product [15] as is made clear from this discussion thread, an attempt has been made to mitigate by deploying a cut-down PKI-like solution, where every credential is digitally signed by the server management (analogous to the "pepper" idea in Username/Password). But the difficulties and shortcomings of this idea are also emphasised. And a quick search through some other open source FIDO server implementations makes it clear that the problem is not widely recognised, and countermeasures are non-existent.

Our experience with standards is that unless certain actions are mandated, they will not be implemented. The reality is that many existing FIDO credential databases are wide open to this attack.

4.3   Is FIDO Truly Two-Factor?

As stated above the client's experience is undoubtedly two-factor. But on closer inspection we can see that in reality it is really two-step. The PIN (first factor) must be entered first to release the private key (second factor). Since the second factor is stored in hardware, the hardware itself may also be considered as the second factor (possession of the authentication token, or of the mobile phone).

Clearly an attacker who can penetrate the vault, does not need the PIN. Therefore in reality the process is only protected by a single factor, that being the integrity of the hardware vault.

4.4   Is Secure Hardware Secure?

Secure hardware certainly seems like a good idea, and potentially provides an important resource for cryptographers. It provides a safe place to store keys and other sensitive material (like biometric templates). It also promises to implement the cryptography faster. But there are problems.

- Hardware cannot be easily patched.
- Hardware cannot be easily replaced.
- Can we trust in the security of the hardware?
- Can we trust that any back-doors will not be exploited?
- Hardware is expensive.

Rather than using secure hardware to protect a secret, an alternative idea is to simply split up the secret, an approach very much in tune with the whole idea of multi-factor authentication.

5 FIDO – Our Verdict

Almost all of the benefits of using FIDO arise from the assumption of the existence of secure hardware on the client side. The use of public key cryptography technology from the 1980s is unimaginative, and clearly not fully understood by many of FIDO's proponents. The internal contradiction highlighted above has never been properly addressed. The advantage over a simple shared secret scheme (with secure hardware and peppered credentials) is in fact minimal.

Most shocking of all is that the most common security threat to username/password, an attack on the credential database and the damage that can be done as a result, has not been addressed by FIDO, which washes its hands of responsibility by declaring it as being "outside of the scope" of its efforts.

6 An Alternative Approach

Next we briefly suggest an alternative solution called M-Pin [9][1], based on bilinear maps on elliptic curves, which is a technology which has proven to be a vehicle for many novel cryptographic constructs [10]. First we identify the main problems as being the credential database and the requirement for secure hardware. Let us dispense with both and get ourselves a simple and practical software-only 2-factor replacement for username/password.

6.1 Client Side, Look – No Hardware!

Consider now that a user's identity (username) is hashed and mapped to a point P of large prime order on an elliptic curve, and that the user is issued with a secret sP, by some trusted authority (TA) who possesses a master secret s. The user then chooses a PIN number α and splits their secret into two factors $(s - \alpha)P$ and α. It is then a trivial matter to recombine these factors to restore sP. It should also be clear that an attacker who gains access to one of these factors cannot determine the other. What is not so immediately obvious is that an attacker who also gains access to the extra information xP and xsP for some unknown x, also cannot determine one factor from the other, as each case reduces to finding a solution to the Decisional Diffie-Hellman (DDH) problem, which is known to be hard in prime order elliptic curve groups of a suitable size.

The form of these secrets allows them to split easily into multiple factors, and hence they offer an ideal vehicle for multi-factor authentication. Furthermore the TA that issues these secrets can also split its responsibilities by trivially observing for example that $sP = s_1P + s_2P$, where s_1 and s_2 are generated and maintained by two independent entities under separate control, known as DTAs (Distributed Trusted Authorities). Now the full master secret s does not exist anywhere in the system, and both DTAs would need to be hacked in order to access it.

---

[1] See https://github.com/miracl/core for implementations and benchmarks.

Next let us consider a signature scheme where the signature of a challenge c is created by generating a random x and consists of the tuple {Username, $U = xP$, $V = (x + c)sP$}, which can be verified by the server. Before we describe how such a signature is verified, let us recap on the client's experience. First the client registers with the DTAs and is supplied with its secret, which it immediately splits into a stored blob of data $(s - \alpha)P$ and a memorised password α. We emphasis that the blob of data does not need to be stored in secure hardware as it derives its security from the fact that it is incomplete and useless due to the missing PIN component. This is true two factor authentication, not two step. When required to authenticate by signing a challenge, the PIN part is restored and the signature created. As before we assume that the server is pro-active in blocking PIN guessing attacks (which present as failed authentication attempts).

6.2   Server Side, Look – No Credential Database!

Now we assume that the elliptic curve chosen is actually a type-3 pairing-friendly curve [10]. Recall that a bilinear pairing is a map $e : \mathbb{G}_1 \times \mathbb{G}_2 \to \mathbb{G}_T$, where $\mathbb{G}_1$ and $\mathbb{G}_2$ are elliptic curve groups, and $\mathbb{G}_T$ is a finite extension field [10]. The client side action takes place entirely in $\mathbb{G}_1$ and is unaffected by the pairing-friendly nature of the curve. The server is issued by the DTAs with a single secret sQ, where Q is a fixed point in $\mathbb{G}_2$. This secret point can be considered as a global public key that works for all clients, and replaces the entire FIDO database. Verification of the signature described above is carried out by independently hashing and mapping the username to the point P, and testing if

$$e(V, Q) \overset{?}{=} e(U + cP, sQ)$$

This signature scheme is a provably secure method first described by [1] and [5]. The PIN extraction idea comes from [14]. That the DDH assumption applies in the group $\mathbb{G}_1$ is the well established XDH assumption.

7   Conclusion

FIDO was a noble attempt to come up with a 2-factor replacement for the legacy username/password method of authentication. Unfortunately its internal contradictions were never properly recognised, and since they are intrinisc to the primitive public key cryptography that underpins it, we would predict that its adoption will eventually stall.

Secure hardware, while potentially a great asset, has its own issues. Amongst these there is a major trust issue, highlighted by the current Huawei controversy. Cryptographers are coming to the conclusion that splitting a secret and/or adopting methods from Multi-Party Computation[2] may be a better way to protect a secret. Credential databases are notoriously difficult to defend, and are best done away with. Here we point out that this is indeed possible, and suggest a practical solution.

---

[2] https://www.unboundtech.com/.

References

1. Bellare, M., Namprempre, C., Neven, G.: Security proofs for identity-based identification and signature schemes. In: Cachin, C., Camenisch, J.L. (eds.) EUROCRYPT 2004. LNCS, vol. 3027, pp. 268–286. Springer, Heidelberg (2004). https://doi.org/10.1007/978-3-540-24676-3_17
2. Grassi, P., et al.: NIST special publication 800–63b digital identity guidelines (2017). https://pages.nist.gov/800-63-3/sp800-63b.html
3. Google: OpenSK is an open-source implementation for security keys written in rust that supports both FIDO U2F and FIDO2 standards. https://github.com/google/OpenSK
4. Google Groups: Why are the already registered keys needed during registration? (2016). https://groups.google.com/a/fidoalliance.org/forum/#!forum/fido-dev
5. Kurosawa, K., Heng, S.-H.: From digital signature to ID-based identification/signature. In: Bao, F., Deng, R., Zhou, J. (eds.) PKC 2004. LNCS, vol. 2947, pp. 248–261. Springer, Heidelberg (2004). https://doi.org/10.1007/978-3-540-24632-9_18
6. Lindemann, R.: FIDO security reference (2018). https://fidoalliance.org/specs/fido-v2.0-id-20180227/fido-security-ref-v2.0-id-20180227.pdf
7. Machani, S., Noor, A.: FIDO enterprise adoption best practices (2019). https://fidoalliance.org/white-paper-fido-and-pki-integration-in-the-enterprise/
8. Mani, A.: Life of a password. Real World Cryptography - 2015. https://www.slideshare.net/ArvindMani1/amanirwcpassword
9. MIRACL: MIRACL trust multi-factor authentication (2020). https://miracl.com/miracl-trust-multi-factor-authentication/
10. El Mrabet, N., Joye, M. (eds.): Guide to Pairing-Based Cryptography. Chapman and Hall/CRC (2016). https://www.crcpress.com/Guide-to-Pairing-Based-Cryptography/El-Mrabet-Joye/p/book/9781498729505
11. Muffet, A.: Facebook: password hashing and authentication. Real World Cryptography - 2015. http://bristolcrypto.blogspot.com/2015/01/password-hashing-according-to-facebook.html
12. Noor, A.: FIDO-enabling a web-application using Universal 2nd Factor (U2F) (2015). https://fidoalliance.org/wp-content/uploads/FIDO-enabling-web-applications_Noor.pdf
13. Queralt, M.: Extending public key infrastructure for mobile users via FIDO universal authentication framework (2019). https://medium.com/@caumike/password-less-authentication-extending-public-key-infrastructure-for-mobile-users-via-fido-3a0da263b385
14. Scott, M.: Authenticated ID-based key exchange and remote log-in with simple token and PIN number. Cryptology ePrint Archive, Report 2002/164 (2002). http://eprint.iacr.org/2002/164
15. StrongKey: Open-source FIDO server, featuring the FIDO2 standard. https://github.com/StrongKey/fido2
16. Tzur-David, S.: Your complete guide to FIDO, FIDO2 and webAuthn. https://doubleoctopus.com/blog/your-complete-guide-to-fido-fast-identity-online/
17. Yubico: Webauthn introduction. https://developers.yubico.com/WebAuthn/

Blockchain-Enabled User Authentication in Zero Trust Internet of Things

Shanshan Zhao[1], Shancang Li[1(✉)], Fuzhong Li[2], Wuping Zhang[2],
and Muddesar Iqbal[3]

[1] Department of Engineering Design and Mathematics,
University of the West England, Bristol, UK
shancang.li@uwe.ac.uk
[2] School of Software, Shanxi Agricultural University,
Taigu, Jinzhong 030801, China
[3] London South Bank University, London, UK
m.iqbal@lsbu.ac.uk

Abstract. The Internet of Things (IoT) connects increasing number of smart devices, which makes the central authorities or third parties (e.g., cloud, fog, firewall, etc.) based authentication scheme very challenging. In recent, the blockchain shows great promises in IoT to provide secure and flexible authentication schemes. In this work, a blockchain enabled authentication scheme is proposed for IoT devices, which ensures a more secure and easily interoperable alternative to IoT systems. It makes it possible to switch smart devices from an untrust to a trusted data using blockchain.

Keywords: Internet of Things · Security · Blockchain · Authentication · Public key infrastructure

1 Introduction

The Internet of Things (IoT) is expected to have a strong influence on many areas (*e.g.*, smart city, smart home, healthcare industry, smart manufacturing, *etc.*) by providing interconnection and information exchanging [1,2]. Increasing number of smart devices are connected by the IoT, which makes the IoT become a growing complicated system and leads to many challenges, like security, privacy, authentications, *etc.* In existing IoT systems, the security requirements mainly relies on the type of applications it serves [3,4]. The authentication scheme in IoT ensures a more secure and easily interoperable alternative to IoT systems. However, the unprecedented number of smart devices makes it very challenge to guarantee every device connected to the IoT system is authenticated and certified. In existing systems, such as smart home [5], healthcare, industrial critical system (ICS), industrial IoT [6], supervisory control and data acquisition (SCADA) systems, public key infrastructure-based schemes have been widely used [7,8]. Most existing IoT systems are secured through following techniques:

© ICST Institute for Computer Sciences, Social Informatics and Telecommunications Engineering 2021
Published by Springer Nature Switzerland AG 2021. All Rights Reserved
D. Wang et al. (Eds.): SPNCE 2020, LNICST 344, pp. 265–274, 2021.
https://doi.org/10.1007/978-3-030-66922-5_18

- Authentication, ensure connected devices in the IoT can access the resource.
- Encryption, encrypt the data before transmission and passed to the storage device, ensure to protect the data from eavesdropping.
- Integrity, guarantee the genuine device and operating correctly, together with conformity and malware free for data in IoT.

1.1   IoT Authentication

This work focuses on the authentication of user and devices in IoT, which can help IoT system to manage smart devices that are mutually authenticated and verified. In may existing IoT systems, resource access control techniques are widely used to secure data and resources in the system. However, due to the diversity of devices in the growing IoT system, it is becoming difficult to use access control techniques due to the security principles and levels might be different for devices.

An IoT system usually contains diverse devices with different communication protocols, security levels, *etc.* A smart home system might use multiple connectivity techniques, like Low energy Bluetooth (BLE), RFID (radio frequency identification), WiFi, 3G/4G/5G, *etc.* Connecting devices using above techniques are usually not difficult due to most IoT devices supports one or more connectivity, and devices are usually authenticated using user-password or multi-factor authentication to verify device and access. The current trends suggest that many IoT systems requires IoT devices should be able to support secure socket layer and public key infrastructure (PKI), where the authenticity of users/devices were proven using digital certificates [9].

PKI based solution would ensure a level of trust of a user/device. In recent, the emerging blockchain technologies show promises in device authentication, which makes it possible to manage users/devices authentication and integrity of messaging between them using blockchain. The blockchain technology, can be seen as a decentralised ledger system, could enhance the security and ensure identity and access management. Devices can potentially be considered as an autonomous node in the system and every access attempt is verified and traced automatically.

On the other hand, the diversity of IoT devices makes it very challenging to run secure PKI. Authentication schemes, such as user-password based, multiple factor authentication, or Azure IoT, can be used to manage all devices. The MQTT (Message Queuing Telemetry Transport) is a lightweight messaging protocol for smart devices in IoT using a publish/subscribe model, which is widely used in energy sensitive IoT applications. Different IoT applications require one or more protocols. However, the overhead generated by the protocol could be too much for resource constrained IoT devices, or device might not support the protocol.

In existing IoT systems, a number of authentication solutions have been developed, most of them significantly rely on application scenarios [10]. It is very challenging to support different types of devices in an authentication scheme that is also secure. However, it is possible to ensure devices are designed by following some security standards in design. To provide IoT systems with strong secure authentication, each devices needs to be authenticated before grantee the access.

1.2   Blockchain

The blockchain technology is a decentralised ledger system that maintains a growing chain of verified blocks, which records generated transactions and data. The blockchain technology can provide excellent traceability, immutability, and transparency, which can well help in verification the authenticity of access in IoT.

Basically, a blockchain system computes a cryptographic hash as unique identification for each record or transactions. The one-way hash function cryptographically generates the same hash value when given the same input, which is a perfect way to verify the authenticity of the input. When slight change is made on the input, it will cause dramatically different hash value. The blockchain can be distributed across all participants and each participant keeps an identical copy of the ledger. In each block, the records were verified and immutable.

The blockchain can further enhance the trustworthiness of users/devices, and data in an IoT system. The blockchain can offer user to anonymously perform secure. Meanwhile, the integrity of it is continuously being verified by the entire network as opposed to a central entity such as a central server or authority. This way, each participant do not have to trust a central entity but security is guaranteed by the strength and computing power of the entire system participating in the blockchain.

2   Blockchain-Enabled Device Mutual Authentication

In blockchain system, each participant holds a public key pair as their address. The network will validate the transaction and after that will start to add the transaction to the blockchain. The blockchain technology can be used for many areas, including e-voting, supply chain, *etc.* In zero trust IoT (zIoT) environment, blockchain can be used to authenticate every individual device without a central authority. Each individual entity can use its key information, such as name, unique identity, id, serial number, *etc.*, to register its identity on the blockchain that can be verified based on its hash value earlier registered on the blockchain.

One challenge is that the blockchain requires all participants to reach a consensus result to make sure it is trustworthy and independent of controlling organisations. In zIoT scenarios, users or devices need to be motivated in a blockchain which can be used for authentication tasks.

For dynamic access, such as a device joining/leaving the zIoT system, its identity need to be well dealt with in blockchain. In case of leaving the system, the decentralised authentication provider based on blockchain technology.

As a good counterpart to the identity authentication/verification protocol, the blockchain technology can verify its own data integrity without requiring for a third party. In a blockchain network, the validity and security can be guaranteed using hash values, while the access control can be coordinated using smart contracts.

In a blockchain based authentication system, only the device/user owns the private key, which can be used to verify its ownership when an IoT service/application needs a proof of identity. The generated data can be stored on a blockchain, the IoT service/application do not have to worry about correctness and ownership of the data.

In this work, we propose a blockchain-enabled device authentication solution for zero trust IoT (BazIoT), which provides a new way to manage device/identity authentication with following features:

– BazIoT utilises blockchain in zIoT by enabling user to verify user/devices and their data without the need for a central entity, which can help reduce the use of third-part validation and further reduce the data breaches or incidents.
– It provides impeccable security, by storing data in blockchain and eliminating single points of failure inherent to centralised system.
– The BazIoT follows 'never trust, always verify' principle to stop malicious access to the data, which verifies each attempt of access to the resources.
– Unlike existing solutions, the BazIoT enables user/device owners fully control their data, in which blockchain network helps users to be in charge of their data and control the access right to these data.

However, in BazIoT, there are still a number concerns need to be addressed

– Incentivising the node, the BazIoT requires different participants, which requires extra computation resources;
– Permanent loss of access, in BazIoT the only way to gain access to the data is through a private key. If a user lose the key, there would be no way to get a new private key used for data access.
– Scalability, the BazIoT tends to get rather complex and fact, scalability could be a huge hurdle for public blockchains.
– Alterability of the data, the nature of immutability of blockchain guarantees the data validity but in IoT scenarios, extra option is needed to make the data owner be able to alter their data.

2.1 Self-sovereign Security

As mentioned above, the blockchain-based authentication system could significantly reduce the identity theft and data breaches. Meanwhile, it can also offer self-sovereign identities and security, enable device/user to fully control their own identity and data.

Self-sovereign security means in the sub IoT system (e.g., smart home), the user holds control of their security and all its attributes (including authentication, encryption, integration, privacy, etc.) and is not dependent on any single issuer or versifier to be online or available at the time of using the device. The self-sovereign security can ensure that device/user control their identity and data and share minimised attributes in the system.

The self-sovereign security system puts the users and device at the heart of the centre of security and its own data, in which smart contract can be used for device/user identity registration on blockchain network, and therefore resilient to censorship and server failure. It enable user/device fully control their identity data and access of their owned resources.

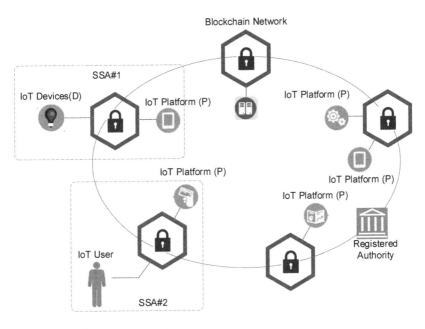

Fig. 1. Blockchain enabled zero trust IoT architecture.

The self-sovereign security can help to give back to the user/device full control on its identity and data, which enable user/device be the central of their own identity and data. There is no central authority needed for the entire IoT system.

2.2 Challenges

The main challenge that authentication of identity by blockchain faces is the involvement of different independent participants to calculate the blockchain to make it trustworthy and decentralised. How to implement consensus of participation is yet to be seen. In recent, the mutual distributed ledger identity schemes have been developed to enable secure the data storage, management, secure access in financial, healthcare, industrial sectors, as shown in Table 1.

New requirements for IoT device is *serverless, paswordless, self-sovereign security*, the decentralised blockchain technology is a perfectly tool to guarantee identity, trust, interoperability. It can also implement decentralised platform that enable secure cryptographic identity management. On the other hand, the blockchain can make the self-sovereign identity/security become a reality.

Table 1. Existing blockchain-based authentication schemes

Solution	Description
Civic [11]	Blockchain enabled biometrics identity manage system on mobile device, supporting multi-factor, without user-password
Helix [12]	Blockchain enabled digital identity system, supporting multiple participants share both self-asserted and verified information
Vida identity [2]	Identity authentication enables distributed key revocation and re-issuance
Spidchain [13]	Self-sovereigh identity, allow individuals to create, recover, revoke identifier, to sign and verify files and claims, etc.
BitID [14]	Bitcoin based authentication protocol, authenticates addresses by signing cryptographic challenge
Clear.me	Verified identity claims are signed by issuers, encrypted, and stored via the blockchain
Digi-ID [15]	DigiByte blockchain based authentication method allowing user to log in to a site, app by simply scanning or tapping on a QR code
CerCoin [16]	Namecoin-based blockchain authentication system that maintains a public ledger of domains and associated public keys
Trusted key [17]	Self-sovereign identity platform based on Ethereum, which offers mobile identity, id verification, password-less login, supports IoT
Ockam [18]	ERC20 based blockchain that registers IoT devices to solve systemic security and interoperability problems

3   Proposed Device Authentication in zIoT

As mentioned in Table 1, a number Blockchain based authentication has been developed and most of them focus on the identity authentication in cloud, blockchain systems. In this work, we will introduce a mutual authentication scheme for devices in the zIoT, in which all devices need to be verified before granting access to the zIoT by following the principle of "never trust, always verify".

In zIoT, the self-sovereign security area (SSA) is dynamic, as shown in Fig. 1, the IoT user/device A belongs SSA #1 and IoT user/device B belongs SSA #2. When A finishes the access session, the SSA #1 will be dismissed and new SSA will be established for new successful access. Actually, the roles of devices can be switched from resource owner to resources requester depends the application scenarios. All key events will be recorded in blockchain and the procedure is coordinated using smart contract.

The proposed protocol verifies to both the authentication participants. The user is the entity attempting to gain access to the protected resource in zIoT [19].

1. The user/device logs with a password-less scheme, with a authentication request ask authentication for accessing resources in an IoT systems.
2. The authentication request would simply encode the {*blockchain address*} using public key of the resource owner in zIoT.
3. The resource owner verifies the request and send back a response.

The blockchain enabled protocol can prevents critical cyber attacks in zIoT, including, spoofing attack, phishing attack, the main-in-the-middle, replay attack, *etc.* In this solution, the verification and encryption (AES) keys are stored on the blockchain and the signing and decryption keys are stored on the device, which can enable the devices can fully control its resources and data.

In a zIoT system, at the IoT device/user side:

(1) The device retrieve the RSA public key P of the verifier in zIoT, which will be used to encrypt blockchain address of the IoT device/user as $(addr, P)$ and then send to the IoT platform;
(2) When the verifier receives the request, it will extract the *addr* using the P, then the zIoT retrieve the RSA public key D of the device/user from the blockchain, and then generate a random string and timestamp *nonce* that then will be sent to the device/user;
(3) The device/user decrypts the received *hash* and then and sends back to verifier in a encrypted envelope.
(4) The verifier verifies the digital signature and confirm the authentic of the access.

As shown in Fig. 2. The identity addresses private key is derived from user keychain phrase that user chooses to use to sign in to the app. It never leaves the user's instance of the browser. This private key signs the authentication response token for an app to indicate that the user approves sign in to that system. The BazIoT is designed to protect enterprise-level zIoT from potential security breaches using blockchain-enabled mutual verification.

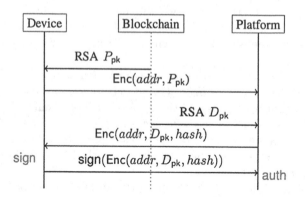

Fig. 2. BazIoT authentication flow

4   Evaluation

Decentralised authentication is more efficient. The BazIoT eliminates the need for reliance on third parties for document authentication and allows users to select their own solution based on their needs. In this work, we use a very simple blockchain to implement the proposed BazIoT.

In BazIoT, the identifier of a device/user could be one or combination of *name, account number, imei number, address, serial number, etc.* In this work, we use device *serial number* as their identifier, as {"device_id": "test-id-z2832"}, the keys generated as shown in Fig. 3.

```
-----BEGIN RSA PUBLIC KEY-----
MIGJAoGBAMP+1J/gX2iidz7WJFFdha9rcyirmEikyEOcaGuZdCFpI+VDnCaZeYLi
X9Yw5MAwELjAGCBGqUYiOFz0h1C8/ttEwzNqDxNMiqKmwCvyAELzwADXAVLXOZ55
Y1dW0SDi6SBqmPqrDNMtS-HLceWjL6xexo+4COj6a11OVtTAYuq5AgMBAAE=
-----END RSA PUBLIC KEY-----

-----BEGIN MESSAGE-----
prTjWJAsyQb+WCr3VXRzgB/cRkF7Hy2zXH+rw30hgDZABOfUXQgF2nBBIJTZJny+
0KKnLrjysn0h7ZoglpmjnERXqpVHq5/Jnzpwab5iUrAvNh511TCbhcb2d8uvJ8Sm
+ipqQ7XBOBHLiueubEj1u0O8njCFl+TbH9/DsTcQIKU=
-----END MESSAGE-----
```

Fig. 3. Keys generated in BazIoT

In the blockchain, we assume IoT platform is the verifier that verifies an IoT device. The blockchain addresses of device and platform are: '0xdf1256d ffa2342ef9da8ed5862ebf732b12972a3', and '0xce1473fdea4235db9ce83d28721bf7 b2d12982c5', respectively. In this work, AES symmetric encryption is used to encrypt data transmission.

In practical systems, the procedure of BazIoT consists three main steps:

Initialise. This stage the BazIoT creates initialisation for a device that wants to access the resource in IoT, key parameters include {*device-id, blockchain address, , , authentication-id, RSA-, RSA-*}, and AES parameters $\mathsf{dpk}, \mathsf{iv}, \mathsf{tag}$;

Authentication. An IoT device first extract the RSA public key of the platform (IoT resources) P, and then encrypts its blockchain address using (P, add), which then will be sent to the resource holder (platform). The platform can decrypts the cipher using (P, add), and then create a 4096-bit *nonce*,

$$H_{nonce} = sha512(nonce \parallel timestamp) \tag{1}$$

The hash then will be encrypted using the public key of device D as (D, H_{nonce}). When receives the encrytped message, the device is able to decrypt and then sign the hash value using its private key. Together with the blockchain address and hash value, the digital signature then will be enveloped into a encrypted envelope using P, which then is sent back to platform.

Verification. When receives the encrypted envelope, the platform can decrypt it using P and extract the {*address, hash, signature*}, if the signature is valid, then returns 'accept', else returns 'reject'.

5   Conclusion

This work investigated the device/user authentication in zero trust IoT environments. A passwordless device authentication scheme is proposed that can provide enhanced security for IoT, meanwhile, it supports the dynamic authentication of IoT devices like joining/leaving. All access to the resource in the IoT system will be verified before granting access.

References

1. Sharma, V.: An energy-efficient transaction model for the blockchain-enabled Internet of vehicles (IoV). IEEE Commun. Lett. **23**(2), 246–249 (2019)
2. Li, S., Choo, K.R., Sun, Q., Buchanan, W.J., Cao, J.: IoT forensics: Amazon echo as a use case. IEEE Internet Things J. **6**(4), 6487–6497 (2019)
3. Xu, L., He, W., Li, S.: Internet of Things in industries: a survey. IEEE Trans. Ind. Inform. **10**(4), 2233–2243 (2014)
4. Yazdinejad, A., Parizi, R.M., Dehghantanha, A., Choo, K.R.: Blockchain-enabled authentication handover with efficient privacy protection in SDN-based 5G networks. IEEE Trans. Netw. Sci. Eng. 1 (2019, in press)

5. Jangirala, S., Das, A.K., Vasilakos, A.V.: Designing secure lightweight blockchain-enabled RFID-based authentication protocol for supply chains in 5G mobile edge computing environment. IEEE Trans. Ind. Inform. **16**, 7081–7093 (2019)
6. Li, S., Xu, L.D., Zhao, S.: The Internet of Things: a survey. Inf. Syst. Front. **17**(2), 243–259 (2015)
7. Yang, D., Jeon, S., Doh, I., Chae, K.: Randomly elected blockchain system based on grouping verifiers for efficiency and security. In: 2020 22nd International Conference on Advanced Communication Technology (ICACT), pp. 159–165 (2020)
8. Haddad, Z., Fouda, M.M., Mahmoud, M., Abdallah, M.: Blockchain-based authentication for 5G networks. In: 2020 IEEE International Conference on Informatics, IoT, and Enabling Technologies (ICIoT), pp. 189–194 (2020)
9. Li, S., Qin, T., Min, G.: Blockchain-based digital forensics investigation framework in the Internet of Things and social systems. IEEE Trans. Comput. Soc. Syst. **6**(6), 1433–1441 (2019)
10. Li, S., Zhao, S., Yang, P., Andriotis, P., Xu, L., Sun, Q.: Distributed consensus algorithm for events detection in cyber-physical systems. IEEE Internet Things J. **6**(2), 2299–2308 (2019)
11. Civic: Flexibility without oversharing your identity (2020). https://www.civic.com/wallet/
12. Helix: Building trusted digital identity (2020). https://blockchain-helix.com/
13. Spidchan: Spidchain (2020). https://gomedici.com/companies/spidchain
14. BitID: Bitcoin authentication open protocol (2020). https://github.com/bitid/bitid
15. Digi-ID: Authentication at its best (2020). https://www.digi-id.io/
16. Feng, T., Chen, W., Zhang, D., Liu, C.: One-stop efficient PKI authentication service model based on blockchain. In: Si, X., et al. (eds.) CBCC 2019. CCIS, vol. 1176, pp. 31–47. Springer, Singapore (2020). https://doi.org/10.1007/978-981-15-3278-8_3
17. Wood, M.: Blockchain digital identity startup trusted key acquired by workday (2020). https://www.ledgerinsights.com/workday-blockchain-digital-identity-trusted-key/
18. Simone, S.D.: Ockam brings blockchain serverless identification to IoT devices (2020). https://www.infoq.com/news/2019/01/ockam-blockchain-iot-identity/
19. Nagpal, R.: Blockchain-based authentication of devices and people (2018). https://medium.com/blockchain-blog/blockchain-based-authentication-of-devices-and-people-c7efcfcf0b32

Cloud Security

Security Analysis and Improvement of a Dynamic-Hash-Table Based Auditing Scheme for Cloud Storage

Qiang Ma[1], Ti Guan[1], Yujie Geng[1], Jing Wang[2], and Min Luo[2(✉)]

[1] State Grid Shandong Electric Power Company, Jinan, China
[2] School of Cyber Science and Engineering, Wuhan University, Wuhan, China
mluo@whu.edu.cn

Abstract. Cloud storage has emerged as a promising solution to the scalability problem of massive data management for both individuals and organizations, but it still faces some serious limitations in reliability and security. Recently, Tian et al. proposed a novel public auditing scheme for cloud storage (DHT-PA) based on dynamic hash table (DHT), with which their scheme achieves higher efficiency in dynamic auditing than the state-of-the-art schemes. They claimed that their scheme is provably secure against forging data signatures under the CDH assumption. Unfortunately, by presenting a concrete attack, we demonstrate that their scheme is vulnerable to the signature forgery attack, i.e., the cloud service provider (CSP) can forge a valid signature of an arbitrary data block. Thus, a malicious cloud service provider can pass the audit without correct data storage. The cryptanalysis shows that DHT-PA is not secure for public data verification. The purposed of our work is to help cryptographers and engineers design/implement more secure and efficient identity-based public auditing schemes for cloud storage by avoiding such kind of attacks.

Keywords: Cloud storage · Public auditing · Dynamic hash table · Auditing security

1 Introduction

With the explosive growth of data in today's world, the significance of cloud storage service is more and more highlighted [1]. Taking the advantages of elastic storage, ubiquitous access and affordable management, cloud storage providers have attracted an increasing number of individuals and organizations to enjoy this service, such as Microsoft Skydrive, Amazon S3 and Google cloud storage [2–4]. By shifting the data from their local storage system to the remote cloud server, individuals and organizations can greatly relieve themselves from the burden of data management and maintenance. Regardless of these benefits, outsourcing the local data to a remote cloud server still faces some security and privacy challenges. For example, the cloud infrastructure may suffer from some inevitable

D. Wang et al. (Eds.): SPNCE 2020, LNICST 344, pp. 277–285, 2021.
https://doi.org/10.1007/978-3-030-66922-5_19

failures that leads to a data corruption, but the cloud service provider (CSP) may hide the accident to avoid financial loss [5]. Therefore, maintaining the integrity and privacy of cloud data is a key point for prompting the serviceability of cloud storage.

To address the security issues, many public auditing schemes have been proposed to verify the integrity of cloud data, which allow an honest-but-curious public auditor (also called trusted public auditor, TPA) verify the integrity of outsourced data periodically without downloading the entire data file from the remote cloud server. Ateniese et al. [6] first presented the notion of Provable Data Possession (PDP) to check the storage correctness of cloud data without downloading the whole file. On the basis of Ateniese et al.'s conception, Shacham and Waters [7] proposed an improved PDP scheme with Boneh-Lynn-Shacham (BLS) signature, which is widely adopted to construct auditing schemes with additional requirements, such as privacy preserving [8,9] and efficient dynamic auditing [10,11].

Note that a secure public auditing scheme should enable an external auditor to check the storage correctness of cloud data without learning any content of the data, as the introduced TPA is credible but curious. Otherwise, the TPA can reconstruct the while file by collecting all data blocks after several auditing procedures, so that the data copyright of the owner may be violated. Wang et al. [12] is the first to come up with a privacy-preserving auditing scheme by using the random masking technique. Later, there are many other improved privacy-preserving public auditing protocols have been proposed for higher efficiency, such as [13–16].

As for the dynamic data auditing, Erway et al. [17] first came up with a dynamic provable data possession (DPDP) scheme by utilizing a ranked-based skip list, but it cannot support public auditing. Then, Wang et al. [18] proposed a dynamic public auditing scheme with Merkle Hash Tree (MHT). However, both the two dynamic auditing schemes would arouse heavy computation and communication costs during the verification and updating processes. In view of these problems, Zhu et al. [19] came up with an efficient dynamic public auditing scheme (IHT-PA) based on an index-hash table (IHT) by storing the auditing metadata in the side of TPA rather than CSP. However, Tian et al. [20] pointed out that IHT-PA is still inefficient in updating procedure, although it can efficiently support dynamic auditing to some degree.

To get a better tradeoff between the dynamic properties and auditing efficiency, Tian et al. [20] presented a new public auditing scheme (DHT-PA) by exploiting the dynamic hash table (DHT) and Boneh-Lynn-Shacham (BLS) signature to achieve dynamic auditing and batch auditing. Tian et al. proved that DHT-PA is much more efficient than IHT-PA at the time of updating data blocks and files. They also claimed that DHT-PA is secure in terms of resisting the signature forgery attack and proof forgery attack. However, we demonstrate that their scheme is vulnerable to signature forgery attack, i.e., the CSP can forge a valid signature of any data block, with which the CSP can further generate a forged auditing proof to pass the TPA's verification. By providing a new math-

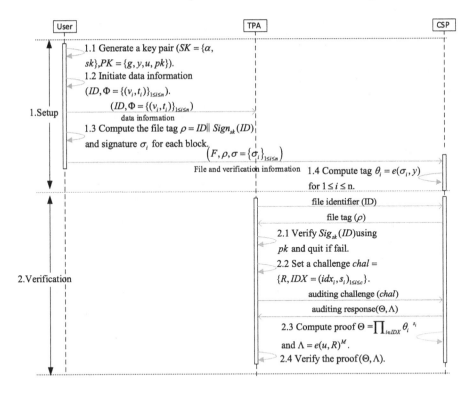

Fig. 1. The auditing process of Tian et al.'s DHT-PA scheme

ematical attack, our work is helpful for cryptographers and engineers to design and implement more secure and efficient identity-based public auditing schemes for cloud storage.

The remainder of this paper is organized as follows: In Sect. 2, we concisely review the scheme proposed by Tian et al. [20]. In Sect. 3, we demonstrate that Tian et al.'s scheme is vulnerable to signature forgery attack, and propose a probable fix to this weakness in Sect. 4. At last, we draw some conclusions for this paper in Sect. 5.

2   Review of DHT-PA

In this section, we give a brief review on Tian et al.'s scheme (DHT-PA) about achieving public dynamic data auditing for cloud storage.

To start with, some definition are presented. $e : G_1 \times G_1 \to G_2$ is viewed as a bilinear map, where G_1 and G_2 are two additive cyclic groups with the same prime order p. $H : \{0,1\}^* \to G_1$ is a secure hash function. Let $F = \{m_1, m_2, \cdots, m_n\}$ denote the outsourced file, which is divided into n blocks.

For the sake of simplicity, we will only describe the first auditing part of DHT-PA with setup phase and verification phase as shown in Fig. 1. And the more details for readers can be referred to [20].

2.1   Setup Phase

1) **Key initiation:** $(SK = \{\alpha, sk\}, PK = \{g, u, y, pk\})$ is a key pair generated by the user, where g and u are two different elements in G_1, (sk, pk) generated for computing file tags.

2) **Data information initiation:** Let ID be the unique identifier of F. And $\Phi = \{v_i, t_i\}_{1 \leq i \leq n}$ denotes the latest version information of data blocks, where v_i, t_i are the version and timestamp of block m_i respectively. Then, the user sends (ID, Φ) to the TPA as a delegation of data auditing.

3) **Signature Generation:** The user first computes the signature for each data block m_i as follows:

$$\sigma_i = H(v_i \| t_i) \cdot u^{m_i + H(v_i \| t_i)}, 1 \leq i \leq n \tag{1}$$

Then, the user calculates $\rho = ID \| Sig_{sk}(ID)$ as the file tag, where $Sig_{sk}(ID)$ is the signature of ID under the secret key sk. Finally, the user outsources (F, ρ, σ) to the CSP before deleting them from the local storage, where $\sigma = \{\sigma_i\}_{1 \leq i \leq n}$.

4) **Tag Generation:** Upon receiving the signatures σ_i, the CSP computes a tag for each data block as follows:

$$\theta_i = e(\sigma_i, y), 1 \leq i \leq n \tag{2}$$

After that, the CSP will store $(\rho, \theta_i)_{1 \leq i \leq n}$ along with the file $F = \{m_1, m_2, \cdots, m_n\}$.

2.2   Verification Phase

1) **File identifier check:** The TPA first verifies the file signature $Sig_{sk}(ID)$ using the public key pk after receiving the tag ρ. If the verification fails, TPA refuse the user's delegation; otherwise, the TPA launches a challenge for data auditing on behalf of the user.

2) **Challenge:** The TPA randomly selects a c-element subset $I = \{idx_1, idx_2, \cdots, idx_c\}$ from the set $\{1, 2, \cdots, n\}$ as the index set of the blocks to be checked. Then it sets $chal = \{R, (idx_i, s_i)\}_{i \in I}$ as the auditing challenge and sends it to the CSP, where s_i is a random number from Z_p, $R = y^r$ ($r \in Z_p$ is also a random number).

3) **Proof generation:** Upon receiving the challenge, the CSP starts to compute the corresponding proof: $\Theta = \prod_{i \in I} \theta_i^{s_i}$, $M = \sum_{i \in I} s_i m_i$ and $\Lambda = e(u, R)^M$. Next, it sends $\{\Theta, \Lambda\}$ back to TPA as the auditing proof.

4) **Proof check:** To perform the verification, the TPA first computes the value of $H = \prod_{i \in I} H(v_i, t_i)^{s_i}$, then it verifies the proof by checking the following equation:

$$\Lambda \cdot e(H \cdot u^{\sum_{i \in I} s_i H(v_i, t_i)}, R) \overset{?}{=} \Theta^r. \tag{3}$$

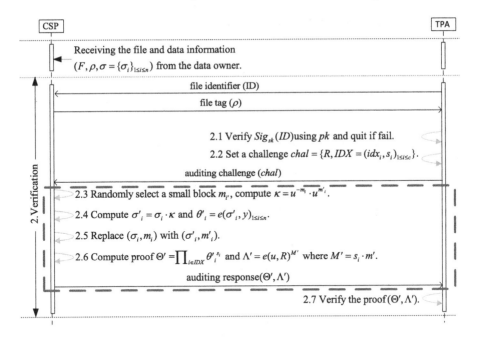

Fig. 2. The forgery attack of Tian et al.'s DHT-PA scheme

If holds, the cloud data is stored correctly; otherwise, the data loses its integrity on the remote node.

3   Cryptanalysis of Li et al.'s Scheme

Tian et al. claimed that their DHT-PA is secure because the CSP cannot keep m'_i instead of m_i to pass the audit. However, in this section, we will analyze the security of DHT-PA on verifying the integrity of the outsourced data, and demonstrate that DHT-PA is insecure against the signature forgery attack, i.e., the CSP can create a legal signature of an arbitrary data block m'_i. In other words, the CSP can keep m'_i instead of m_i to pass the audit successfully as shown in Fig. 2. Some details about the attack are presented as below.

Assume that the file F to be outsourced is divided into n blocks, i.e., $F = m_1 \| m_2 \| \cdots \| m_n$. The signature of each data block m_i is denoted as σ_i. Let $\mathscr{A}$ denote the malicious CSP, and it can pass the verification even if it does not correctly store the data by executing the following steps:

1) $\mathscr{A}$ randomly retrieves a signature σ_i of the data block m_i. As the messages transmitted from a user to the CSP is over public channel, thus the step is easily to for an network adversary $\mathscr{A}$ as the way referred in [21,22].
2) $\mathscr{A}$ randomly selects another data block m'_i $(m'_i \neq m_i)$, and computes the value of $\kappa = u^{-m_i} u^{m'_i}$ due to the fact that m_i, m'_i and u are public to the CSP.

3) $\mathscr{A}$ computes $\sigma_i' = \sigma_i \cdot \kappa$, and outputs it as the signature of data block m_i'. Since $\sigma_i = H(v_i\|t_i) \cdot u^{m_i + H(v_i\|t_i)}$, we would get

$$\sigma_i \cdot \kappa = H(v_i\|t_i) \cdot u^{m_i + H(v_i\|t_i)} \cdot u^{-m_i} u^{m_i'}$$

$$= H(v_i\|t_i) \cdot u^{m_i' + H(v_i\|t_i)}$$

$$= \sigma_i'$$

Obviously, σ_i' is a valid signature on m_i' according to the above equation.

4) Replace $(m_i, \sigma_i)_{1 \le i \le n}$ with $(m_i', \sigma_i')_{1 \le i \le n}$.
5) Upon receiving the auditing challenge, $\mathscr{A}$ computes the forged response proof: $\Theta' = \prod_{i \in I}(\theta_i')^{s_i}$, $M' = \sum_{i \in I} s_i m_i'$ and $\Lambda' = e(u, R)^{M'}$, where $\theta_i' = e(\sigma_i', y)$.
6) $\mathscr{A}$ returns (Θ', Λ') as auditing proof.

$\mathscr{A}$'s response can surely pass the TPA's verification, we prove it as below:

$$\Lambda' \cdot e(H \cdot u^{\sum_{i \in I} s_i H(v_i, t_i)}, R)$$

$$= e(u, R)^{\sum_{i \in I} s_i m_i'} e(\prod_{i \in I} H(v_i\|t_i)^{s_i} \cdot u^{\sum_{i \in I} s_i H(v_i, t_i)}, R)$$

$$= e(u^{\sum_{i \in I} s_i m_i'}, R) e(\prod_{i \in I} H(v_i\|t_i)^{s_i} \cdot u^{\sum_{i \in I} s_i H(v_i, t_i)}, R)$$

$$= e(\prod_{i \in I} H(v_i\|t_i)^{s_i} \cdot u^{\sum_{i \in I} s_i(m_i' + H(v_i, t_i))}, R)$$

$$= e(\prod_{i \in I} (H(v_i\|t_i) \cdot u^{m_i' + H(v_i, t_i)})^{s_i}, g^{r \cdot \alpha})$$

$$= e(\prod_{i \in I} (\sigma_i')^{s_i}, g^{\alpha})^r$$

$$= \prod_{i \in I} e(\sigma_i', y)^{s_i r}$$

$$= (\Theta')^r$$

Hence, the proof (Θ', Λ') provided by $\mathscr{A}$ can certainly pass the verification of TPA without being detected when it does not store the user's data correctly. In other words, a malicious CSP can hide the corrupted data blocks caused by hardware/software failures; and it also can replace the large data blocks with smaller ones or directly deletes the unfrequently accessed data for space saving. So DHT-PA is not secure as an auditing scheme.

4   Possible Countermeasure

In the above attack, $\mathscr{A}$ just uses the value of $\kappa = u^{-m_i} \cdot u^{m_i'}$ to compute a legal signature σ_i' for another data block m_i', and then constructs a legal proof to pass the TPA's audit. Therefore, to withstand this attack, we should prevent $\mathscr{A}$ from computing $\kappa = u^{-m_i} \cdot u^{m_i'}$ to derive a valid signature. To achieve this goal, we can modify the *Signature Generation* step and *Tag Generation* step as follows.

Signature Generation: Given each data block m_i and public key u, the user generates a corresponding signature σ_i by following equation:

$$\sigma_i = (H(v_i\|t_i) \cdot u^{m_i+H(v_i\|t_i)})^\alpha \tag{4}$$

where α is the user's private key generated in *Key Initiation* step. Next, the user sends $(F, \rho, \boldsymbol{\sigma})$ to the CSP, where $\rho = ID\|Sig_{sk}(ID)$, $\boldsymbol{\sigma} = \{\sigma_i | 1 \le i \le n\}$.

Compared to the Eq. (1), we exploit the private key to sign the data block, with which $\mathscr{A}$ is not able to obtain the forged signature σ_i', because nobody knows the private key α except the data owner.

Tag Generation: Based on the received signature σ_i, the CSP generates a tag θ_i for each data block m_i, namely,

$$\theta_i = e(\sigma_i, g) \tag{5}$$

Compared to the Eq. (2), we replace the public key y with g which are both generated in *Key Initiation* step.

As for the verification phase, it does not need to have any modification. Now, we verify the correctness of Eq. (3) based on the Eq. (4) and Eq. (5) as follows:

$$\Lambda \cdot e(H \cdot u^{\sum_{i \in I} s_i H(v_i,t_i)}, R)$$

$$= e(u, R)^{\sum_{i \in I} s_i m_i} e(\prod_{i \in I} H(v_i\|t_i)^{s_i} \cdot u^{\sum_{i \in I} s_i H(v_i,t_i)}, R)$$

$$= e(u^{\sum_{i \in I} s_i m_i}, R) e(\prod_{i \in I} H(v_i\|t_i)^{s_i} \cdot u^{\sum_{i \in I} s_i H(v_i,t_i)}, R)$$

$$= e(\prod_{i \in I} H(v_i\|t_i)^{s_i} \cdot u^{\sum_{i \in I} s_i(m_i+H(v_i,t_i))}, R)$$

$$= e(\prod_{i \in I} (H(v_i\|t_i) \cdot u^{m_i+H(v_i,t_i)})^{s_i}, g^{r \cdot \alpha})$$

$$= e(\prod_{i \in I} (H(v_i\|t_i) \cdot u^{m_i+H(v_i,t_i)})^{\alpha \cdot s_i}, g^r)$$

$$= e(\prod_{i \in I} (\sigma_i)^{s_i}, g)^r$$

$$= \prod_{i \in I} e(\sigma_i, g)^{s_i r}$$

$$= \prod_{i \in I} \theta_i^{s_i r}$$

$$= (\varTheta)^r$$

Remark. From the above correctness analysis, we can see that the proposed countermeasure can be used to audit the cloud data at the cost of only small performance loss in computing block signatures. By adding a random exponent to the original tag, it will break the linear relationship between different message tags. And from this point, it may improve the security level of the original DHT-PA scheme by avoiding the attack described in Sect. 3.

5   Conclusion

In this paper, we reviewed the scheme DHT-PA proposed by Tian et al. [20], which is a public auditing scheme using the dynamic hash table to support dynamic auditing. Tian et al. claimed that DHT-PA is secure due to the unforgeability of data signatures and auditing proofs. However, the cryptanalysis of their DHT-PA scheme demonstrates that a malicious CSP can create a valid signature of any data block, so that it can pass the audit of TPA without correct data storage. Therefore, DHT-PA is not secure for practical application. To address the problem, we come up with a possible countermeasure to enhance the security of DHT-PA. And in the near future, we will be devoted ourselves to design a more secure and efficient public auditing scheme.

Acknowledgment. The work was supported by the National Key Research and Development Program of China (No. 2018YFC1604000) and the National Natural Science Foundation of China (Nos. 61972294, 61932016).

References

1. Wang, C., Wang, Q., Ren, K., Cao, N., Lou, W.: Toward secure and dependable storage services in cloud computing. IEEE Trans. Serv. Comput. **5**(2), 220–232 (2012)
2. Buyya, R., Yeo, C.S., Venugopal, S., Broberg, J., Brandic, I.: Cloud computing and emerging IT platforms: vision, hype, and reality for delivering computing as the 5th utility. Future Gener. Comput. Syst. **25**(6), 599–616 (2009)
3. Liu, J.K., Au, M.H., Huang, X., Lu, R., Li, J.: Fine-grained two-factor access control for web-based cloud computing services. IEEE Trans. Inf. Forensics Secur. **11**(3), 484–497 (2016)
4. Li, Y., Yu, Y., Yang, B., Min, G., Wu, H.: Privacy preserving cloud data auditing with efficient key update. Future Gener. Comput. Syst. **78**, 789–798 (2016)
5. Libing, W., Wang, J., Zeadally, S., He, D.: Privacy-preserving auditing scheme for shared data in public clouds. J. Supercomput. **74**(11), 6156–6183 (2018)
6. Ateniese, G., et al.: Provable data possession at untrusted stores. In: Proceedings of the 14th ACM Conference on Computer and Communications Security, pp. 598–609. ACM (2007)
7. Shacham, H., Waters, B.: Compact proofs of retrievability. In: Pieprzyk, J. (ed.) ASIACRYPT 2008. LNCS, vol. 5350, pp. 90–107. Springer, Heidelberg (2008). https://doi.org/10.1007/978-3-540-89255-7_7
8. Cui, H., Mu, Y., Au, M.H.: Proof of retrievability with public verifiability resilient against related-key attacks. IET Inf. Secur. **9**(1), 43–49 (2015)
9. Yu, Y., Zhang, Y., Ni, J., Au, M.H., Chen, L., Liu, H.: Remote data possession checking with enhanced security for cloud storage. Future Gener. Comput. Syst. **52**, 77–85 (2015)

10. Barsoum, A.F., Hasan, M.A.: Provable multicopy dynamic data possession in cloud computing systems. IEEE Trans. Inf. Forensics Secur. **10**(3), 485–497 (2015)
11. Zhang, Y., Ni, J., Tao, X., Wang, Y., Yong, Yu.: Provable multiple replication data possession with full dynamics for secure cloud storage. Concurr. Comput.: Pract. Exp. **28**(4), 1161–1173 (2016)
12. Wang, C., Chow, S.S.M., Wang, Q., Ren, K., Lou, W.: Privacy-preserving public auditing for secure cloud storage. IEEE Trans. Comput. **62**(2), 362–375 (2013)
13. Zhao, H., Yao, X., Zheng, X.: Privacy-preserving TPA auditing scheme based on skip list for cloud storage. IJ Netw. Secur. **21**(3), 451–461 (2019)
14. Yang, Z., Wang, W., Huang, Y., Li, X.: Privacy-preserving public auditing scheme for data confidentiality and accountability in cloud storage. Chin. J. Electron. **28**(1), 179–187 (2019)
15. Zhang, X., Zhao, J., Xu, C., Li, H., Wang, H., Zhang, Y.: CIPPPA: conditional identity privacy-preserving public auditing for cloud-based WBANs against malicious auditors. IEEE Trans. Cloud Comput. (2019)
16. Tian, H., Nan, F., Chang, C.-C., Huang, Y., Jing, L., Yongqian, D.: Privacy-preserving public auditing for secure data storage in fog-to-cloud computing. J. Netw. Comput. Appl. **127**, 59–69 (2019)
17. Erway, C., Küpçü, A., Papamanthou, C., Tamassia, R.: Dynamic provable data possession. In: Proceedings of the 16th ACM Conference on Computer and Communications Security-CCS 2009, p. 213. ACM Press (2009)
18. Wang, Q., Wang, C., Ren, K., Lou, W., Li, J.: Enabling public auditability and data dynamics for storage security in cloud computing. IEEE Trans. Parallel Distrib. Syst. **22**(5), 847–859 (2011)
19. Zhu, Y., Ahn, G.-J., Hu, H., Yau, S.S., An, H.G., Hu, C.-J.: Dynamic audit services for outsourced storages in clouds. IEEE Trans. Serv. Comput. **6**(2), 227–238 (2013)
20. Tian, H., et al.: Dynamic-hash-table based public auditing for secure cloud storage. IEEE Trans. Serv. Comput. **10**(5), 701–714 (2015)
21. Libing, W., Wang, J., Kumar, N., He, D.: Secure public data auditing scheme for cloud storage in smart city. Pers. Ubiquitous Comput. **21**(5), 949–962 (2017)
22. Xu, Z., Wu, L., Khan, M.K., Choo, K.-K.R., He, D.: A secure and efficient public auditing scheme using RSA algorithm for cloud storage. J. Supercomput. **73**(12), 5285–5309 (2017)

A Public Auditing Framework Against Malicious Auditors for Cloud Storage Based on Blockchain

Song Li[✉], Jian Liu, and Guannan Yang

College of Information Engineering,
Nanjing University of Finance and Economics, Nanjing, China
lisong@nufe.edu.cn

Abstract. In the cloud storage applications, the cloud service provider (*CSP*) may delete or damage the user's data. In order to avoid the responsibility, *CSP* will not actively inform the users after the data damage, which brings the loss to the user. Therefore, increasing research focuses on the public auditing technology recently. However, most of the current auditing schemes rely on the trusted third public auditor (*TPA*). Although the *TPA* brings the advantages of fairness and efficiency, it cannot get rid of the possibility of malicious auditors, because there is no fully trusted third party in the real world. As an emerging technology, blockchain technology can effectively solve the trust problem among multiple individuals, which is suitable to solve the security bottleneck in the *TPA* based public auditing scheme. This paper proposed a public auditing scheme with the blockchain technology to resist the malicious auditors. In addition, through the experimental analysis, we demonstrate that our scheme is feasible and efficient.

Keywords: Cloud storage · Pubic auditing · Blockchain

1 Introduction

With the rapid development of the cloud computing, users can access the cloud services more economically and conveniently today: for example, the cloud users can outsource the numerous computing tasks to the *CSP* and reduce the purchase of local hardware resources [1]; besides, with the help of cloud storage service such as, Amazon, icloud, Dropbox, etc. [2], users can put aside the geographical restrictions and upload the local data to the *CSP*, with only a small amount of payment but a greatly reduction of local storage resources and more convenience of the data sharing with others. For the enterprise users, due to the explosive growth of business data, enterprises need to spend high cost to purchase software/hardware resources to build an IT system and maintain a professional technical team to manage this system, which causes extra burden to enterprises. Hence, the "pay as you go" service mode of the cloud storage is more convenient and practical. Users can dynamically apply for the storage space according to their data volume from the *CSP*, so as to avoid resource waste through the elastic resource allocation mechanism.

Although the cloud storage service has a broad market prospect, there are still many data security problems to be solved. Many famous *CSP* have experienced information

D. Wang et al. (Eds.): SPNCE 2020, LNICST 344, pp. 286–300, 2021.
https://doi.org/10.1007/978-3-030-66922-5_20

disclosure and service termination [3], such as icloud's information disclosure, Amazon cloud's storage outage, Intuit's power failure, sidekick's cloud disaster, Gmail's email deletion, etc. In August 6, 2018, Tencent cloud admitted to the user's silent error caused by the firmware version of the physical hard disk, i.e. the data written is inconsistent with the data read, which damages the system metadata [4]; therefore, solving the data integrity problem not only can enhance the user's confidence in the cloud storage services, but also can effectively promote the development of the cloud storage services industry. Since the cloud computing has become the basic infrastructure at the era of big data, the data security is the primary concern of cloud users.

However, in the practical applications, due to system vulnerabilities, hacker attacks, hardware damage, human operation errors or even to maximize the interests, *CSP* may delete or damage some user's data [5–7]. For example, the hospital outsourced all the electrical disease records to the *CSP*, but *CSP* may lose part of the stored data. It will cause a great loss to the users when these records cannot be retrieved. In order to avoid responsibility, the *CSP* may not actively inform the data owners after the data is damaged; in addition, in some special service models, *CSP* claims to provide multi-backup storage service, but in the actual process, they only provide ordinary single-backup storage service, and cheat the consumers to obtain additional service fees. All of these factors will cause the cloud users unable to trust the *CSP* fully.

The traditional method of checking the integrity of remotely stored files is to download all the data from the *CSP* to the local machine, then data owner checks it locally by computing the message authentication code or signature [8–11]. However, if the large amount of data has been stored in the remote cloud server, such as for the online retailer

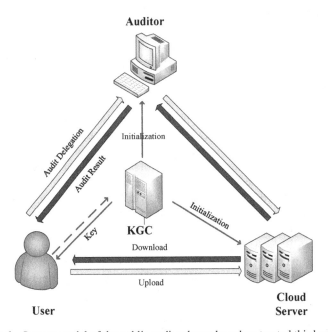

Fig. 1. System model of the public audit scheme based on trusted third party

like Amazon that produced the hundreds of *PB* data every day, it is unrealistic to download all these data to the local machines every time when checking the integrity, because this will cause a lot of bandwidth/storage resources waste; on the other hand, the integrity checking is a periodicity task, it is expensive for mobile devices with limited resources to execute locally [12]; for the fairness at last, it is not reasonable to let either part of the *CSP* or data owners to audit after the data corruption, so it is an ideal choice to introduce a trusted third party to replace *CSP* or data owners to check the data integrity [13] (Fig. 1). However, after the third-party auditor (*TPA*) has been introduced, the problem of privacy disclosure is also produced. For example, the malicious auditor obtains the data owner's identity information in the auditing process, so as to know which part of the stored data is more valuable to the user [14]; in addition, it is possible for the *TPA* to know the content of the stored data block in the interaction with *CSP* [15].

2 Related Works

In 2003, Deswarte et al. [8] proposed a remote data integrity checking scheme based on the challenge-response protocol for the distributed system. Although their scheme does not need to download all the data when checking the remotely stored data, their scheme causes a large number of modular exponential operations on the server side resulting in large computing overhead, besides, the client needs to maintain all the data backup locally. In 2004, Sebe et al. [9] proposed a remote integrity checking scheme based on the Diffel-Hellman protocol. In their scheme, the client needs to store n-bits data for each data block to be stored, that is to say, only when the size of the data block is much larger than n, their scheme has practical significance (otherwise, it is not better than storing all the data locally); in 2005, Opera et al. [10] proposed a scheme based on the tweakable encryption. However, the client needs to download all the files in the checking phase, and their scheme aims at data retrieval, which is not suitable for the scenario of data integrity checking. In 2006, Schwarz et al. [11] solved the data security problem of remote storage across multiple servers based on algebraic signature. However, the security cost in the client side increases dramatically with the increasing of the data blocks to be checked.

The proposed schemes introduced above have the same problem: the client needs to access the complete data back-up, however, it is not suitable in practice obviously as mentioned before. Many scholars have carried out research on this issue later. In 2007, Ateniese et al. [16] proposed the concept of provable data possession (PDP) firstly based on RSA homomorphic linear authenticator and random sampling technology. User can check the data stored in the remote server without downloading all the data to the local thus solve the defect existed in the early proposed schemes, however, their scheme only supports of the static data. In 2008, Shacham and Water proposed two improved schemes based on BLS short signature [17]: the first scheme based on BLS signature supports infinite times public verifications on the data; the second scheme calculates the authenticators using pseudo-random function but does not support public verification.

Except of the static data, users may also add, delete, or modify the remote data, these dynamic operations will change the index of the data block resulting in the invalid of the original authenticators, as shown in Fig. 2. If all the authenticators will be recalculated each time when the data owner performs dynamic operations, a lot of computing and communication cost will be produced. Therefore, many scholars studied on the dynamic data supported schemes. In 2008, Ateniese et al. [18] proposed the dynamic PDP scheme based on symmetric key firstly. However, for the reason that their scheme is based on symmetric encryption, it does not support public auditing. In reference [19], Erway et al. introduced a dynamic PDP scheme that can support dynamic data using rank-based skip lists technology. In reference [20], Zhu et al. proposed a scheme with indexing-hash table to support the effective update of the dynamic data.

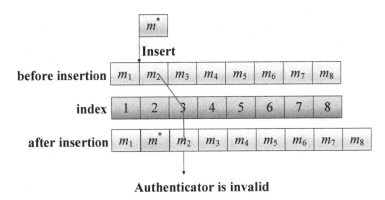

Fig. 2. The invalid of authenticator caused by the data dynamic operation (insertion)

In 2011, Hao et al. [21] expanded the scheme of Sebe et al.'s scheme [9], and proposed a dynamic auditing scheme in block-level based on RSA homomorphic tag. The so-called block level dynamic means that the data owners can insert, delete or update data blocks, but after the update, they still need to recalculate the authenticators which is not flexible.

In the practical applications, the integrity checking task is performed by *TPA* and most of the schemes proposed later support public auditing. In 2009, Wang et al. [13] proposed the integrity checking scheme with *TPA* firstly based on BLS short signature and MHT (Merkel hash tree). In this scheme, any entities in the network can challenge the *CSP* to check the integrity of the data stored on the cloud server, but this scheme does not support the full dynamic operations on the data.

Although the introduction of the *TPA* brings many benefits, it also brings new security and privacy issues. Therefore, the public auditing scheme supporting privacy preserving has become a hotspot recent years. In 2010, Wang et al. [14] proposed a public auditing scheme supporting content privacy preserving based on random mask technology. This scheme supports batch verification of multi-user tasks. However, due to the large number of verification tags generated on the server side, the system suffers a large storage burden. In 2012, Wang et al. [15] proposed a public auditing scheme to protect the identity privacy of the group users based on group signature technology, but the group signature produced huge computing cost in the data owner side, and their scheme did not consider the situation that the users can leave and join the group dynamically. In their scheme, users need to recalculate the authenticators of all the stored data block when the group key has changed; in 2014, Wang et al. [22] proposed an auditing scheme based on ring signature technology, which can protect the identity privacy of group membership and support group members to join/leave the group dynamically, but the efficiency of their scheme will be decreased with the increasing number of the group members, and the malicious users cannot be tracked in their scheme.

In the process of authenticator generation phase, a large number of signature operations are involved, however, many of the existing terminal equipment are embedded devices with low-power capacity such as mobile phones or sensors in IoT applications, therefore, public auditing schemes for low-power equipment have also been studied: in 2015, He et al. [23] proposed a public auditing scheme based on the certificateless cryptosystem, and applied it into the cloud-assisted wireless body area networks. Based on their certificateless mechanism, certificates do not need to be transferred and stored comparing with the previous proposals thus reduced the bandwidth resources; the users does not need to do the CRL(certificate revocation list) querying which greatly saves the computing resources. In 2016, Li et al. [12] proposed two auditing schemes for low-performance equipments based on online-offline signature technology. In the first basic scheme, the *TPA* needs to store some offline signature information, so it is only suitable for users to upload some short data (such as phone number, etc.) in the cloud; in the second scheme, the author solved the problem that the *TPA* needs to store a large number of offline signatures.

In 2017, Li et al. [24] pointed out that most of the existing schemes are based on the PKI infrastructure and the security of these schemes depend on the security of the key, then Li et al. proposed a public auditing scheme based on fuzzy identity signature technology. In this scheme, the user's identity (ID) is the public key, which improves the security of the system. However, Xue et al. [25] pointed out that Li's scheme can't resist malicious auditor's attack; Yu et al. put forwarded a scheme to resist key disclosure attack in the literature [26], which guarantees the forward security of the system by supporting the key updating mechanism, and the updated keys can still audit the previous data block tagged with the old keys.

In 2013, Liu et al. [27] proposed a public auditing scheme based on the rank-based Melkel-hash tree to improve the efficiency of the traditional hash tree algorithm. However, this algorithm causes a lot of computation cost to the *TPA*. If there are a large number of data blocks, the *TPA* needs to spend a lot of time to calculate the path of the Melkel tree. Yang et al. [28] proposed a scheme based on index table structure and BLS Signature algorithm, which supports the PDP mechanism of full dynamic data operation. In their scheme, because the index table is used to store the metadata of block file through continuous storage space, the deletion and insertion move a large number of data. With the expansion of user data scale and the increase of the number of block files, the time cost of deletion and insertion will increasing dramatically, which directly leads to the increasing of verification time cost after dynamic operation and reduces the auditing efficiency. In 2016, Li et al. [29] proposed a PDP auditing model based on LBT structure (large branching tree proofs of data possession, LPDP) to solve the problem of the authentication path is too long in building the MHT. LBT adopts a multi-branch path structure, and the depth of the LBT to be constructed decreases with the increasing of out-degree, thus reducing the auxiliary information in the process of data integrity checking, simplifying the process of data dynamic update, and reducing the calculation overhead between entities in the system. In 2017, Garg et al. [30] added indexes and timestamps to the MHT structure introduced in the scheme of reference [13] and proposed rist-MHT (relative indexed and time staged Merkle hash tree) structure, based on this structure, they proposed a PDP mode. Compared with the MHT structure, rist-MHT structure shortens the authentication length in MHT, thus reduces the time cost of node query. On the other hand, time stamp attribute gives the authenticator data freshness. However, although these algorithms based on MHT hash tree [13, 27, 30] avoid downloading all the data in the auditing process, but the correct verification results can only prove that the cloud server stores the hash tree but not the uploaded data.

In recent years, many scholars have carried out researches on the other issues such as group user revocation, data de-duplication, sensitive information sharing and antiquantum attack etc.

In 2018, Zhang et al. [31] pointed out that in the existed group sharing schemes, user revocation results in the large computational cost of the authenticator associated with the revoked users, so they proposed an identity-based public auditing scheme that can support user revocation, in which the revoking of malicious user does not affect the auditing of the previous data blocks.

Taek-Young Youn etc. [32] combined the ciphertext de-duplication technology with public auditing scheme. Because a large number of data uploading work are transferred to the *CSP*, the client only needs to carry out a single tag calculation step, which is suitable for low-performance client environment.

Shen et al. [33] proposed a public auditing scheme that can hide sensitive information when data owner sharing the data with other users based on IBE (identity-based-encryption). In this scheme, a role of data transfer (sanitizer) is added to transfer the sensitive data and its signature to realize the privacy preserving of the sensitive information in shared medical record.

In 2019, Tian et al. [34] pointed out that up to now, none of the schemes above can meet all the security properties and put forward a new scheme. In the process of tagging, the user's signatures will be converted into group signature, thus protecting the identity privacy of the users; in the auditing process, the content privacy is protected by using mask technology; all data operations will be recorded in the operation history table so that all illegal activities can be tracked.

Xue et al. [25] proposed a public auditing scheme based on blockchain to resist malicious auditors. In their scheme, the challenge verification information is generated based on bit-coin algorithm. However, the final auditing result of their scheme still relies on *TPA* uploading to the blockchain, which does not eliminate the threat of malicious *TPA* fundamentally.

Through the analysis above, we can see that the proposed schemes have the following defect presently: the security of these schemes relies on the trusted third party - *TPA*. Although the *TPA* brings advantages of the fairness and efficiency to the auditing process, it cannot get rid of the possibility of the malicious auditor, because there is no completely trusted third party in the real world. Although some scholars have conducted research on privacy protection problem in *TPA* based public auditing schemes with group signature, ring signature and other privacy protection technologies, the *TPA* needs to be treated as a semi-trusted entity and the risk of malicious auditor have not be eliminated fundamentally. As a new technology, blockchain technology can effectively solve the trust problem among multiple individuals, which is suitable to solve the security bottleneck problem in the *TPA* based public auditing scheme. This paper intends to solve the malicious auditor problem in the public auditing schemes combined with blockchain technology. The main contributions are summarized as follows:

1) We proposed a framework of public auditing scheme without trusted third part based on blockchain and given a basic work-flow;
2) We proposed a certificateless public auditing scheme based on the proposed framework to resist the malicious auditor and key escrow problems;
3) We gave a proven security analysis on our proposed schemes, the efficiency and security properties comparison shows that our scheme is better than previous schemes.

3 Preliminaries

In this section, we introduce the cryptographic techniques used to construct our scheme.

Definition 1: Bilinear Map

Given a cyclic multiplicative group G with order q and another multiplicative cyclic group G_T with the same order q. A bilinear pairing refers to a map $e: G \times G \to G_T$ should satisfy the following properties:

1) **Bilinearity**: For all $P, Q \in_R G$ and $a, b \in_R \mathbf{Z}_q^*$, $e(a \cdot P, b \cdot Q) = e(P, Q)^{ab}$.
2) **Non-degeneracy**: There exist $P, Q \in_R G$ such that $e(a \cdot P, b \cdot Q) \neq 1_{GT}$.
3) **Computability**: For all $P, Q \in_R G$, there exists an efficient algorithm to compute $e(a \cdot P, b \cdot Q)$.

Definition 2: Elliptic Curve Discrete Logarithm Problem (ECDLP)

Suppose that P and Q are two points over elliptic curve $E_p(a, b)$, and we know that P and Q has the relationship of $Q = s \cdot P$, it is difficult to find out the integer $s \in \mathbf{Z}_q^*$ only with P and Q.

Definition 3: Computational Diffel Hellman (CDHP)

Suppose that P, $a \cdot P$ and $b \cdot P$ are three points over elliptic curve $E_p(a, b)$, it is difficult to compute the result $a \cdot b \cdot P$ only with P, $a \cdot P$ and $b \cdot P$.

4 The Framework of Public Auditing Scheme Based on Blockchain

4.1 System Model

In our proposed framework, there are four roles: cloud server provider (*CSP*), client, key generating center (*KGC*) and auditors.

(1) *Cloud Service Provider*: In our scheme, the *CSP* is a semi-trusted entity with strong computing/storage resources, and the client uploads the local data to the remote CSP for storage. The CSP will faithfully complete the whole process of our auditing protocol with the other entities, however, he/she will attempt to cover up the fact of data corruption.

(2) *Client*: The client is a cloud storage service user. He/she could store his/her data in the *CSP* to reduce the storage burden locally. To ensure the integrity of the remotely stored data, the client can delegate the auditor to execute the interactive protocol with *CSP* and get the auditing result from auditor.

(3) *KGC*: The *KGC* is a trusted entity in our proposal, which is mainly used to generate the public parameters of the whole system and the client's partial secret key in certificateless cryptosystem.

(4) *Auditor*: auditors are distributed nodes deployed on the blockchain nodes, the *ProofVerify* algorithm are deployed on the auditors as the form of smart contract. After get the proof generated by the *CSP*, the auditors calculate the checking result and store them into the storage layer of blockchain.

The relationship among these entities is shown in Fig. 3.

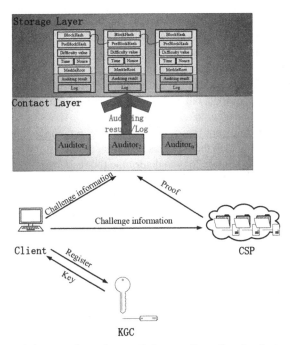

Fig. 3. The proposed framework against malicious auditors for cloud storage based on the blockchain

4.2 The Proposed Framework

In this section, we proposed a basic framework of public auditing scheme based on blockchain technology, and give a general work flow. In our framework, in order to avoid the problem of malicious attackers in the traditional *TPA* based schemes, we use the distributed nodes in the blockchain network as auditors to check the integrity.

Before the client uploads the data to the *CSP*, it uses the private key issued by the *KGC* to calculate the linear authenticator of the file. The calculation process divides the file into data blocks for calculation firstly, and then the user uploads the data and the corresponding linear authenticator to the *CSP* for storage. When the clients wants to check the integrity of the stored data in the cloud, the client sends the challenge information (randomly generated integers) and sends it to the auditors and *CSP*; the *CSP* calculates the proof according to the challenge information and returns proof to the auditors.

Auditors are smart contracts deployed on the blockchain nodes, the function of which mainly includes of two parts: processing client auditing request and execute the *ProofVerify* algorithm (the main part of the auditing scheme). The distributed auditors calculates the auditing results according to the proof returned by the *CSP*, stores the results into the storage layer of the blockchain, and maintain a history that cannot be tampered.

Secondly, when the client performs the data updating operations (such as adding, deleting, querying and modifying) on the stored data, the *CSP* generates the client's operation log of this time and compute the multiple–signatures on this log by client and *CSP* which indicate that all members agree with this result. It should be noted that auditing is a periodic process, it can be arranged every day at a certain fixed period of every day such as after zero clock, but each time the user performs an updating operation, an auditing action will be triggered automatically.

If the client or *CSP* finds out the stored data has been damaged, they can compare the current auditing results with the previous historical records stored in the blockchain, and combine the signed operation logs to determine the responsibility for data damage; because these data are stored in the distributed ledger with non-repudiation and non-tampering, neither party can refuse to admit it.

5 The Detailed Scheme

In this section, we give a detail proposal based on the framework we introduced above. Our scheme is constructed based on He et al.'s CLPA [24] scheme and Xue et al.'s scheme IDBA [26].

Setup: Input the security parameter κ, The *KGC* generates the system parameters and the master key executes the following steps:

1) The *KGC* selects a large prime number q, an additive group G_1, and uses the bilinear group generator to generate the bilinear group G_2. The *KGC* chooses a bilinear pairing e: $G_1 \times G_1 \to G_2$.

2) Let P be a generator of group G_1. The KGC selects a big integer $s \in Z_q^*$ randomly as the master key, keeps s secretly, computes the public key $P_{pub} = s \cdot P$

3) The KGC publishes the system parameters $\textbf{Para} = \{q, G_1, G_2, P, e, h_1(\cdot), h_2(\cdot), h_3(\cdot), H_1(\cdot), H_2(\cdot), P_{pub}\}$.

PartialPrivateKeyExtract: The client registers with the KGC to extract the partial private key with the following steps:

1) Client submits his/her identity ID_U to the KGC.

2) After received the client's identity ID_U, the KGC chooses a random big integer $t_U \in Z_q^*$ and computes $T_U = t_U \cdot P$, $h_U = h_1(ID_U, T_U)$, and $s_U = t_U + s \cdot h_U \bmod q$.

3) The KGC sends the partial private key $D_U = \{s_U, T_U\}$ to the user secretly.

SetSecretValue: The client sets his/her secret value as follows.

1) The client chooses a big integer x_U randomly as his/her secret value.

2) The client keeps x_U secretly.

SetPublicKey: The client sets his/her public key as follows.

1) The clients computes $P_U = x_U \cdot P$.

2) The clients sets $pk_U = \{T_U, P_U\}$ as his/her public key.

SetPrivateKey: The client sets $ssk_U = \{s_U, x_U\}$ as his/her private key.

Store: The client O with identity ID_O, private key $sk_O = \{s_O, x_O\}$, and public key $pk_O = \{T_O, P_O\}$ runs this algorithm to generate the tags for the data file F. Firstly, the data file F should be divided into n blocks $\{m_1, m_2, ..., m_n\}$, for every data blocks m_i, the client compute the tags with the following steps, where $i \in \{1, 2, ..., n\}$.

1) The client computes $k_O = h_2(ID_O, pk_O, P_{pub})$ and $Q = H_1(P_{pub})$.

2) The client computes $S_i = (s_O + k_O \cdot x_O)(r \cdot H_2(m_i) + H_2(id_i) + m_i \cdot Q)$ and sends $\{m_i, id_i, S_i, R\}$ to the CSP, where id_i is the unique identity of m_i, r is a random number,

$$R = r \cdot (T_O + h_O \cdot P_{pub} + k_O \cdot P_O) \tag{1}$$

Audit: To check the integrity of the uploaded data, the user executes the follows challenge-response protocol with CSP:

1) *Challen:*

The client generates challenge information as follows:

- Selects a random l-element subset $J = \{a_1, a_2, ..., a_l\}$ of the set $[1, n]$
- Selects a random $v_j \in Z_q^*$ for each $j \in J$.

- Generate the challenge information: $Chall = \{j, v_j\}_{j \in J}$, and broadcast it in the network, CSP and all the auditors can get it.

2) *ProofGen*:

After receiving the challenge information $Chall = \{j, v_j\}_{j \in J}$ from the client, the CSP generates a proof which prove of the correctly possession of selected blocks as follows:

- Choose a big integer $x \in \mathbf{Z}_q^*$ randomly
- Computes:

$$u = x^{-1}(\sum_{j=a_1}^{a_l} m_j \cdot v_j + h3(\sigma)) \tag{2}$$

$$\sigma = x \cdot Q \in G_1 \tag{3}$$

$$\delta = \sum_{j=a_1}^{a_l} v_j \cdot S_j \tag{4}$$

- Broadcast the proof information $Prof = \{\delta, u, \sigma, R\}$ to the auditors; if the auditing client choose to use in the more efficient model, the CSP divides the data blocks into k parts, and send them to the k auditors separately, k means the number of auditors.

ProofVerify: Upon receiving the $Prof = \{\delta, u, \sigma, R\}$, the auditors execute this algorithm to check the integrity of the data stored in the CSP. Here, the $Prof$ indicates the proof generated by CSP; in the secure model, the $Prof$ is the proof information of all the data blocks; in the efficient model, $Prof$ is the partial proof information, we use the same express as $Prof$ here.

1) The auditors computes $h_O = h_1(ID_O, T_O)$, $k_O = h_2(ID_O, pk_O, P_{pub})$, and $Q = H_1(P_{pub})$.
2) The auditors checks whether equation:

$$e(\delta, P) = e(\sum_{j=a_1}^{a_1} v_j \cdot H_2(id_j), T_O + h_O \cdot P_{pub} + k_O \cdot P_O)$$

$$\cdot e(\sum_{j=a_1}^{a_1} v_j \cdot H_2(m_j), R) \cdot e(u\sigma - h_3(\sigma)Q, T_O + h_O \cdot P_{pub} + k_O \cdot P_O) \tag{5}$$

holds, and output 1 if the equation holds that represents the correct storage of the data File F; otherwise, output 0 to indicate data corruption.
3) Create an **entry**(t, *nonce*, *Chall*, *Prof*, 0/1) and broadcast it in the network, all the auditors get the full auditing result and store the result; in the secure model, for the reason that each auditor can calculating the auditing result by themselves, the broadcast operation is not needed.

DataUpdate: when the user updates the file in the cloud, a recording log *Log* is generated by *CSP* to record the details of the user's operation. *CSP* and user execute the *MultiSign(Log)* and broadcast it in the blockchain network for storage. After each data update operation, the system automatically triggers the *Audit* phase.

6 Security Requirements Discussions

This section discussed that our proposed scheme satisfies the security requirements of auditing schemes.

1) *Publicly verifiability*: through the correctness proof part, we can see that as long as the client correctly calculates the data tags before uploading the data file, the auditor can perform interactive algorithm with the *CSP*, and get the real storage situation of the data blocks without the help of the client. Therefore, we say that our scheme has achieved the property of publicly verifiability.
2) *Privacy preserving*: We can see that in the process of the data auditing, the auditors can only get the aggregated data blocks and the tags, but through these information, auditors can not get any available information about stored data. Therefore, we say that our scheme achieves the goal of privacy protection.
3) *Batch auditing*: through the derivation of the correctness analysis, we can see that in the process of the auditing phase, multiple data blocks can be sampled at one time, and multiple data auditing tasks can be batch verified to improve the auditing efficiency. Therefore, we say that our scheme achieves the goal of the batch auditing.
4) *Key escrow resistant:* similar to the scheme IDBA [26], our scheme is based on the certificateless cryptography, the secret key to generate the authenticator has two parts which is derived from the KGC and client respectively. Therefore, the KGC cannot get the full of user's secret key like the scheme CLPA [24] based on the identity cryptosystem.
5) *Malicious auditor resistant:* in our auditing scheme, the auditing result is calculated by the distributed nodes, none of them can tamper the auditing result only if the attacker controls 51% nodes in the network; compare to the existing blockchain based public auditing scheme [26], the *ProofVerify* phase is transferred to the blockchain as the form of smart contract, instead of relying on the third-party auditor to upload the auditing result to the blockchain, thus, the possibility of the auditor creates the false result is eliminated fundamentally; besides, for the reason that the data blocks are confused with the mask code and the auditors can get nothing about the auditing data, the privacy of the data content has been protected.

7 Experimental Analysis

This section compares the performance of our proposed scheme with that of He et al.'s CLPA [24] scheme and the scheme IDBA [26]. Table 1 is the notation list we used in Table 2.

Table 1. The notations for operations

Symbol	The time cost of corresponding operation
T_M	The point multiplication operation in G_1
T_p	The pairing operation
T_H	Hash to point function
T_h	Hash function
k	The number of auditors

Table 2. The computation cost comparison of our scheme with CLPA and IDBA

Scheme	User's computational cost	Auditing computational cost	Communication cost
CLPA [24]	$2nT_M + (n+1)T_H + T_h$	$2T_p + (n+3)T_M + (n+1)T_H + 2T_h$	$|Z_q| + |G_1|$
IDBA [26]	$3nT_M + nT_H + nT_h$	$3T_p + (2n+3)T_M + nT_H + (n+1)T_h$	$|Z_q| + 3|G_1|$
Ours	$(3n+3)T_M + (2n+1)T_H + T_h$	$4T_p + ((2n+4)T_M + 2nT_H + T_h)/k$	$|Z_q| + 3|G_1|$

Table 2 shows the security overhead of these schemes in *Store* phase on the client side and the *ProofVerify* phase on the auditors' side. From the Table 2, we can see that in the *Store* phase, the time consumption of the authenticator calculation in our scheme is slightly higher than the other two schemes.

In the *ProofVerify* stage, because we used the distributed auditors to audit the data blocks, we get the better efficiency than the other schemes. We can see that if we do not use distributed auditors for auditing tasks, the computing cost of our scheme is still the highest, but after using the distributed processing mechanism in the efficient model, the efficient has been improved greatly.

Communication Cost: In the three schemes, the challenge information is the same; in the response phase, the proof returned by our scenario is: $Prof = \{\delta, u, \sigma, R\} = |Z_q| + 3|G_1|$. Through the comparison of Table 2, we can find that our scheme has the same communication cost with IBDA and slightly higher than CLPA.

8 Conclusion

In this paper, we pointed out that most of the *TPA* based public auditing schemes cannot resist the malicious auditor. To solve this problem, we proposed a public auditing framework with blockchain technology and certificateless crptography. In this framework, we used the distributed nodes in the blockchain network as auditors to check the integrity and the checking results will be stored into the storage layer of the blockchain with the tamper-resistant manner; the client operations on the data will be recorded as log signed by the data owners and *CSP* which indicate that all members agree with this result. Anyone can check the historical records stored in the blockchain nodes, and combine with the signed operation logs to determine the responsibility for data damage. We gave a detailed proven security proof of our scheme. A comprehensive performance evaluation shows that our scheme is more feasible and efficient than similar schemes.

References

1. Armbrust, M., et al.: A view of cloud computing. Commun. ACM **53**(4), 50–58 (2010)
2. Feng, D.-G., Zhang, M., Zhang, Y., et al.: Study on cloud computing security. J. Softw. **22**(1), 71–83 (2011)
3. Shen, W., Yu, J., Xia, H., et al.: Light-weight and privacy-preserving secure cloud auditing scheme for group users via the third party medium. J. Netw. Comput. Appl. **82**, 56–64 (2017)
4. http://www.sohu.com/a/245553016_671058
5. Ren, K., Wang, C., Wang, Q.: Security challenges for the public cloud. IEEE Internet Comput. **16**(1), 69–73 (2012)
6. Song, D., Shi, E., Fischer, I., Shankar, U.: Cloud data protection for the masses. IEEE Comput. **45**(1), 39–45 (2012)
7. Juels, A., Oprea, A.: New approaches to security and availability for cloud data. Commun. ACM **56**(2), 64–73 (2013)
8. Deswarte, Y., Quisquater, J.-J., Saïdane, A.: Remote integrity checking. In: Jajodia, S., Strous, L. (eds.) Integrity and Internal Control in Information Systems VI. IIFIP, vol. 140, pp. 1–11. Springer, Boston, MA (2004). https://doi.org/10.1007/1-4020-7901-X_1
9. Sebe, F., Martinez-Balleste, A., Deswarte, Y., et al.: Time-bounded remote file integrity checking. Technical report 04429 (2004)
10. Oprea, A., Reiter, M.K.: Space-efficient block storage integrity. In: Network and Distributed System Security Symposium, NDSS 2005, San Diego, California, USA. DBLP (2005)
11. Schwarz, T.S.J., Miller, E.L.: Store, forget, and check: using algebraic signatures to check remotely administered storage. In: IEEE International Conference on Distributed Computing Systems. IEEE (2006)
12. Li, J., Zhang, L., Liu, J.K., et al.: Privacy-preserving public auditing protocol for low-performance end devices in cloud. IEEE Trans. Inf. Forensics Secur. **11**(11), 2572–2583 (2016)
13. Wang, Q., Wang, C., Li, J., Ren, K., Lou, W.: Enabling public verifiability and data dynamics for storage security in cloud computing. In: Backes, M., Ning, P. (eds.) ESORICS 2009. LNCS, vol. 5789, pp. 355–370. Springer, Heidelberg (2009). https://doi.org/10.1007/978-3-642-04444-1_22
14. Wang, C., Wang, Q., Ren, K., et al.: Privacy-preserving public auditing for data storage security in cloud computing. In: 29th IEEE International Conference on Computer Communications, Joint Conference of the IEEE Computer and Communications Societies, INFOCOM 2010, 15–19 March 2010, San Diego, CA, USA. IEEE (2010)
15. Wang, B., Li, B., Li, H.: Knox: privacy-preserving auditing for shared data with large groups in the cloud. In: Bao, F., Samarati, P., Zhou, J. (eds.) ACNS 2012. LNCS, vol. 7341, pp. 507–525. Springer, Heidelberg (2012). https://doi.org/10.1007/978-3-642-31284-7_30
16. Ateniese, G., Bums, R., Curtmola, R., et al.: Provable data possession at untrusted stores. In: Proceedings of the 14th ACM Conference on Computer and Communications Security, pp. 598–609. ACM (2007)

17. Shacham, H., Waters, B.: Compact proofs of retrievability. In: Pieprzyk, J. (ed.) ASIACRYPT 2008. LNCS, vol. 5350, pp. 90–107. Springer, Heidelberg (2008). https://doi.org/10.1007/978-3-540-89255-7_7

18. Ateniese, G., Pietro, R.D., Mancini, L.V., et al.: Scalable and efficient provable data possession. In: Proceedings of the 4th International Conference on Security and Privacy in Communication Networks. ACM (2008)

19. Erway, C.C., Küpçü, A., Papamanthou, C., et al.: Dynamic provable data possession. ACM Trans. Inf. Syst. Secur. (TISSEC) 17(4), 1–29 (2015)

20. Zhu, Y., Hu, H., Ahn, G., et al.: Cooperative provable data possession for integrity verification in multicloud storage. IEEE Trans. Parallel Distrib. Syst. 23(12), 2231–2244 (2012)

21. Hao, Z., Zhong, S., Yu, N.: A privacy-preserving remote data integrity checking protocol with data dynamics and public verifiability. IEEE Trans. Knowl. Data Eng. 23(9), 1432–1437 (2011)

22. Wang, B., Li, B., Li, H.: Oruta: privacy-preserving public auditing for shared data in the cloud. IEEE Trans. Cloud Comput. 2(1), 43–56 (2014)

23. He, D., Zeadally, S., Wu, L.: Certificateless public auditing scheme for cloud-assisted wireless body area networks. IEEE Syst. J. 12(1), 64–73 (2015)

24. Li, Y., Yu, Y., Min, G., et al.: Fuzzy identity-based data integrity auditing for reliable cloud storage systems. IEEE Trans. Dependable Secur. Comput. 16(1), 72–83 (2017)

25. Xue, J., Xu, C., Zhao, J., Ma, J.: Identity-based public auditing for cloud storage systems against malicious auditors via blockchain. Sci. China Inf. Sci. 62(3), 32104 (2019)

26. Yu, J., Wang, H.: Strong key-exposure resilient auditing for secure cloud storage. IEEE Trans. Inf. Forensics Secur. 12(8), 1931–1940 (2017)

27. Liu, C., Chen, J., Yang, L.T., et al.: Authorized public auditing of dynamic big data storage on cloud with efficient verifiable fine-grained updates. IEEE Trans. Parallel Distrib. Syst. 25(9), 2234–2244 (2013)

28. Yang, K., Jia, X.: An efficient and secure dynamic auditing protocol for data storage in cloud computing. IEEE Trans. Parallel Distrib. Syst. 24(9), 1717–1726 (2013)

29. Yong, L., Ge, Y., Linan, L., Xiaofei, Z., Kun, Y.: LBT-based cloud data integrity verification scheme. J. Tsinghua Univ. (Sci. Technol.) 56(5), 504–510 (2016)

30. Garg, N., Bawa, S.: RITS-MHT: relative indexed and time stamped Merkle hash tree based data auditing protocol for cloud computing. J. Netw. Comput. Appl. 84, 1–13 (2017)

31. Zhang, Y., Yu, J., Hao, R., et al.: Enabling efficient user revocation in identity-based cloud storage auditing for shared big data. IEEE Trans. Dependable Secur. Comput. 17, 608–619 (2018)

32. Youn, T.Y., Chang, K.Y., Rhee, K.H., et al.: Efficient client-side deduplication of encrypted data with public auditing in cloud storage. IEEE Access 6, 26578–26587 (2018)

33. Shen, W., Qin, J., Yu, J., et al.: Enabling identity-based integrity auditing and data sharing with sensitive information hiding for secure cloud storage. IEEE Trans. Inf. Forensics Secur. 14(2), 331–346 (2018)

34. Tian, H., Nan, F., Jiang, H., et al.: Public auditing for shared cloud data with efficient and secure group management. Inf. Sci. 472, 107–125 (2019)

A Secure and Verifiable Outsourcing Scheme for Machine Learning Data

Cheng Li, Li Yang$^{(\boxtimes)}$, and Jianfeng Ma

Xidian University, Xi'an 710071, Shaanxi, China
15991727802@sina.cn, yangli@xidian.edu.cn, jfma@mail.xidian.edu.cn

Abstract. In smart applications, such as smart medical devices, in order to prevent privacy leaks, more data needs to be processed and trained locally or near the local end. However, the storage and computing capabilities of smart devices are limited, so some computing tasks need to be outsourced; concurrently, the prevention of malicious nodes from accessing user data during outsourcing computing is required. Therefore, this paper proposes EVPP (efficient, verifiable, and privacy-preserving), a machine learning method based on a collaboration of edge computing devices. In this solution, the computationally intensive part of the model training process is outsourced. Meanwhile, a random encryption perturbation is performed on the outsourced training matrix, and verification factors are introduced to ensure the verifiability of the results. In addition, when a malicious service node is found, verifiable evidence can be generated to build a trust mechanism. Through the analysis of theoretical and experimental data, it can be shown that the scheme proposed in this paper can effectively use the computing power of the equipment.

Keywords: Machine learning · Edge computing · Privacy-preserving · Mobile devices · Outsourced computing

1 Introduction

With the development of the Internet of Things, 5G communication networks, AI technology and the construction of intelligent facilities that promote the development of mobile devices, connected cars and smart wearable devices have been developed. Concurrently, a large amount of data has also been generated that is processed by different companies and different servers. Data are collected on various cloud computing platforms for various data analyses and mining. It is expected that by 2020, an average person will generate approximately 250 million bytes of data per day [1], which may come from mobile phone sensors, smart wearable devices, and so on.

We would like to thank the anonymous reviewers for their careful reading and useful comments. This work was supported by the National Key Research and Development Project (2017YFB0801805), the National Natural Science Foundation of China (61671360).

Abundant data require intelligent terminal processing, calculations, storage, etc. [2]; however, the storage and computing capabilities of smart devices are limited, and more data is being continuously collected, transmitted, and calculated. The transmission capabilities and data storage capabilities have become increasingly powerful, but in the face of a geometrically increasing amount of data, it is also difficult to meet users' requirements for data processing capabilities and transmission quality. Furthermore, the transmission of these data in the network will definitely apply a great pressure to the network.

The traditional centralized computing architecture based on a cloud centre [3–5] has been unable to meet the requirements of modern devices and applications for low latency, high efficiency, and low cost applications. In some special scenarios, such as smart healthcare [6,7], identity recognition [9], smart homes [10], all have high requirements on time and accuracy. Transferring data to cloud servers will raise latency, but running artificial intelligence algorithms such as machine learning and deep learning locally will bring an additional consumption of computing and power to the device.

Therefore, the application of edge computing [11] technology is used to outsource the calculation of data to edge nodes that are close and satisfy the computing power to reduce the computing and processing pressure of the device and reduce the delay in data transmission. At the same time, to reduce the pressure of network transmission, some data needs to be processed locally, such as the basic operations including simple data cleaning and partial data processing; simultaneously, in order to avoid a lengthy time, it is necessary to seek auxiliary computing nodes in the model training process on the near device side due to the limitation of the network with a high delay and high network pressure.

When applying edge computing to model training for local devices and nodes, data security and privacy issues cannot be ignored [12,13]. For example, in the application scenarios of user data collection such as smart medical devices and smart bracelets, the local device continuously accesses the user's geographic location, physical characteristics (including heart rate, stride, voiceprint, and other characteristics) or medical characteristics, which is apart from the collection and processing data by these devices that include a large number of user's privacy characteristics. As mentioned earlier, the local device's computing and processing capabilities cannot process and return results in a timely manner. To avoid the leakage of users' private data and to ensure that the calculation results are obtained in a timely and effective manner, advancements are needed.

Therefore, as shown in Fig. 1, this paper uses edge computing to solve data processing and computing problems in the construction of intelligent facilities, such as the Internet of Things, ensuring a high availability of data, effectively reducing network pressure and network delays. Concurrently, it will combine existing artificial intelligence and machine learning algorithms; the machine learning algorithm training process is "local + edge" for effective and safe training, and finally, the machine learning algorithm model is obtained. EVPP (efficient, verifiable, and privacy-preserving) is proposed: an outsourcing algorithm for device-to-edge machine learning model training. This algorithm is a good compromise between privacy-preserving and execution efficiency. For example, deep learning is adjusted and compressed to reduce complexity, and high-complexity computing tasks are deployed at the edge of the network. The device only needs to perform some relatively simple operations to complete the entire model training process to appropriately reduce the network time delay via the effective use of computing resources, While to ensure the security of the data and the correctness of the calculation results, the outsourced data is encrypted and replaced, and the existence of malicious service nodes is taken into consideration to ensure the correctness of the calculation results of rational computing nodes. This paper also adds a trust mechanism to further increase the security of the system.

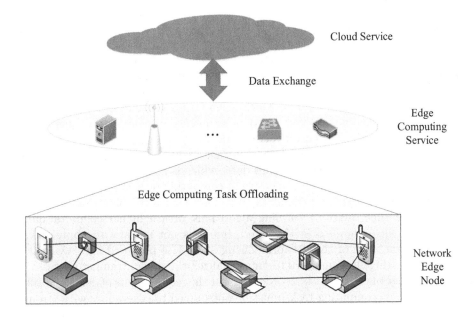

Fig. 1. Schematic diagram of data processing and calculation using edge computing.

The contributions of this article are as follows:

1. To solve the high calculation and high storage pressure caused by local machine learning algorithms on the device (especially mobile devices), a method called EVPP is proposed to outsource the computing part of the training process.
2. To solve the problems of high latency and network transmission pressure in outsourced computing, a near-local outsourcing algorithm is proposed in conjunction with edge computing, and concurrently, a cryptographic device is designed to solve the privacy and security problems brought by data outsourcing, a random matrix calculation scheme is introduced to randomly perturb the calculation data.

3. To prevent the dishonest outsourced computing nodes from affecting the training process, a trust mechanism with the arbitration function is proposed, which can guarantee the correctness of the calculation results of rational outsourced computing nodes.

The organizational structure of this paper is as follows: the first section will briefly introduce the related research, the second section will further describe the problems and challenges studied in this paper, and the third section will discuss the scheme and its algorithms in detail. The fourth section will focus on the security and performance proofs of the proposed goals in this paper. In the last section, the scheme will be summarized, and future research directions will be discussed.

2   Related Work

Smarter healthcare [6–8], urban transportation [14,15], connected cars [16], social networks [17] and other scenarios are increasingly applying machine learning algorithms for prediction and analysis. In these application scenarios, the data are outsourced to the cloud, which has a strong computing power, so that resource-constrained devices can use the cloud centre to complete various complex computing tasks and better serve users [18].

However, with the advent of the Internet of Everything, the edge of the network no longer generates abundant data, which is not suitable for processing in the cloud centre from the perspective of computing or network transmission. Applying edge computing technology is a good choice to solve this problem. At the same time, machine learning algorithms with a higher computational complexity can be applied in the cloud, but it is not realistic to apply them directly on the edge of the network. Zhang et al. [19] proposed the "OpenEI" open-edge intelligent framework to "marginalize" the model training process.

At the edge of the network, in order to meet the requirements of delay and efficiency, complex computing tasks must be outsourced to edge nodes with strong computing capabilities, but these nodes are often untrustworthy. To ensure the security and privacy of the data, the data must be encrypted and calculated in the ciphertext domain. This work has become more mature in cloud computing. For example, secure multiparty computing can be applied [20–22], including homomorphic encryption [23], differential privacy [24], and attribute-based encryption [12], but these solutions are not friendly to edge devices with low computing and storage capabilities.

Since determinant and matrix calculations are widely used in the fields of science and engineering, especially in various AI algorithms, there have been many studies on outsourcing calculations of determinants and matrixes [25–28]. Salinas et al. [25], a large-scale deterministic secure outsourcing computing solution was proposed. The client can effectively verify the correctness of the calculation results of the outsourced data. Chen et al. [29], a scheme for scrambling the original matrix data using diagonal matrix multiplication was proposed to ensure the security of the data; subsequently, it was improved in Zhou et al. [30], and it must be further improved in terms of security and result verification. Hu et al. [31], a matrix outsourcing inversion matrix scheme that can be applied to cloud computing and other scenarios was proposed, which effectively reduces the computational complexity of the client.

3   Problem Description and Research Goals

3.1   Research Goals and Challenges

In this paper, to better solve the computing and privacy issues of edge devices and ensure the security and accuracy of computing, four research goals are proposed.

- Privacy: These data contain tremendous user identity information, privacy data, etc. The collection and processing of these data are extremely prone to leakage of information. Therefore, the outsourcing of data computing needs to ensure the privacy of the data.
- Verifiability: Due to the instability of the system, network or computing nodes, the nodes to which the data are outsourced should be assumed to be incompletely trusted or even malicious. They may steal or peek at the user's data; furthermore, the operation may not be performed in accordance with the protocol at all, and the wrong calculation result for the user is returned, leading to the failure of the entire data training. Therefore, the calculation result of the data should be verifiable.
- High efficiency: In the whole process, the user's calculation amount in the outsourced calculation process should be lower than the entire operation performed by the device itself, otherwise the outsourced operation will be useless.
- Accuracy: This requires the design of the entire system to ensure that the calculation results can be guaranteed under the premise of the correct operation at each stage.

3.2   System Model

To ensure the security and availability of the system and achieve the research goals proposed in Sect. 3.1, this paper designs an outsourcing model training scheme based on edge computing, as shown in Fig. 2:

In the solution, the system is divided into three layers (the cloud computing problem is not considered here), which are the sensor node layer (or data acquisition layer), the edge node layer, and the edge service layer.

The sensor node layer is responsible for collecting data, but because of its poor computing and storage capabilities, the collected data cannot be calculated, organized, and stored. The collected data must be transmitted to the edge node layer of other networks for processing. For example, smart phones are implanted with several sensors, and these sensors transmit data to the mobile phone's computing unit for processing and storage. To ensure the availability of data and the security of users' data, the edge node layer mainly guarantees the data cleaning,

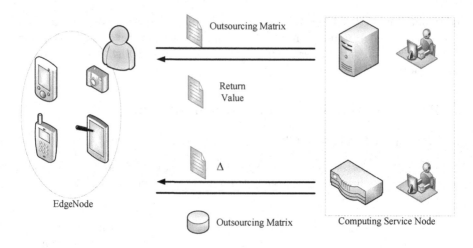

Fig. 2. Training model of the outsourcing model based on edge computing.

calculation, and storage tasks of sensor devices. Concurrently, in order to ensure the timeliness and accuracy of calculations, it also performs some calculation tasks to offload the edge service layer. The main task of the edge service layer is to assist the devices in the edge node layer to perform collaborative computing. However, due to the difficulty in ensuring security, there are the following risks: (1) the layer may peep and steal data, leading to information leakage; (2) it may not complete the computing task as agreed, causing the computing task of the edge node to fail.

As shown in Fig. 2, the solution proposed in this paper includes edge service nodes that need to outsource computing tasks and edge service nodes that assist edge nodes to outsource the computing tasks. Edge service nodes include two parts that assist users in key generation calculations, and an edge server that assists users in computing tasks.

3.3 Linear Regression and Gradient Descent

There are many optimization and learning algorithms in machine learning and deep learning. Most of these algorithms are based on matrix calculations and training models. During the training phase, the device performs a large number of matrix multiplications and additions. Through the analysis of the corresponding algorithm, it is not difficult to find that the number of multiplication operations is higher than that of addition operations. At the same time, the computational complexity of multiplication is higher than addition. Linear regression and gradient descent methods are more common methods. Therefore, in order to facilitate the description of the scheme, this paper uses gradient descent to optimize the model in the linear regression problem and finally obtain the training model. At the same time, describe the main ideas of the scheme.

Given the n sample set (X, Y) where the i-th sample X_i contains d features, that is, $X_i = (x_1, x_2, ..., x_d)$, adjust the objective function $h(X_i) = x_1 \cdot w_1 + x_2 \cdot w_2 + ... + x_d \cdot w_d = X_i \cdot w^T$, where $w = (w_1, w_2, ..., w_d)$.

Adjust the parameters through the training process to yield the appropriate to make $J_i(w) = (\frac{1}{2n} \sum_{i=1}^{n} (h(X_i - Y_i)^2))$ the smallest, that is, $Y_i \approx h(X_i)$.

The gradient descent method is widely used in machine learning to solve optimization problems. When targeting linear regression problems, for the j-th feature of X_i, the weight is $w_j = w_j - \alpha \frac{\partial J_i(w)}{\partial w_i} = w_j - \alpha \frac{1}{n} \sum_{i=1}^{n} (f(X_i) - Y_i) X_i^j$, and by representing vectors as a matrix, w can be expressed as:

$$w := w - \alpha X^T \times (X \times w - Y) \text{ [21]}.$$

Among them, α is the learning rate or step size, which is a fixed value, and this parameter determines the convergence degree of the algorithm; Y is a vector of $n \times l$ dimensions, that is, a given data tag set.

In the gradient descent method, because all the samples are used for training at one time, it will cause pressure on the memory and calculations, so $|B|$ samples are selected for small batch training. In the formula, $|B|$ is the amount of data, and $w := w - \alpha \frac{1}{|B|} X_B^T \times (X_B \times w - Y_B)$ [21].

Therefore, this paper decomposes the matrix calculations with abundant calculations. Among them, suppose

$$\begin{aligned}
\Delta &= X_B^T \times (X_B \times x - Y_B) \\
&= X_B^T \times (X_B \times x + (-Y_B)) \\
&= X_B^T \times (X_B \times x) + X_B^T \times (-Y_B)
\end{aligned} \tag{1}$$

That is, the update formula can be expressed as:

$$w := w - \alpha \frac{1}{|B|} \Delta \tag{2}$$

3.4 System Framework

The main idea of our solution is shown in Fig. 3, which includes the following five parts:

Step 1 (outsourced data generation algorithm): The client constructs a reversible matrix D for scrambling the data matrix, generating a random matrix, and randomly generating a verification matrix. Concurrently, the client has a training data set. The sample set X contains n samples x_i. Each sample set can be represented as an m-dimension vector, and the tag set is represented as Y.

The client calculates the confusion matrix forms C_1, C_2, C_3, and C_4 corresponding to the sample set X and its transposed matrix X^T, tag set Y, and initialization weight matrix w, and sends the data and corresponding calculation rules $f(C')$ to the edge service layer for calculation. At the same time, the matrix verification block is calculated and saved to facilitate subsequent result verification work.

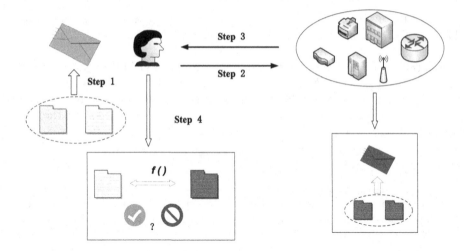

Fig. 3. The main design ideas of the scheme.

Step 2 (outsourcing data calculation algorithm): The edge service layer node outputs the calculation result Δ^* according to the outsourcing calculation rule $f(C')$ sent by the client and sends the calculated result back to the client.

Step 3 (training result generation algorithm): The client receives the calculation result Δ^* sent back by the edge server layer node and performs a recovery operation based on the information held locally to obtain the calculation result Δ. Then, the result is brought into Formula (1) and calculated, obtaining w_t. By comparison, $w_t < w_{t-1}$ indicates that the function has not reached the convergence value, the scrambling operation is performed, and return to Step 2 to continue training; $w_t > w_{t-1}$ indicates that the function has reached the convergence value, which terminates this calculation task and enters Step 5.

Step 4 (data verification algorithm): The client receives the calculation result Δ^* returned by the edge server layer node, extracts the verification matrix block V^* from it, and compares it with $V? = V^*$. When they are equal, the result indicates that the edge service layer node has performed the calculation operation correctly, otherwise it indicates that it has not faithfully calculated the outsourcing task; the client retains the test evidence, publishes the evidence and the identity of the edge service layer node, and executes Step 5 to find new computing-nodes.

Step 5 (End the calculation task): When the function reaches the convergence value, the client sends W^0 to the edge service layer node. When the edge service layer node receives the message, it knows that the calculation task is terminated, and the edge service layer node clears all relevant data. (This step may be performed in the following two situations: (1) The protocol execution process is normally completed, and (2) it is found that the edge server does not faithfully execute the calculation protocol. Therefore, it is not identified in Fig. 3).

4 System Solutions

In this section, the application scenario of EVPP is introduced in detail. The solution takes the gradient descent method as an example to achieve the four goals required by the previously described.

4.1 Encryption and Decryption Methods for Outsourced Data

In this section, we will describe the construction, encryption, and decryption processes of Formula (1) and Formula (2) in the scheme.

To ensure the security of the data and the simplicity of the result verification, the training data is encrypted:

The edge node encrypts $m \times n$ data X, $m \times l$ data Y, and $n \times l$ data w to ensure data security and then operates the matrix according to the following methods:

The edge nodes randomly generate $m \times t$ order matrixes M_1 and M_3; $n \times t$ order matrixes M_2 and M_4 ; four randomly generated t order matrixes V_1, V_2, V_3, and V_4; randomly select the diagonal matrix R; and construct the reversible matrix D. Finally, we can obtain the outsourcing matrix:

$$(X^T)'_{(n+t) \times (m+t)} = \begin{bmatrix} X^T D^{-1} M_2 \\ 0 \quad V_2 \end{bmatrix} \tag{3}$$

$$w'_{(n+t) \times (i+t)} = \begin{bmatrix} R(w)^T M_3 \\ 0 \quad V_3 \end{bmatrix} \tag{4}$$

$$Y'_{(n+t) \times (i+t)} = \begin{bmatrix} R(Y)^T M_4 \\ 0 \quad V_4 \end{bmatrix} \tag{5}$$

$$X'_{(m+t) \times (n+t)} = \begin{bmatrix} DX \quad M_1 \\ 0 \quad V_1 \end{bmatrix} \tag{6}$$

The construction process of the invertible matrix D will be described in detail [30]:

The invertible matrix $D = D_1 + D_2$ and the matrixes D_1, D_2 and D are all square matrixes of order $m \times m$. The matrix D_1 is a random diagonal matrix, in which $a_{11} = 0$, and other diagonal position element values are randomly selected from the edge nodes. In the matrix $D_2 = P^T Q$, where $P = (p_1, p_2, ..., p_m)$ and $Q = (1, q_1, q_2, ..., q_{m-1})$. D will be shown below as an invertible matrix.

Proof. First, the matrixes D_1, D_2, and D are all square matrixes of $m \times m$.

Second, D_1 is a diagonal matrix, $D_2 = P^T Q$, and $D = D_1 + D_2$. Therefore,

$$D = \begin{bmatrix} p_1 & q_1 p_1 & q_2 p_1 & \cdots & q_{m-1} p_1 \\ p_2 & q_1 p_2 + d_2 & q_2 p_2 & \cdots & q_{m-1} p_2 \\ p_3 & q_1 p_3 & q_2 p_3 + d_3 & \cdots & q_{m-1} p_3 \\ \vdots & \vdots & \vdots & \ddots & \vdots \\ p_m & q_1 p_m & q_2 p_m & \cdots & q_{m-1} p_m + d_m \end{bmatrix} \tag{7}$$

Subtract the first row from 2 to m of the matrix D to yield

$$D = \begin{bmatrix} p_1 & q_1p_1 & q_2p_1 & \cdots & q_{m-1}p_1 \\ 0 & d_2 & 0 & \cdots & 0 \\ 0 & 0 & d_3 & \cdots & 0 \\ \vdots & \vdots & \vdots & \ddots & \vdots \\ 0 & 0 & 0 & \cdots & d_m \end{bmatrix} \tag{8}$$

Finally, the unit matrix I can be obtained from the above matrix through a further elementary transformation, so it can be easily obtained that the matrix D must be an invertible matrix.

The matrix inversion process can be performed with reference to [31].

Next, we will demonstrate how to ensure that the calculation results are recoverable and verifiable in the matrix.

For the formula $\Delta = X_B^T \times (X_B \times w) + X_B^T \times (-Y_B)$, the method of matrix block construction can be obtained: $\Delta' = (X^T)' \times ((X)' \times (w)') + (X^T)' \times (-Y')$. Multiply two block matrixes, for example:

$$\begin{aligned} C_2 \times C_1 &= \begin{bmatrix} X^T D^{-1} & M_2 \\ 0 & V_2 \end{bmatrix}^T \times \begin{bmatrix} DX & M_1 \\ 0 & V_1 \end{bmatrix} \\ &= \begin{bmatrix} X^T D^{-1} DX & X^T D^{-1} M_1 + M_2 V_1 \\ 0 & V_2 V_1 \end{bmatrix} \end{aligned} \tag{9}$$

It is not difficult to see from the result of the matrix multiplication that the upper left part of the matrix is equivalent to solving $X^T \times X$; the edge node can calculate $\det(V) = \det(V_1 V_2 V_3 - V_2 V_4)$ as the verification factor for the outsourced calculation, which indirectly proves the correctness of the calculation result.

4.2   Description of Scheme

In this section, Algorithm 1 and Algorithm 2 will be described according to the algorithm described in Sect. 4.1, and the initialization process of the scheme has been completed. Use $f()$ to represent the calculation rules for outsourced data. In this paper, $f()$ uses the Δ of Sect. 3.4 to define the calculation rules.

Step 1 (outsourced data generation algorithm): The client constructs a reversible matrix D for scrambling the data matrix, generates a random matrix M_1, M_2, M_3, M_4, and randomly generates a verification matrix V_1, V_2, V_3, V_4; R is a diagonal matrix. Concurrently, the client has a training data set. The sample set X contains n samples x_i. Each sample set can be represented as a d-dimension vector, and the tag set is represented as Y.

Outsource the invertible matrix D to obtain D^{-1}.

The client can obtain the sample set X and its transposed matrix X^T, the tag set Y, and the initial w confusion matrixes C_2, C_1, C_3, and C_4 through

---

Algorithm 1. Outsourced data generation algorithm

Input: Key k, Invertible matrix D_1, P^T, Q, Random matrixes (M_1, M_2, M_3, and M_4), Random verification matrixes (V_1, V_2, V_3, and V_4), Sample set X, Tag set Y, and the initialization vector w.
Output: Outsourcing matrix C_1, C_2, C_3, C_4.

1: Calculate $D = D_1 + D_2$;
2: Goto the inverse matrix algorithm is outsourced to obtain the inverse matrix D^{-1};
3: The edge computing node calculates the key matrix $K \leftarrow kI$;
4: Initialize the vector for scrambling (perturbation) $w' \leftarrow Kw$ and tag set $Y' \leftarrow KY$;
5: Design calculation rules $f()$;
6: According to the calculation rules, solve the verification factor $\det V$;
7: Construct the sample set $X' \leftarrow DX$ and its transpose matrix $(X^T)' \leftarrow X^T D^{-1}$;
8: **return** $C_2 \leftarrow ((X^T)'||M_2||V_2)$, $C_1 \leftarrow (X'||M_1||V_1)$, $C_3 \leftarrow (w'||M_3||V_3)$,
   $C_4 \leftarrow (Y'||M_4||V_4)$;

Algorithm 2. Outsourcing data calculation algorithm

Input: Key k, Matrix C_1, C_2, C_3, C_4, and Calculation Rule $f()$.
Output: Calculation result Δ^*.

1: Calculation $\Delta^* = C_2 C_1 G_3 + C_2(-C_4)$
2: Send the calculated result Δ^* to the edge computing node.

Algorithm 1. The constructed calculation rules and confusion matrix is sent to the edge service node for outsourced calculation.

Step 2 (outsourcing data calculation algorithm): The edge service layer node outputs the calculation result Δ^* according to the outsourcing calculation rule $f(C')$ sent by the client (the calculation rule is based on the gradient descent method for linear regression) and send the calculation result to the client.

Step 3 (training result generation algorithm): The client receives the calculation result δ^* returned by the edge server layer node and executes Algorithm 3. In this section, to better explain the application process of the algorithm, Formula (1) will be taken as an example.

Step 4 (data verification algorithm): After the client receives the calculation result Δ^* returned by the edge server layer node, it verifies the calculation result.

Step 5 (end the calculation task) When the function reaches the convergence value or the edge service node calculation task fails, the edge calculation node executes this step algorithm.

When the edge computing node checks and finds that there is an error in the calculation result returned by the edge service node, Algorithm 6 is executed. This algorithm is used to generate evidence that the edge service node has not faithfully performed the model training task according to the protocol, thereby announcing that the node is untrustworthy and building a system trust mechanism.

When other nodes verify the security of the edge service node, the verification matrix is extracted from the evidence $E_{u \rightarrow s}$, and the corresponding results are obtained according to the calculation rules to determine whether the evidence is

---

Algorithm 3. Training result generation algorithm

1: $w_t^* \leftarrow \Delta^*$, extract submatrix part of the matrix w_t^*;
2: Goto Algorithm 4;
3: **if** $w_t' = w_t^*$ **then**
4:      The results returned by outsourced calculations are true;
5:      **if** $w_t' < w_{t-1}'$ **then**
6:          The function does not reach the convergence value and generates a new validation factor V_t to replace the validation factor in Δ^*;
7:          Perform scrambling operation to generate $\Delta^* \leftarrow w_t'$;
8:          Goto Step2;
9:      **else**
10:          The function reaches a convergence value;
11:      **end if**
12: **end if**
13: $w_t = K^{(-1)}W_t'$;
14: Goto Step 5.

Algorithm 4. Data verification algorithm

Input:Key: k, Calculation result Δ^*.
Output: Validation result $V? = V^*$.

1: Extract validation matrix blocks $V^* \leftarrow \Delta^*$;
2: Calculation $\det(V^*)$;
3: **if** $\det(V) = \det(V^*)$ **then**
4:      **return** "*True.*"
5: **else**
6:      Goto Step5 and Step6;
7:      Find new computing nodes and perform model training tasks;
8: **end if**

---

valid. When the verification node records the verification result, a trust record is built locally.

5 System Analysis

5.1 Security

The security of the solution is considered from the following aspects: data security and privacy, the correctness of the calculation results, and the trust mechanism of edge service nodes.

1. The correctness of calculation results

When calculating a block matrix, it can be divided into two parts: matrix addition and matrix multiplication for analysis and consideration.

Algorithm 5. End the calculation task

1: The client will send W^0 to the edge server node;
2: When the edge service layer node receives W^0, it learns that the computing task is terminated;
3: The edge service layer node deletes the computing data related to the training task.

Algorithm 6. Evidence generation and adjudication algorithm

Input:Key: k, Validation matrixes V_1, V_2, V_3, V_4, and Calculation Rule $f()$.
Output: Evidence E.

1: Generate evidence's signature $S \leftarrow H(V_1 \| V_2 \| V_3 \| V_4 \| V^* \| H_u(ID) \| H_s(ID))$;
2: Generate evidence $E_{u \rightarrow s} \leftarrow (V_1 \| V_2 \| V_3 \| V_4 \| V^* \| f() \| S)$.

First, verify the correctness of matrix multiplication:
In the calculation of $(X^T)' \times X'$, the result is

$$\begin{bmatrix} X^T D^{-1} DX & X^T D^{-1} M_1 + M_2 V_1 \\ 0 & V_2 V_1 \end{bmatrix}. \tag{10}$$

And the result in the upper right corner of the matrix is not difficult to see as $X^T \times X$. For a further description of the calculation with w, it can be simplified to $\begin{bmatrix} X^T D^{-1} DX(Rw^T) & \cdots \\ 0 & V_2 V_1 V_3 \end{bmatrix}$.

Since R is a diagonal matrix, it is assumed that k is randomly selected as the diagonal element value of the matrix. The matrix element of $X^T \times X$ is x_{ij}. The calculation result of Formula (1) indicates

$$\begin{bmatrix} \sum_{j=1}^n \sum_{i=1}^n x_{1j} k w_{i1} & \cdots & \sum_{j=1}^n \sum_{i=1}^n x_{1j} k w_{il} \\ \vdots & \ddots & \vdots \\ \sum_{j=1}^n \sum_{i=1}^n x_{nj} k w_{i1} & \cdots & \sum_{j=1}^n \sum_{i=1}^n x_{nj} k w_{il} \end{bmatrix} \tag{11}$$

.

It is not difficult to see that multiplying the matrix by $\frac{1}{k}I$ can restore the original data.

The correctness of the matrix addition is relatively simple and will not be repeated; the reader can prove it by himself.

2. Proof of algorithm security

In this section, the security of this scheme is demonstrated under five assumptions of insecurity.

– Hypothesis 1: Malicious users obtain data through intermediate parameters.

In the embodiment, the data used for the matrixes X, Y, and w are used for the matrixes randomly perturbing invertible matrix operation, and the elements in the invertible matrix D are randomly generated and have no correlation.

Therefore, the service edge node cannot guess any information data from the matrix, to ensure the privacy of the data. At the same time, both the target matrix and the intermediate parameters are the result of adding K to the disturbance, and only users who master K can restore the data.

For $\Delta = X_B^T \times (X_B \times w) + X_B^T \times (-Y_B)$, there is a transformation of $X^T D^{-1} DX$, which guarantees data security and recovery.

Meanwhile because of the learning rate (step), use the data to set the size of the initial parameters, and other related parameters of the system are performed locally. Therefore, the program also ensures the safety system model in the training process, the iterative training process is completed, and the edge service node only grasps the intermediate value of, thus, ensuring the safety training model.

This paper only describes the solution with a linear regression model. When other machine learning algorithms are needed, only the calculation process needs to be improved.

- Hypothesis 2: The malicious edge server can recover the intermediate parameter through the inverse matrix.

In the solution described in this paper, all user data is added with disturbance and encapsulation, and edge servers cannot know the true meaning of their calculation data. Here, we assume a more powerful enemy that can conspire with the edge server that computes the inverse matrix $(D^{-1})'$ (this is not truly an inverse matrix). In this case, although the malicious server obtains the intermediate value of the relevant inverse matrix D^{-1}, it cannot recover the accurate inverse matrix information, and it also cannot restore the original data. After generating the inverse matrix, the inverse matrix returned by the server contains the data disturbance. Therefore, except for the owner of the data, it is difficult for others to restore the original matrix and its inverse matrix.

- Hypothesis 3: The two edge servers conspire to get user data.

In this method, the biggest threat is that the two edge service providers recover the data by obtaining the inverse matrix D^{-1} information of the user, and other methods cannot recover the data. Similar to Hypothesis 2, due to a matrix calculated by one of the servers. The matrix contains inverse matrix and disturbance information, and to restore the true inverse matrix, adversary need to grasp the parameters of its disturbance information. Therefore, the final inverse matrix D^{-1} cannot be obtained between these two edge servers, so this scheme is already safe under this assumption.

- Hypothesis 4: Trust mechanism of edge service nodes.

In the system, this paper introduces an arbitration mechanism. When the edge node detects that the calculation result returned by the edge service node is abnormal, that is, the returned result is inconsistent with the verification result, then the edge node considers labelling the edge service node as malicious.

The edge node verifies the information, including the returned result v, the raw data used to generate the verification results (V_1, V_2, V_3, and V_4), and the relevant identity information $H(ID_u)|H(ID_s)$ of the edge node and the service node, and then generates evidence and publishes it to other nodes. When each node in the system receives the evidence, it performs a test. If it is indeed proved that the edge service node has not performed calculation tasks according to the agreement, it will be added to the blacklist, and the data will no longer be outsourced to the node. Concurrently, to ensure the long-term effectiveness of the system, you can look for trusted authorities (for example, government agencies) to store evidence and maintain node information in the system.

– Hypothesis 5: Adversary obtains enough ciphertext for analysis.

The scheme in this paper is to protect the security and privacy of the sample data X, tag data Y and parameter (weight) information w. Ensure that the adversary can assist the user in the calculation task, and does not obtain any data information.

In the scheme, the invertible matrix D and its inverse matrix D^{-1} are the key to ensure the computability and safety of X and X^T. It has been proved in Hypothesis 1 and 2 that it can guarantee the security of the data. However, long-term use of the same set of matrices (including the invertible matrix D and its inverse matrix D^{-1}) is a security risk. The adversary may recover the original data contained in the disturbance data by collecting a large number of matrices (for example, C_1, C_2). In turn, threatens the security of the program. Therefore, the frequency of use of D and D^{-1} can be adjusted to achieve better data security. The safest solution is to use different D and D^{-1} each time. Of course, this will increase the computational cost of the initialization phase (Algorithm 1 and Algorithm 2). The computational cost of the initialization phase will be discussed in the experimental part.

5.2   System Performance Analysis

For the performance of the system, this paper uses a real data set for comparison experiments (It is not difficult to see through a theoretical analysis that the calculation results generated by using this solution are consistent with the original calculation results, so the comparison of the calculation results will not be performed here).

The experimental parameters are as follows: CPU I5-2450M (2.5 GHz), 8 GB of memory, and Windows 10 64-bit operating system. The system is implemented using Python language, and because some edge devices may not be equipped with a GPU, this paper does not use a GPU to accelerate processing.

In this paper, we first use the Boston house price prediction data set for experiments. The data set is 507 * 13. During the experiment, 400 pieces of data are selected for training. The average of 15 test results is used to determine the final experimental data results. The experimental test results are shown in Fig. 4: the method proposed in this paper takes 24.76 ms, and the general training

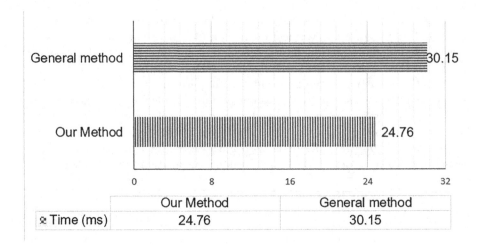

Fig. 4. Comparison of time consumptions for the Boston house price prediction dataset.

process takes 30.15 ms. From the results, it can be seen that the method in this paper is approximately 18% faster than the general method (In this paper, it refers to the general machine learning algorithm, that is, linear regression algorithm).

The feature dimensions of the Boston house price prediction data set are small, as there are only 13 features. Afterwards, this paper randomly generates a data set of size 500, 600, 700, etc., which has 300 features and 200 training rounds. The experimental results obtained are shown in Fig. 5:

It can be seen from Fig. 5 that the solution proposed in this paper can effectively reduce the calculation amount of edge computing nodes. Among devices with limited resources, the system will effectively reduce the computing pressure of the device and alleviate the consumption of local resources.

From the experimental results, the scheme in this paper saves time by approximately 20%, when compared with the unoptimized scheme. The machine learning algorithm used in this paper is relatively basic and has a small number of calculations. The results have a certain advantage in terms of time consumption. Since the main computing tasks in the training process are outsourced to other service nodes, appropriate adjustments can be applied to machine learning algorithms with more complex training processes. In general, the solution in this article is suitable for application scenarios where the edge device cannot execute or is difficult to handle.

At the same time, as shown in Fig. 6, this article also compares with homomorphic encryption. By using a homomorphic encryption scheme to perform addition operations and multiplication operations of the same order of magnitude, this method can be seen to be more suitable for mobile devices in terms of efficiency. It is worth noting that, because the addition homomorphic encryption

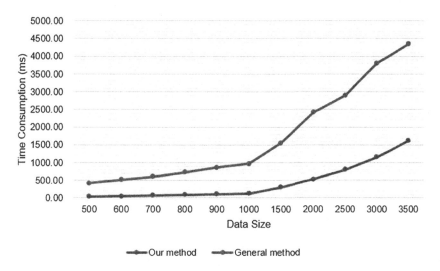

Fig. 5. Comparison of the efficiencies of the proposed scheme and the general method.

time based on Paillier cryptography runs far longer than does the scheme in this paper during the addition operation of the same order of magnitude (data is not shown in the figure).

In the initialization phase of the algorithm, the initialization time of the multiplicative homomorphic encryption based on RSA is approximately 45.4 ms, while the initialization time of the additive homomorphic encryption based on Paillier is approximately 1687.8 ms; the initialization time of the method in this paper varies with the amount of data formed by the matrix. When processing the same amount of data, the scheme in this paper takes the shortest time of time at this stage.

Through the experimental comparison and theoretical analysis, it can be seen that, by comparing with the scheme without any encryption and data perturbation, the scheme in this paper significantly improves the execution efficiency and other aspects, as well as ensured the security and accuracy of the data. Compared with the homomorphic encryption system, the security of this solution is weaker than the homomorphic encryption technology, but the execution efficiency is more suitable for devices with lower computing capabilities. This solution can enable low computing power equipment to process larger machine learning algorithms, while ensuring their safety and accuracy.

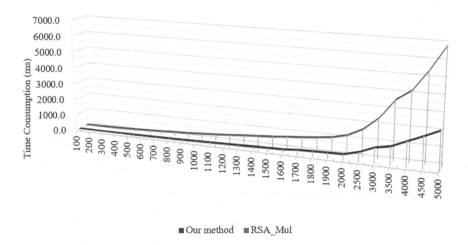

Fig. 6. Comparison of the efficiencies of this scheme and RSA multiplication homomorphic encryption.

6 Conclusion

With the continuous improvement in mobile device capabilities and the need for low-latency applications, computing tasks will increasingly be migrated locally, but this also brings issues such as device energy consumption, occupied device computing, and storage. Therefore, outsourcing data to a near local end can reduce network transmission delays and reduce pressure on mobile devices. However, the security and privacy issues arising during the outsourcing process cannot be ignored. Therefore, this paper proposes EVPP: a secure data outsourcing computing solution based on matrix operations. The outsourcing matrix adds a lightweight verification factor so that the verification process does not place excessive computing pressure on the mobile device. In terms of the trust mechanism, when the device finds a malicious service node in the system, it can verify the data generates arbitration evidence so that other nodes in the system can verify and avoid sending outsourced tasks to malicious nodes. Through a theoretical analysis and experimental comparison, it can be verified that the scheme exhibits certain improvements in efficiency, safety and correctness and can be applied to practical applications. Because the solution in this article only optimizes the training process, it has certain limitations. To better reduce the complexity of machine learning models for equipment training in edge environments, the next step will be combining with federated learning [32] to conduct a further study in the distributed system.

References

1. Petrov, C.: Big data statistics (2019). https://techjury.net/stats-about/big-data-statistics/. Accessed 22 Mar 2019

2. Zhang, X., Qiao, M., Liu, L., et al.: Collaborative cloud-edge computation for personalized driving behavior modeling. In: Proceedings of the fourth ACM/IEEE Symposium on Edge Computing (SEC). ACM/IEEE, Washington (2019)
3. Jia, K., Li, H., Liu, D., et al.: Enabling efficient and secure outsourcing of large matrix multiplications. In: 2015 IEEE Global Communications Conference (GLOBECOM), pp. 1–6. IEEE, San Diego (2015)
4. Lei, X., Liao, X., Huang, T., et al.: Achieving security, robust cheating resistance, and high-efficiency for outsourcing large matrix multiplication computation to a malicious cloud. Inf. Sci. **280**, 205–217 (2014)
5. Li, P., Li, J., Huang, Z., Gao, C.-Z., Chen, W.-B., Chen, K.: Privacy-preserving outsourced classification in cloud computing. Cluster Comput. 1–10 (2017). https://doi.org/10.1007/s10586-017-0849-9
6. Abdellatif, A.A., Mohamed, A., Chiasserini, C.F., et al.: Edge computing for smart health: context-aware approaches, opportunities, and challenges. IEEE Netw. **33**(3), 196–203 (2019)
7. Pathinarupothi, R.K., Durga, P., Rangan, E.S.: IoT-Based smart edge for global health: remote monitoring with severity detection and alerts transmission. IEEE Internet Things J. **6**(2), 2449–2462 (2018)
8. Liu, X., Deng, R.H., Choo, K.R., et al.: Privacy-preserving reinforcement learning design for patient-centric dynamic treatment regimes. IEEE Trans. Emerg. Top. Comput. 1 (2019). (Early Access)
9. Chen, S., Wen, H., Wu, J., et al.: Radio frequency fingerprint-based intelligent mobile edge computing for internet of things authentication. Sensors **19**(16), 3610 (2019)
10. Froiz-Míguez, I., Fernández-Caramés, T., Fraga-Lamas, P., et al.: Design, implementation and practical evaluation of an IoT home automation system for fog computing applications based on MQTT and ZigBee-WiFi sensor nodes. Sensors **18**(8), 2660 (2018)
11. Shi, W., Cao, J., Zhang, Q., et al.: Edge computing: vision and challenges. IEEE Internet Things J. **3**(5), 637–646 (2016)
12. Li, Q., Zhu, H., Xiong, J., Mo, R., Ying, Z., Wang, H.: Fine-grained multi-authority access control in IoT-enabled mHealth. Ann. Telecommun. **74**, 389–400 (2019). https://doi.org/10.1007/s12243-018-00702-6
13. Chui, K.T., Liu, R.W., Lytras, M.D., et al.: Big data and IoT solution for patient behaviour monitoring. Behav. Inf. Technol. **38**, 1–10 (2019)
14. Liang, X., Du, X., Wang, G., et al.: A deep reinforcement learning network for traffic light cycle control. IEEE Trans. Veh. Technol. **68**(2), 1243–1253 (2019)
15. Zhou, P., Braud, T., Alhilal, A., et al.: ERL: edge based reinforcement learning for optimized urban traffic light control. In:2019 IEEE International Conference on Pervasive Computing and Communications Workshops (PerCom Workshops), pp: 849–854. IEEE, Mannheim (2019)
16. Joo, J., Park, M.C., Han, D.S., et al.: Deep learning-based channel prediction in realistic vehicular communications. IEEE Access **7**, 27846–27858 (2019)
17. Feng, B., Fu, Q., Dong, M., et al.: Multistage and elastic spam detection in mobile social networks through deep learning. IEEE Netw. **32**(4), 15–21 (2018)
18. Liu, X., Deng, R.H., Choo, K.R., et al.: Privacy-preserving outsourced support vector machine design for secure drug discovery. IEEE Trans. Cloud Comput. (2018). (Early Access)
19. Zhang, X., Wang, Y., Lu, S., et al.: OpenEI: an open framework for edge intelligence. arXiv preprint arXiv:1906.01864 (2019)

20. Sun, Y., Wen, Q., Zhang, Y., et al.: Two-cloud-servers-assisted secure outsourcing multiparty computation. Sci. World J. **2014** (2014)
21. Mohassel, P., Zhang, Y.: SecureML: a system for scalable privacy-preserving machine learning, In: 2017 IEEE Symposium on Security and Privacy (SP), pp. 19–38. IEEE, San Jose (2017)
22. Huang, K., Liu, X., Fu, S., et al.: A lightweight privacy-preserving CNN feature extraction framework for mobile sensing. IEEE Trans. Dependable Secure Comput. 1 (2019). (Early Access)
23. Vengadapurvaja, A.M., Nisha, G., Aarthy, R., et al.: An efficient homomorphic medical image encryption algorithm for cloud storage security. Proc. Comput. Sci. **115**, 643–650 (2017)
24. Piao, C., Shi, Y., Yan, J., et al.: Privacy-preserving governmental data publishing: a fog-computing-based differential privacy approach. Future Gener. Comput. Syst. **90**, 158–174 (2019)
25. Salinas, S., Luo, C., Chen, X., et al.: Efficient secure outsourcing of large-scale sparse linear systems of equations. IEEE Trans. Big Data **4**(1), 26–39 (2018)
26. Salinas, S., Luo, C., Chen, X., et al.: Efficient secure outsourcing of large-scale linear systems of equations. In: 2015 IEEE Conference on Computer Communications (INFOCOM), pp. 1035–1043. IEEE (2015)
27. Yu, Y., Luo, Y., Wang, D., et al.: Efficient, secure and non-iterative outsourcing of large-scale systems of linear equations. In: 2016 IEEE International Conference on Communications (ICC), pp. 1–6. IEEE (2016)
28. Lei, X., Liao, X., Huang, T., et al.: Cloud computing service: the caseof large matrix determinant computation. IEEE Trans. Serv. Comput. **8**(5), 688–700 (2014)
29. Chen, F., Xiang, T., Lei, X., et al.: Highly efficient linear regression outsourcing to a cloud. IEEE Trans. Cloud Comput. **2**(4), 499–508 (2014)
30. Zhou, L., Zhu, Y., Choo, K.K.R.: Efficiently and securely harnessing cloud to solve linear regression and other matrix operations. Future Gener. Comput. Syst. **81**, 404–413 (2018)
31. Hu, C., Alhothaily, A., Alrawais, A., et al.: A secure and verifiable outsourcing scheme for matrix inverse computation. In: IEEE INFOCOM 2017 IEEE Conference on Computer Communications, pp. 1–9. IEEE (2017)
32. Cheng, K., Fan, T., Jin, Y., et al.: SecureBoost: a lossless federated learning framework. arXiv preprint arXiv:1901.08755 (2019)

Support Vector Machine Intrusion Detection Scheme Based on Cloud-Fog Collaboration

Ruizhong Du[1,2], Yun Li[1(✉)], Xiaoyan Liang[1,2], and Junfeng Tian[1,2]

[1] Cyberspace Security and Computer College, Hebei University,
Baoding 071002, China
drzh@hbu.edu.cn, 15232045203@163.com
[2] Key Laboratory on High Trusted Information System in Hebei Province,
Baoding 071002, China

Abstract. Fog computing is a new computing paradigm in the era of the Internet of Things. Aiming at the problem that fog nodes are closer to user equipment, with heterogeneous nodes, limited storage capacity resources, and greater vulnerability to intrusion, a lightweight support vector machine intrusion detection model based on Cloud-Fog Collaboration (CFC-SVM) is proposed. Due to the high dimensionality of network data, first, Principal Component Analysis (PCA) is used to reduce the dimensionality of the data, eliminate the correlation between attributes and reduce the training time. Then, in the cloud server, a support vector machine (SVM) optimized by the particle swarm algorithm is used to complete the training of the dataset, obtain the optimal SVM intrusion-detection classifier, send it to the fog node, and carry out attack detection at the fog node. Experiments with the classic KDD CUP 99 dataset show that the model in this paper is better than other similar algorithms in regard to detection time, detection rate and accuracy, which can effectively solve the problem of intrusion detection in the fog environment.

Keywords: Cloud-fog collaboration · Intrusion detection · Support vector machine · Particle swarm optimization

1 Introduction

The rapid development of the Internet of Things technology, the popularity of 5G networks, the emergence of a large number of industrial Internets, and augmented reality/virtual reality have increased network transmission capacity requirements, data distribution processing capabilities and real-time performance [1]. A single cloud-computing architecture cannot meet heterogeneous, low-latency, dense network access and services. Fog computing [2] allocates computing, communication, control and storage resources and services to users or devices and systems close to users to extend the cloud-computing mode to the edge of the network and provide network users with low latency, more flexible access and more secure network communication services. Fog and cloud computing do not have a mutual replacement relationship but have complementary infrastructure.

© ICST Institute for Computer Sciences, Social Informatics and Telecommunications Engineering 2021
Published by Springer Nature Switzerland AG 2021. All Rights Reserved
D. Wang et al. (Eds.): SPNCE 2020, LNICST 344, pp. 321–334, 2021.
https://doi.org/10.1007/978-3-030-66922-5_22

Together, the two form a computing model that is more suitable for the application scenario of the Internet of Things.

Compared to traditional cloud computing, the new features of fog computing [3] also create entirely new security and privacy protection issues. First, due to the heterogeneity of fog nodes, the whole fog computing system, including network facilities, service facilities, virtualization facilities and user terminals, is jointly owned by multiple owners. The result of this situation is that any part of the whole fog environment can be the target of network attack and privacy theft. Second, since the nodes providing computing and storage capabilities are deployed on the side of the network that is closer to the user, the nodes mainly provide services to nearby users [4]. This deployment and service method is a double-edged sword. On the one hand, it limits the scope of the network attack to a smaller range. On the other hand, once the attacker successfully controls the node, the whole region that the node serves will face risks. Common network threats include denial-of-service attacks [5], man-in-the-middle attacks, and probe attacks. Intrusion detection is the second security barrier behind the firewall, which can quickly detect security risks in the network. Traditional single-intrusion detection schemes have problems such as slow detection speed and low accuracy and are no longer adapted to the requirements of fog computing and edge computing environments [6] in response speed and high real-time performance. Aiming at this problem, this paper proposes a lightweight intrusion-detection model based on cloud-fog collaboration.

We will first analyze the security issues of the fog environment. Based on the resource constraints of fog nodes, a cloud-fog collaboration support vector machine intrusion detection model is proposed. This design makes full use of the advantages of the fog node and the cloud server and trains the classifier for detection in the cloud. After training, the training model is sent to the fog node. Our contributions in this study can be summarized as follows.

(1)   Based on the classic three-layer architecture of fog computing, a lightweight intrusion detection model based on cloud-fog cooperative support vector machines is proposed.
(2)   The principal component analysis process is added in the data processing stage to reduce the complexity of high-dimensional data by eliminating the correlation between features to reduce the dimension, making the data more lightweight and adapting to the fog nodes with limited resource storage capabilities.
(3)   Particle swarm optimization is added to select SVM parameters to speed up training.
(4)   By comparing the performance of the proposed scheme with the ELM, AdaBoost, and decision tree classifiers, the superiority of the proposed scheme is verified.

The rest of the paper is organized as follows. In Sect. 2, we will review related works on fog computing security and intrusion detection. In Sect. 3, we propose a support vector machine intrusion detection scheme based on cloud-fog collaboration. In Sect. 4, we introduce the SVM-based intrusion detection algorithm for fog computing. Section 5 will describe the simulation to verify the algorithm. The paper is concluded in Sect. 6.

2   Related Works

Currently, there are many studies analyzing the security threats and development directions of fog computing and mobile edge computing. Among them, Elazhary et al. [7] analyzed the intersections between the prominent areas of cloud computing and mobile computing research in detail to make their definitions more standardized. Parikh et al. [8] compared the structural advantages of cloud computing, edge computing, and fog computing with the possible security threats. Saad et al. [9] pointed out that the fog computing system can process a large amount of data locally, is completely portable, and can be installed on heterogeneous hardware. These characteristics make the fog computing platform very suitable for processing time- and location-sensitive applications. [10, 11] proposed fog computing as a new computing paradigm that can provide computing, storage and network services between users and traditional cloud platforms and described in detail the security issues in several different scenarios of fog computing, such as intelligent instrument certification and MIM attacks.

In recent years, there have been many studies on the security of fog computing systems. Among them, Razouk et al. [12] proposed a secure middleware architecture suitable for fog computing that provided computing power and secure communication links. This architecture focuses on ensuring the security of the communication process, and threats from outside the system are not considered. Hosseinpour et al. [13] proposed a distributed lightweight intrusion detection system based on an artificial immune system (AIS) suitable for fog computing, but the research did not perform in-depth analysis on the detection rate and accuracy of the system. Peng et al. [14] proposed a decision tree intrusion-detection system based on the characteristics of a large amount of data generated in a fog environment. The system did not consider the real-time requirements of fog computing and the limited storage capacity of fog nodes. Zhou et al. [15] applied the concept of fog computing to DDoS mitigation, distributing traffic monitoring and analysis work to local devices, and coordinating and merging work to the cloud center server to achieve rapid response in the case of a low false alarm rate.

Many scholars have proposed intrusion-detection schemes based on the characteristics of fog nodes. Among them, An et al. [16] fully considered the advantages of cloud servers and fog nodes and proposed an intrusion-detection scheme based on sample selection. This scheme shortened the training time but did not consider the problem of the limited storage capacity of fog nodes. S. Prabavathy et al. [17] proposed an intrusion-detection scheme based on an extreme learning machine (ELM) for the IoT environment of fog computing, which improved the accuracy and detection rate of intrusion detection but lacked consideration of real-time detection.

The support vector machine (SVM) is a small sample of machine learning methods based on statistical learning theory that has the characteristics of fast training speed and strong generalization ability and is suitable for intrusion detection in fog environments. From the current research results [18–21], the SVM algorithm has shown good results in various types of classification systems.

3   Support Vector Machine Intrusion Detection Scheme Based on Cloud-Fog Collaboration

The intrusion behavior of fog computing mainly comes from the terminal device layer. Therefore, the intrusion detection of fog computing is the process of discriminating and classifying the network data from the terminal device layer to determine whether it is an intrusion behavior. The fog node has fast calculation speed and limited resource storage capacity, and it is difficult to store a large number of training samples. To make reasonable use of the resources in the fog computing system and effectively perform intrusion detection, a collaborative intrusion-detection scheme for cloud servers and fog nodes is designed. The process is shown in Fig. 1:

1) Establish a communication link: Due to the heterogeneity of user terminals, the fog node provides network connections to each terminal device through different protocols and collects the data generated by each terminal device in real time.
2) The cloud server preprocesses and trains the original data: The entire training set is stored in the cloud server, and the entire training process is completed in the cloud server to generate a training model and save it.
3) The fog node sends a detection instruction: After the fog node establishes a communication link with the terminal device, it collects a large amount of network data generated by the terminal device and sends a detection instruction to the cloud server.
4) The cloud server sends the training model: After receiving the detection instruction, the cloud server sends the data processing model and the trained classification detection model to the fog node.
5) Fog node detection process: After receiving the model, the fog node uses the model for data processing and detection and generates detection results.
6) Intrusion response: The detected abnormal data are sent to the intrusion response module, and the intrusion response module performs the corresponding processing.

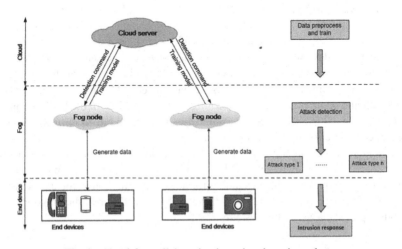

Fig. 1. Cloud-fog collaboration intrusion detection scheme

4 PCA and PSO-Optimized SVM for Intrusion Detection

4.1 PCA Data Dimensionality Reduction

Network data generally have high dimensional characteristics, resulting in a longer training time required for intrusion detection. Therefore, before training, dimension-reduction processing is performed on the training data to eliminate the correlation between attributes and maximize the retention of original data information. The core idea of principal component analysis (PCA) is to transform a set of variables that may be related to a set of linearly uncorrelated variables through orthogonal transformation. This paper uses PCA to make dimensional reduction on the training data.

This method can reduce the storage cost while removing the correlation of different dimensions. The contribution rate and cumulative contribution rate data of each eigen-value are shown in Fig. 2. The value of the main ordinate axis represents the proportion of the variance value of each principal component in the total square difference after dimension reduction. The larger the proportion is, the more important the principal component is. The value of the secondary ordinate axis represents the cumulative value of the proportion of the variance value of each principal component to the total square difference, that is, the cumulative contribution rate.

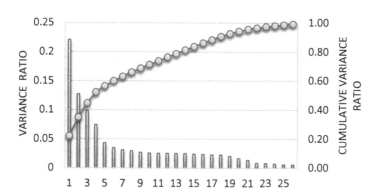

Fig. 2. Principal component contribution rate and cumulative contribution rate

4.2 SVM Algorithm

SVM (Support Vector Machine) was first proposed by Vapnik et al. The second kind is the problem that does not have linear separability. The SVM algorithm can use a kernel function to map the nonlinear separable dataset to high-dimensional space so that it has separability in high-dimensional space to achieve classification.

The calculation flow of the SVM algorithm is as follows: for linearly separable sample set $T = \{(x_1, y_1), (x_2, y_2), \cdots, (x_l, y_l)\}$, construct and solve the optimization problem of variables w and b:

$$\min J = \frac{\|w\|^2}{2} + C \sum_{i=1}^{l} \xi_i$$

$$\text{s.t. } y_i(wx_i + b) \geqslant 1 - \xi_i$$

$$\xi > 0_i, i = 1, 2, \cdots, l \tag{1}$$

where w is the weight vector, b is the threshold value, C is the penalty factor, which is used to punish the wrong samples, and ξ_i is the slack variable. Equation (7) is transformed into its dual problem by using Lagrange dual theory:

$$\max Q(\alpha) = \sum_{i=1}^{l} \alpha_i - \frac{1}{2} \sum_{i=1}^{l} \sum_{j=1}^{l} \alpha_i \alpha_j y_i y_j (x_i x_j)$$

$$s.t. \sum_{i=1}^{l} \alpha_i y_i = 0, \quad \alpha_i > 0, i = 1, 2, \cdots, l \tag{2}$$

where α_i, α_j are the LaGrange multipliers. Using the quadratic programming method in operational research to solve Eq. (8), the optimal classification function is obtained as follows:

$$f(x) = \text{sgn}\left[\sum_{i=1}^{l} \alpha_i^* y_i (x^* x_i) + b^*\right] \tag{3}$$

where the sgn($\cdot$) function is used to identify the sample category, α_i^* is the solution of formula (8), and $\alpha_i^* \neq 0$, b^* is the classification threshold. For the linear nonseparable type in this paper, the kernel function $K(x_i, x_j)$ satisfying the Mercer condition should be used to map the sample to the high-dimensional space to transform it into the linear separable problem. After mapping to the high-dimensional space, the corresponding dual problem becomes:

$$\max Q(\alpha) = \sum_{i=1}^{l} \alpha_i - \frac{1}{2} \sum_{i=1}^{l} \sum_{j=1}^{l} \alpha_i \alpha_j y_i y_j K(x_i, x_j)$$

$$s. t. \sum_{i=1}^{l} \alpha_i y_i = 0, \quad 0 \leqslant \alpha_i \leqslant C, i = 1, 2, \cdots, l \tag{4}$$

Finally, the optimal classification function is obtained as follows:

$$f(x) = \text{sgn}\left[\sum_{i=1}^{l} \alpha_i^* y_i K(x^*, x_i) + b^*\right] \tag{5}$$

In this paper, the radial basis kernel function RBF is used, where g is the kernel parameter.

$$K(x_i, x_j) = \exp\left(-\frac{\|x_i - x_j\|^2}{2\sigma^2}\right) = \exp\left(-g\|x_i - x_j\|^2\right) \quad g > 0 \tag{6}$$

4.3 PCA and PSO-Optimized Support Vector Machine for Intrusion Detection

The specific steps of PCA and PSO-optimized support vector machine for intrusion detection are as follows:

(1) Use principal component analysis to analyze the original data, and the first k principal components with a cumulative contribution rate greater than 99% are used as input variables, which are imported into the support vector machine model for simulation and prediction.

(2) Initialize the particle swarm position and velocity.

(3) Calculate the fitness value of all particle swarms. The fitness function is $F = \sum_{i=1}^{n} (y_i - \overline{y_i})^2$, where y_i and $\overline{y_i}$ are the actual output value and the expected output value of SVM model training, respectively.

(4) Evaluate population U_{id}, find the individual extreme value and global extreme value, and update the particle velocity and position.

(5) Determine whether the end condition of optimization is met. If it is, the optimization is ended, and the optimal solution is found. If not, return to step (3).

(6) Give the optimal parameters are given to the SVM model. The trained model is used to detect the fog nodes.

5 Experimental Evaluation

This simulation environment is performed under a Windows 10 64-bit operating system, i5 3.20 GHz processor, 16 GB memory environment, and the SVM algorithm is implemented in MATLAB (R2013a).

There is currently no dataset specifically for fog computing intrusion detection. This article uses the public intrusion detection dataset KDD CUP 99 23 to verify the effectiveness of the proposed scheme.

The dataset mainly includes four types of attack types: denial-of-service (DOS), remote-to-local (R2L), user-to-root (U2R), and port monitoring and scanning attacks (probing). The public dataset provides multiple datasets for training and testing. This paper uses the corrected dataset, which contains a total of 303,736 datasets. The distribution of various attack types in the corrected dataset is shown in Fig. 3. In experiment 1, this dataset was split into two mutually exclusive sets using the "hold-out" method, where the set containing 80% of the data was used as the training set S, and the other set containing 20% of the data was used as the test set T. T is used to evaluate the test error as an estimate of the generalization error.

5.1 Data Preprocessing

The corrected dataset includes 41 attributes, and the data format is as follows: 2, tcp, smtp, SF, 1684, 363, 0, 0, 0, 0, 0, 1, 0, 0, 0, 0, 0, 0, 0, 0, 0, 0, 1, 1, 0.00, 0.00, 0.00, 0.00, 1.00, 0.00, 0.00, 104, 66, 0.63, 0.03, 0.01, 0.00, 0.00, 0.00, 0.00, 0.00, and normal. Before application, data preprocessing is required. The process of data preprocessing is as follows:

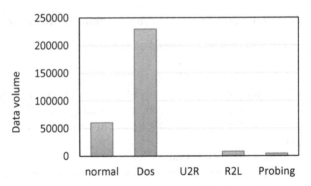

Fig. 3. Various types of data distribution maps corrected

Numeralization

Transform the symbol attributes protocol_type, service, flag in the training set and test set into numeric types. The category labels are converted into numerical representations. The specific assignment of the digitized category labels is shown in Table 1.

Table 1. Four kinds of attack classification are marked

Normal = 0	Normal = 0
DOS = 1	back = 1 land = 1 neptune = 1 pod = 1 smurf = 1 teardrop = 1
Probe = 2	ipsweep = 2 nmap = 2 portsweep = 2 satan = 2
R2L = 3	ftp_write = 3 guess_passwd = 3 imap = 3 multihop = 3 phf = 3 spy = 3 warezclient = 3 warezmaster = 3
U2R = 4	buffer overflow = 4 loadmodule = 4 perl = 4 rootkit = 4

Standardization

Standardize the data obtained in (1) to eliminate the influence of different dimensions on the calculation results.

$$new\_data = \frac{ori\_data - ori\_avg}{ori\_std} \tag{7}$$

The *new_data* in the above formula represents the standardized data, *ori_data* represents the original data, *ori_avg* represents the vector of the mean of each dimension on the original dataset, and *ori_std* represents the vector of the variance of each dimension.

Normalization

Normalize each standardized data to [0, 1] with the min-max method, as follows:

$$v' = \frac{v - min_i}{max_i - min_i} \tag{8}$$

where v is a value of the ith attribute column, min_i is the minimum value of the ith attribute column, and max_i is the maximum value of the ith attribute column.

5.2   Experimental Evaluation Index

As shown in Table 2, TP (True Positive) indicates the number of abnormal samples correctly classified as abnormal samples, TN (True Negative) indicates the number of normal samples classified as normal samples, FP (False Positive) indicates the number of normal samples that are incorrectly classified as abnormal samples, and FN (False Negative) indicates the number of incorrect classifications of abnormal samples into normal samples.

Table 2. Confusion matrix of classification results

Confusion matrix		Predict	
		Attack (positive)	Normal (negative)
True value	Attack (positive)	TP	FN
	Normal (negative)	FP	TN

To verify the effectiveness of the scheme, this paper uses calculation time, precision, detection rate (DR), accuracy (Acc), and false alarm rate (FAR) as the measurement algorithm performance indicators as follows:

$$\text{Precision} = \frac{TP}{TP + FP} \tag{9}$$

$$DR = \frac{TP}{TP + FN} \tag{10}$$

$$Acc = \frac{TP + TN}{TP + TN + FP + FN} \tag{11}$$

$$FAR = \frac{FP}{TN + FP} \tag{12}$$

5.3   Experimental Analysis

To verify the validity of the CFC-SVM intrusion detection model, two experiments are designed in this paper.

Experiment 1: compare the effect of SVM and other classifiers after PCA dimensionality reduction;

Experiment 2: verify the effect of cloud server and fog node collaborative detection on detection time.

The SVM algorithm uses radial basis kernel functions. Penalty parameter C and kernel function parameter g are obtained by the iterative optimization of particle swarm optimization. The parameters of the particle swarm algorithm are set as follows: the number of iterations is 100, and the number of populations is 50. The final parameter optimization results are C = 19.528 and g = 1.238.

The first experiment compares a single SVM algorithm, an extreme learning machine (ELM) algorithm that retains 99% of the information through PCA dimensionality reduction, an Adaboost algorithm, and a decision tree algorithm. To prove the superiority of the PCA-SVM classifier, the experiment was conducted 10 times, and 20% of the data in the dataset was randomly selected as the test set each time. The average of the experimental results was 10 times. Table 3 shows the average value of the data of various attack types in the second experimental test set. Table 4 shows the comparison of PCA-SVM and SVM detection effects. It can be seen from the figure that the dimensionality reduction operation does not reduce the SVM detection effect while retaining the original dataset information and greatly accelerates the training time.

Table 3. Average value of various attack types in 10 experimental test sets

Records	Test
Normal	12,050
DOS	44,772
Probe	465
R2L	1,165
U2R	8
Total	58,460

Table 4. Comparison of SVM and PCA-SVM

	Running time (s)	Normal		DOS		R2L		U2R		Probe	
		DR (%)	FAR (%)	DR (%)	FAR (%)	DR (%)	FAR (%)	DR (%)	FAR (%)	DR (%)	FAR (%)
SVM	9.12	99.8	0.12	99.5	1.57	80.2	2.59	21.2	5.89	91.2	3.56
PCA-SVM	6.53	99.7	0.12	99.5	1.55	81.3	2.58	20.6	5.88	92.3	3.12

Figure 4 shows the comparison of the detection precision between the SVM classifier and other classifiers when PCA dimension reduction is used to retain 99% of the original dataset information. As seen from Fig. 4, after PCA dimension reduction, SVM classification has better detection accuracy for various attack types than Adaboost, ELM, and decision tree.

Figure 5 shows the comparison of the detection rate between the SVM classifier and other classifiers when PCA is also used to reduce the dimensionality and retain 99% of the original dataset information. The PCA-SVM classifier has a high detection rate for

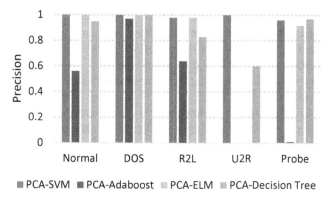

Fig. 4. Precision contrast of four kinds of attacks in four modes.

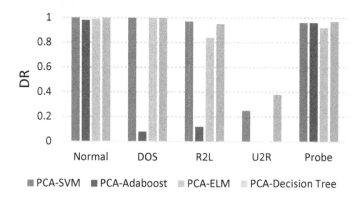

Fig. 5. Detection rate contrast of four kinds of attacks in four modes.

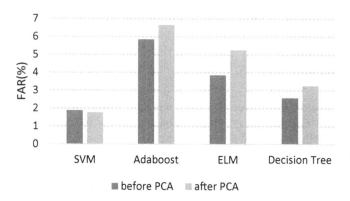

Fig. 6. Comparison of false alarm rate of four classifiers before and after PCA

the R2L, DOS, probe and Normal types. Because the content of U2R attack type data in the dataset is very low, the detection rate of U2R is low.

Figure 6 shows the comparison of the false alarm rate of four classifiers before and after dimension reduction. It can be seen from the figure that the false alarm rate of the SVM classifier decreases after dimension reduction by PCA, while that of the other three classifiers increases after dimension reduction. After PCA dimensionality reduction, SVM is better than the other algorithms in terms of precision, detection rate and false alarm rate.

Experiment 2 is used to verify the impact of the intrusion-detection scheme proposed in this paper, which is collaborative between the cloud server and fog node, on the detection time. MATLAB is used for simulation. SVM is deployed in the cloud server, and 10000 pieces of data are randomly selected as the detection samples. Ten experiments are conducted, and the comparison parameters are the average value of detection time and detection accuracy.

Table 5. Comparison of different algorithms in accuracy and detection time

Algorithm	Accuracy (%)	Detection time (s)
CFC-SVM	98.5	4.1
SVM	96.5	4.5
ELM	96.1	8.8
Decision tree	84.5	12.7

According to the results in Table 5, the PCA-SVM algorithm has relatively high accuracy and less detection time when applied to intrusion detection in a fog environment. The main reason is that the dimensionality reduction of data makes the calculation speed faster. The particle swarm optimization algorithm is applied to the selection of SVM parameters, which increases the training speed. In the fog environment, data training in the cloud server and detection in the fog end are adopted. The layered model can greatly reduce the detection time.

6   Conclusions

The network where the fog nodes are located is highly dynamic and highly real time, which presents a new challenge to the intrusion detection of fog computing. This paper proposes a cloud-fog cooperative intrusion detection model, which is added during pre-processing PCA dimensionality reduction, eliminating the correlation between various dimensions of network data and reducing the storage and calculation overhead. The dataset is trained in the cloud server with an SVM classifier (where the selection of SVM parameters is optimized by particle swarm optimization) and saving the training model greatly saves the detection time at the fog nodes. The proposed model is verified on the KDD CUP 99 dataset, and several experiments have demonstrated the advantages of this model from different perspectives and have contributed to solving the problems caused by resource constraints in fog computing.

Acknowledgment. This project is supported by Natural Science Foundation of China (No. 61572170, No.61170254), The Key Projects of Natural Science Foundation of Hebei Provience (No. F2019201290), The Natural Science Foundation of Hebei Province (No. F2018201153) and Hebei Natural Science Foundation under grant No. F2018201197, Research on Impreciseness Analysis Key Technology of Network Access Control Intention. We hereby express our thanks.

References

1. Shi, W., Cao, J., Zhang, Q., Li, Y., Xu, L.: Edge computing: vision and challenges. IEEE Internet Things J. **3**(5), 637–646 (2016)
2. Liu, C., Xiang, F., Wang, P., Sun, Z.: A review of issues and challenges in fog computing environment. In: DASC/PiCom/DataCom/CyberSciTech, pp. 232–237 (2019)
3. Puliafito, C., Mingozzi, E., Longo, F., Puliafito, A., Rana, O.: Fog computing for the internet of things: a survey. ACM Trans. Internet Tech. **19**(2), 18:1–18:41 (2019)
4. Oma, R., Nakamura, S., Duolikun, D., Enokido, T., Takizawa, M.: An energy-efficient model for fog computing in the Internet of Things (IoT). Internet of Things **1–2**, 14–26 (2018)
5. Puthal, D., Mohanty, S.P., Bhavake, S.A., Morgan, G., Ranjan, R.: Fog computing security challenges and future directions [energy and security]. IEEE Consum. Electron. Mag. **8**(3), 92–96 (2019)
6. Noura, H.N., Salman, O., Chehab, A., Couturier, R.: Preserving data security in distributed fog computing. Ad Hoc Netw. **94**, 101937 (2019)
7. Elazhary, H.: Internet of Things (IoT), mobile cloud, cloudlet, mobile IoT, IoT cloud, fog, mobile edge, and edge emerging computing paradigms: disambiguation and research directions. J. Netw. Comput. Appl. **128**, 105–140 (2019)
8. Parikh, S., Dave, D., Patel, R., Doshi, N.: Security and privacy issues in cloud, fog and edge computing. In: EUSPN/ICTH 2019, pp. 734–739 (2019)
9. Khan, S., Parkinson, S., Qin, Y.: Fog computing security: a review of current applications and security solutions. J. Cloud Comput. **6**, 19 (2017). https://doi.org/10.1186/s13677-017-0090-3
10. D'Souza, C., Ahn, G.-J., Taguinod, M.: Policy-driven security management for fog computing: Preliminary framework and a case study. In: IRI 2014, pp. 16–23 (2014)
11. Ficco, M.: Internet-of-Things and fog-computing as enablers of new security and privacy threats. Internet Things **8** (2019)
12. Razouk, W., Sgandurra, D., Sakurai, K.: A new security middleware architecture based on fog computing and cloud to support IoT constrained devices. In: IML 2017, pp. 35:1–35:8 (2017)
13. Hosseinpour, F., Amoli, P.V., Plosila, J., et al. An intrusion detection system for fog computing and iot based logistic systems using a smart data approach. Int. J. Digit. Content Technol. Appl. **10**(5) (2016)
14. Peng, K., Leung, V.C.M., Zheng, L., Wang, S., Huang, C., Lin, T.: Intrusion detection system based on decision tree over big data in fog environment. Wirel. Commun. Mob. Comput. **2018** (2018)
15. Zhou, L., Guo, H., Deng, G.: A fog computing based approach to DDoS mitigation in IIoT systems. Comput. Secur. **85**, 51–62 (2019)
16. An, X., Zhou, X., Lü, X., Lin, F., Yang, L.: Sample selected extreme learning machine based intrusion detection in fog computing and MEC. Wirel. Commun. Mob. Comput. **2018** (2018)
17. Prabavathy, S., Sundarakantham, K., Shalinie, S.M.: Design of cognitive fog computing for intrusion detection in internet of things. J. Commun. Netw. **20**(3), 291–298 (2018)

18. Liu, Y., Bi, J.-W., Fan, Z.-P.: A method for multi-class sentiment classification based on an improved one-vs-one (OVO) strategy and the support vector machine (SVM) algorithm. Inf. Sci. **394**, 38–52 (2017)
19. Cui, J., Shi, G., Gong, C.: A fast classification method of faults in power electronic circuits based on support vector machines. Nephron Clin. Pract. **24**(4), 701–720 (2017)
20. Li, L.: Analysis and data mining of intellectual property using GRNN and SVM. Pers. Ubiquit. Comput. **24**(1), 139–150 (2019). https://doi.org/10.1007/s00779-019-01344-8
21. Gu, J., Wang, L., Wang, H., Wang, S.: A novel approach to intrusion detection using SVM ensemble with feature augmentation. Comput. Secur. **86**, 53–62 (2019)
22. Subba, B., Biswas, S., Karmakar, S.: Enhancing performance of anomaly based intrusion detection systems through dimensionality reduction using principal component analysis. In: ANTS 2016, pp. 1–6 (2016)
23. KDD CUP 99 data set. http://kdd.ics.uci.edu/databases/kddcup99/kddcup99.html

An Multi-feature Fusion Object Detection System for Mobile IoT Devices and Edge Computing

Xingyu Feng[1], Han Cao[1], and Qindong Sun[1,2(✉)]

[1] School of Computer Science and Engineering, Xi'an University of Technology, Xi'an 710048, Shaanxi, China
sqd@xaut.edu.cn
[2] Shaanxi Key Laboratory of Network Computing and Security, Xi'an University of Technology, Xi'an 710048, Shaanxi, China

Abstract. With the increase of data scale and computing power, deep learning algorithm has made a prominent breakthrough in computer vision and other complex problems. However, its high complexity and large memory requirements make it very difficult to run in real time on the Internet of things terminal mobile devices. There is still delay the employing of cloud services cannot meet the real-time requirement. With the popularity of mobile terminal devices and the development of Internet of things, it is of great significance to design a real-time deep learning algorithm on IOT edge mobile devices with limited computing and memory resources. This paper proposes a new object detection method based on the current state-of-the-art object detection deep network model RetinaNet and traditional feature extraction method SIFT. RetinaNet is a one-stage detector with excellent detection speed and accuracy. We use RetinaNet as the object location method, then extract the CNN features and SIFT features of the fixed position image and combine them to train a new classifier. The object classification result will be based on the final classifier.

Keywords: Object detection · Deep learning · SIFT · IoT

1 Introduction

It is a great challenge to deploy applications based on deep learning technology on mobile devices. How to reduce the time complexity and memory requirements on the premise of ensuring the accuracy of the algorithm, so that it can run efficiently on mobile devices is an urgent problem to be solved.

Object detection aims at locating and classifying objects accurately from an image, which has practical application value, such as intelligent video surveillance [1], robot perception [2] and so on. Traditional object detection is roughly divided into three steps: (1) region selection, (2) feature extraction, and (3)

© ICST Institute for Computer Sciences, Social Informatics and Telecommunications Engineering 2021
Published by Springer Nature Switzerland AG 2021. All Rights Reserved
D. Wang et al. (Eds.): SPNCE 2020, LNICST 344, pp. 335–340, 2021.
https://doi.org/10.1007/978-3-030-66922-5_23

object classification. Feature extraction can be performed by Haar, HOG, LBP and SIFT methods. Compared with traditional methods, object detection based on deep learning has greatly improved the accuracy. At present, there are R-CNN [3], SSD [4], YOLOV1-YOLOV4 [5–8], and others.

The development of artificial intelligence has set off a wave of unmanned retail. The unrestricted placement of unmanned retail scenes poses challenges for object detection: (1) The same model shows different accuracy in different scenarios. (2) In this scenario, mobile devices need to be able to process images in real time. In order to solve this problem, this paper proposes an embedded object detection system for internet of things and mobile edge computing.

The structure of this paper is as follows, the next section introduces some work related to object detection. The Sect. 3 systematic elaborates the method proposed in this paper, and then shows the experimental results. Finally, we summarize the work done in this paper.

2   Related Work

At present, industry and academia have been committed to running deep learning applications on mobile devices. Srinivasan et al. [9] had trained lightweight object detection networks to help visually impaired users identify and locate household items through wearable devices. Blanco-Filgueira et al. [10] proposed an end-to-end solution for multi-object tracking based on deep learning for embedded and low-power IoT platforms.

Prior to 2012, object detection mainly uses SIFT, HOG, DPMs and other traditional features. In 2012, Krizhevsky et al. used 1.2 million data containing 1,000 different categories to train a deep CNN, which became the most advanced technology at the time. The methods based on CNN breaks through the bottleneck of traditional methods.

R-CNN [3] is an object detection method based on the candidate region method. Later, improved versions Fast R-CNN [11] and Faster R-CNN [12] appeared. OverFeat [13] is the first one-stage detectors based on deep learning and won the ILSVRC2013. After that, YOLO and SSD appeared one after another. At present, YOLO has four versions. Lin et al. [14] pointed out that one-stage detector detection accuracy is limited by class imbalance, thereby focal loss is proposed to solve this problem, and the network structure is named RetinaNet.

Therefore, this paper uses RetinaNet for object location, integrates CNN and SIFT features to classify, so that the final model has strong robustness to changes in the external environment.

3   Approach

The location of the object is obtained by using RetinaNet, and then the CNN features and SIFT features of the object image are fused, finally the category of object is obtained. The overall process of the framework is as follows (Fig. 1):

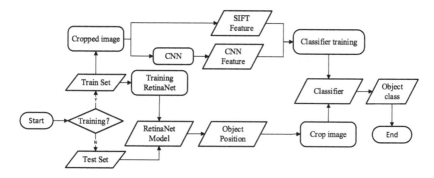

Fig. 1. The overall process of the object detection framework.

3.1 RetinaNet for Object Location

The main network of RetinaNet uses ResNet and FPN to generate feature pyramid model. After that, two sub-networks are connected: Classification Subnet and Box Regression Subnet, which are used for classifying and location respectively. The last layer of two sub-networks has the same outer structure but does not share parameters. The RetinaNet network structure is shown in Fig. 2.

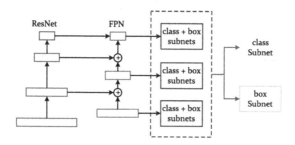

Fig. 2. Structure of RetinaNet.

Focal loss as a loss function to solve the problem that is class imbalance and distinguishing difficult instances:

$$FL(P_t) = -\alpha_t(1 - p_t)^\gamma \log(p_t) \tag{1}$$

Where α is used to adjust the class imbalance, the range of value is $[0, 1]$, $(1 - p_t)^\gamma$ is the modulation factor, which can differentiate the easy example and hard example, γ is tunable focusing parameter and $\gamma \in [0, 5]$.

3.2 Fusion of SIFT Features and CNN Features to Classification

SIFT features have scale and illumination invariance. The extraction process takes grayscale images as input and scale space is used to detect image features of different scales. According to the scale space theory, the scale space L of an image can be defined as:

$$L(x, y, \sigma) = G(x, y, \sigma) \times I(x, y) \tag{2}$$

$$G(x, y, \sigma) = \frac{1}{2\pi\sigma^2} \exp\left(-\frac{(x - x_i)^2 + (y - y_i)^2}{2\sigma^2}\right) \tag{3}$$

(x, y) is the spatial coordinate, σ is scale coordinate, and $G(x, y, \sigma)$ is a scale variable Gauss function. SIFT algorithm suggests that the feature detection on a certain scale can be obtained by subtracting the images in two adjacent Gaussian scale spaces to obtain the responsive image $D(x, y, \sigma)$. The local feature points are located in the spatial position and scale space by searching the local maximum value of the $D(x, y, \sigma)$. The mathematical description is:

$$G(x, y, \sigma) = (G(x, y, k\sigma) - G(x, y, \sigma)) \times I(x, y) = L(x, y, k\sigma) - L(x, y, \sigma) \tag{4}$$

SIFT uses local extremum points in Gauss difference scale space as candidate feature points, determines the main direction of feature points to generate feature descriptors, and normalizes them to form SIFT feature vectors. Because the number of SIFT feature points extracted from each image is different, BOW algorithm is used to construct a uniform dimension feature vector for each image.

CNN neural network is usually used in image processing. It is composed of multiple convolution layers. Each convolution layer contains multiple convolution kernels. These convolution kernels are used to scan the entire image from left to right and from top to bottom to obtain output data called a feature map. This paper uses VGG19 to extract CNN features of the image. The subregion where the object is located is the input image. We use the output of the second complete acquisition layer as the CNN feature of the image, and input it with the SIFT feature into the classifier for training to build the classifier. The classification result is used as the target classification result of target detection.

4   Experimental Results and Analysis

The training set and test set used in this paper are the real-life pictures of Intelligent vending cabinet. We use RetinaNet's ResNet-101 model for training, and crop all object sub-images labeled in the training data set. Extract VGG19 features and SIFT features for fusion, and train the classifier.

The accuracy rate of RetinaNet is 97.93% where test set in the same environment as training set. This is an exciting result. The loss results in the training process of RetinaNet are shown in Fig. 3.

However, when we tested in different environments, we found that the performance of the model was not ideal. The following repeatability issues will occur:

1. Location failure. Sometimes has an object location failure that other items that have not appeared in the training set are displayed.
2. Misclassification. The object item is positioned accurately but the category is wrong.

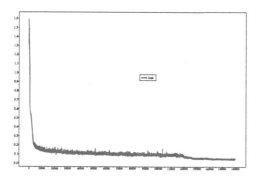

Fig. 3. Curve of loss.

It can be seen that changes in the external environment affect the accuracy of object detection. We also tried to expand the image with brightness, light, etc., and then trained with RetinaNet, but it did not achieve the desired results. Therefore, we try to train new classifiers by means of hybrid SIFT-CNN feature to improve detection accuracy. The output of RetinaNet is $[x_{min}, y_{min}, x_{max}, y_{max}, classes]$, x_{min}, y_{min} are the coordinates of the upper left corner of the object position, and x_{max}, y_{max} are the lower right corner, $classes$ is the confidence matrix, and the maximum value is the predicted class.

We will get the location of the object. It is used to crop the object, and extract the SIFT-CNN features of this area and input it into the classifier. For images whose SIFT features cannot be extracted, the category of the RetinaNet model will be the final category. Some areas with position errors input classifiers can obtain lower confidence, which can be filtered out by threshold. The classifier output can correct wrong label to a certain extent. The test results have been improved in changing scenarios, but there are still misclassifications. However, the speed has also declined. On TITAN XP, using RetinaNet alone, detecting an image takes an average of 180 ms, and our method will add about 23 ms to SIFT-CNN feature extraction and classification.

5   Conclusion and Future Work

In this study, we use RetinaNet to locate objects, then, the use CNN-SIFT hybrid feature of the detection area to predict the category. Although this corrects location errors and classification errors to a certain extent, improving the overall accuracy of object detection. It has not completely solved the accuracy degradation caused by external environmental change. In the future, we will study how to improve the robustness and accuracy of the model.

Acknowledgement. The research presented in this paper is supported in part by the National Natural Science Foundation (No.: 61571360), The Youth Innovation Team of Shaanxi Universities, the Innovation Project of Shaanxi Provincial Department of Education (No.: 17JF023) and the Project of Xi'an Technology Bureau (No.: GXYD14.12).

References

1. Wang, X.: Intelligent multi-camera video surveillance: a review. Pattern Recogn. Lett. **34**(1), 3–19 (2013)
2. Guo, L., Zhang, M., Wang, Y., et al.: Environmental perception of mobile robot, pp. 348–352. IEEE (2006)
3. Girshick, R., Donahue, J., Darrell, T., et al.: Rich feature hierarchies for accurate object detection and semantic segmentation. In: Proceedings of the IEEE Conference on Computer Vision and Pattern Recognition, pp. 580–587. IEEE, Piscataway (2014)
4. Liu, W., et al.: SSD: single shot multibox detector. In: Leibe, B., Matas, J., Sebe, N., Welling, M. (eds.) ECCV 2016. LNCS, vol. 9905, pp. 21–37. Springer, Cham (2016). https://doi.org/10.1007/978-3-319-46448-0_2
5. Redmon, J., Divvala, S., Girshick, R., et al.: You only look once: unified, real-time object detection. In: Proceedings of the IEEE Conference on Computer Vision and Pattern Recognition, pp. 779–788. IEEE, Piscataway (2016)
6. Redmon, J., Farhadi, A.: YOLO9000: better, faster, stronger. In: Proceedings of the IEEE Conference on Computer Vision and Pattern Recognition, pp. 7263–7271. IEEE, Piscataway (2017)
7. Redmon, J., Farhadi, A.: YOLOv3: an incremental improvement. arXiv preprint arXiv:1804.02767 (2018)
8. Bochkovskiy, A., Wang, C.Y., Liao, H.Y.M.: YOLOv4: optimal speed and accuracy of object detection. arXiv preprint arXiv:2004.10934 (2020)
9. Li, P., Li, J., Huang, Z., et al.: Multi-key privacy-preserving deep learning in cloud computing. Future Gener. Comput. Syst. **74**, 76–85 (2017)
10. Blanco-Filgueira, B., García-Lesta, D., Fernández-Sanjurjo, M., et al.: Deep learning-based multiple object visual tracking on embedded system for IoT and mobile edge computing applications. IEEE Internet Things J. **6**(3), 5423–5431 (2019)
11. Girshick, R.: Fast R-CNN. In: Proceedings of the IEEE International Conference on Computer Vision, pp. 1440–1448. IEEE, Piscataway (2015)
12. Ren, S., He, K., Girshick, R., et al.: Faster R-CNN: towards real-time object detection with region proposal networks. In: Advances in Neural Information Processing Systems, pp. 91–99. The MIT Press, Cambridge (2015)
13. Sermanet, P., Eigen, D., Zhang, X., et al.: OverFeat: integrated recognition, localization and detection using convolutional networks. arXiv preprint arXiv:1312.6229 (2013)
14. Lin, T.Y., Goyal, P., Girshick, R., et al.: Focal loss for dense object detection. In: Proceedings of the IEEE International Conference on Computer Vision, pp. 2980–2988. IEEE, Piscataway (2017)

Cryptography

Generative Image Steganography Based on Digital Cardan Grille

Yaojie Wang[1,2(✉)], Xiaoyuan Yang[1,2], and Wenchao Liu[1,2]

[1] Engineering University of PAP, Xi'an 710086, China
wangyaojie0313@163.com
[2] Key Laboratory of Network and Information Security of PAP, Xi'an 710086, China

Abstract. In this paper, a generative image steganography algorithm based on digital Cardan Grille is proposed. Combining the ideas of traditional Cardan Grille and the semantic image inpainting technique, the stego image are driven by secret messages directly. The algorithm first embeds the information based on digital Cardan Grille, and then uses generative adversarial network (GANs) to complete the damaged image. The adversarial game not only reconstruct the corrupted image, but also generate a stego image which contains the logic rationality of image content. The experimental results verify the feasibility of the proposed method.

Keywords: Image steganography · Cardan grille · Semantic completion · Generative adversarial network

1 Introduction

In ancient China, a clever steganography was invented. The sender and the receiver each hold an identical piece of paper with many small holes, and the positions of these holes are randomly selected. The sender covers this piece of paper with a hole, writes the secret information in the position of the small hole, and then removes the paper above, and fills in the empty space to make the whole text logical significance. The receiver can read the secret information left in the small hole as long as the perforated paper is covered with this ordinary text. In the early 16th century, the Italian mathematician Cardan (1501–1576) also invented this method, which is now called Cardan Grille [1]. As an important branch of information security, steganography has made great progress. For a long time, steganography has mostly used images and videos as the cover, And modification of a small number of pixels is the most commonly used method. Although various technologies were adopted to correct and cover up the modification traces, the security and practicability of steganography faced great challenges along with the improvement of computing power [2].

Fortunately, a new technique of deep learning, generative adversarial networks (GANs) [3], has become a new research hotspot in the field of information hiding.

D. Wang et al. (Eds.): SPNCE 2020, LNICST 344, pp. 343–355, 2021.
https://doi.org/10.1007/978-3-030-66922-5_24

The biggest advantage and feature of GANs is the ability to sample real space and generate samples driven by noise, which provides the new possibility for steganography. Combined with deep generative model technology, this paper gives new vitality to the Cardan Grille, and a generative steganography algorithm based on digital Cardan grille is proposed. First, the random digital Cardan grille is generated automatically, which plays the role of a key in cryptography. Secondly, a blank image is taken as the cover, and the secret information is written to the area that needs to be filled by a digital Cardan grille. In the case of keeping the secret message unchanged, the semantic image completion is realized by using the generative adversarial network. The secret message is hidden in the reconstructed image after completion. The experiments on the image database confirms the feasibility of such simple method.

The remainder of this paper is organized as follows: In the following sections we detail the current status of deep generation models in steganography. Section 3 shows how to build digtal Cardan grille by GANs. Experiment results are demonstrated in Sect. 4. Section 5 concludes this research and details our future work.

2   Related Work

Modern steganography [4] entered the world in 1985 with the advent of personal computers being applied to classical steganography problems. Fridrich [5] considers steganographic channel is divided into three categories, cover selection, modification and synthesis. cover modification is a steganography method adopted by most traditional steganography techniques, which is mainly aimed at reducing the difference between covers before and after modification. Due to the high-dimensional characteristics of the cover itself, the modified carrier always leaves traces of modification, which can be attacked from different dimensions. cover selection is essentially a mapping between the original cover and the secret message. For example, Zhou et al. [6] used the bag-of-words model (BOW) to extract the visual words (VW) of the image. The embedding rate of this method is very low, and there is a security risk when it is used many times, so its practical application is less. Cover synthesis attempts to generate a natural cover, which was very difficult before the deep generative model appeared. With the help of texture synthesis, [7, 8] also use the texture sample and a bunch of color points generated by secret messages to construct dense texture images. This kind of texture-based steganography is based on the premise that the cover may not represent the content in real world.

With the development of machine learning, the deep generation model represented by generative adversarial networks has developed rapidly and achieved remarkable achievements. Recently, this new technology has applied adversarial training to steganographic problems. Volkhonskiy etc. [9] first propose a new model for generating image-like containers based on Deep Convolutional Generative Adversarial Networks (DCGAN) [10]. This approach allows to generate more setganalysis-secure message embedding using standard steganography algorithms. Shi etc. [11] introduce a new generative adversarial networks to improve convergence speed, the training stability the image quality. On the basis of [9, 11], Wang et al. [12] introduced a new method of steganography to improve the training effect. However, most of these GAN-based steganographic schemes are still the cover modification techniques for steganography. These methods focus on the adversarial game, but ignore the core aim of the GAN is to build a powerful generator.

Since GAN's biggest advantage is to generate samples, it is a intuitive idea to use GANs generate a semantic cipher from a message directly as the Cardan did. Some researcher made a preliminary attempt on this intuitive idea. Hu et al. [13] proposed a steganographic scheme without cover modification. The core of this method is the extraction of information. The mean square error is used as the evaluation criterion to obtain the noise extractor. Hayes et al. [14] Proposed a steganography method that uses a three-party adversarial network to make the machine automatically learn. Liu et al. [15] proposed a steganographic algorithm for data sampling based on a deep generation model. This method does not depend on a specific cover, the stego image is actually obtained by sampling by the generator. Liu et al. [16] also proposed the Stego-ACGAN method, which uses semi-supervised learning to establish a mapping relationship between information to achieve the generation of a specified carrier. Ke [17] proposed generative steganography method called GSK in which the secret messages are generated by a cover image using a generator rather than embedded into the cover, thus resulting in no modifications in the cover.

In this paper, a generative steganography, based on digital Cardan Grille, is proposed. A mask called digital Cardan grille for determining the hidden location is introduced to hide the message. According to the position corresponding to digital Cardan Grille, the secret information is written into the blank image in advance, and the above information remains unchanged during the entire steganography process. Images with secret messages are then input into a generative adversarial networks to complete semantic completion. The adversarial game not only complement broken images, but also generate a stego image which contains the logic rationality of image content.

3   The Proposed Algorithm

With the help of generative adversarial network technology, the framework of generative steganography based on digital Cardan Grille proposed in this paper is shown in Fig. 1. The sender and receiver define a mask together, called digtal Cardan grille, to determine where the message is hidden, it acts as a key in cryptography. According to the "1" position corresponding to digital Cardan Grille (the "0" position is omitted), the secret information is written into the blank image in advance, and the above information remains unchanged during the entire steganography process. Images with secret messages are then input into a generative adversarial network to complete semantic completion. A stego image is transmitted to the recipient through the public channel. The receiver uses the shared digital Cardan Grille to extract the secret information in a reverse process. The algorithm mainly includes three parts: the design of digital Cardan Grille, the automatic generation of digital Cardan Grille, and the completion of the image after embedding the information. The basic principles will be explained one by one in the next section. In fact, this framework describes a general automated Cardan grille, and the input cover vector can be a piece of text, image, video, and other type of media.

3.1   Design Principles of Digital Cardan Grille

In this paper, digital Cardan Grille still adopts this clever idea of the traditional Cardan Grille, using the Hamard product [18] in matrix multiplication to complete the digital

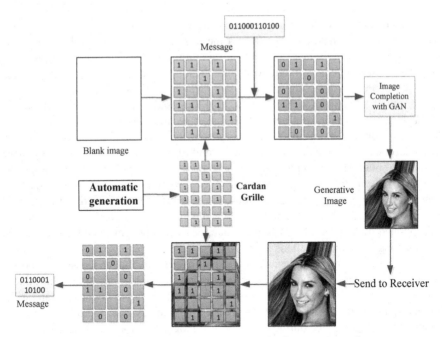

Fig. 1. The structure of the proposed generative steganography

conversion. Hadamard product is a type of matrix operation. If $A = (a_{ij})$ and $B = (b_{ij})$ are two matrices of the same order, if $c_{ij} = a_{ij} \times b_{ij}$, then the matrix $C = (c_{ij})$ is called A and B's Hadamard product. In mathematics, the Hadamard product is expressed as follows:

Definition Suppose A and B, and $A = (a_{ij})$, $B = (b_{ij})$. Then the matrix $C = (c_{ij})$

$$C = (c_{ij}) = \begin{bmatrix} a_{11}b_{11} & a_{12}b_{12} & \dots & a_{1n}b_{1n} \\ a_{21}b_{21} & a_{22}b_{22} & \dots & a_{2n}b_{2n} \\ \vdots & \vdots & & \vdots \\ a_{m1}b_{m1} & a_{m2}b_{m2} & \dots & a_{mn}b_{mn} \end{bmatrix} \tag{1}$$

is the Hadamard product of the matrices A and B, denoted as $A \odot B$.

According to the above, the design process of digital Cardan Grille is described as follows: if A is stego information and B is a Cardan Grille of a binary mask, which can only be represented by two values of 0 and 1, so the secret information can be hidden and extracted by using the Hadamard product. Taking a 2×2 simple image as an example, a blank image is covered with a digital Cardan Grille in the hiding process. The position of 1 indicates a small hole in the traditional Cardan Grille, which means that information can be embedded. A position with a value of 0 indicates that there are no holes, which means that information cannot be embedded, and which needs to be completed later by GANs, as shown in Fig. 2(a).

When extracting information, the same digital Cardan Grille is still used, and the embedded secret information is obtained through the Hadamard product. A value of 1 indicates information that needs to be retained in the image, and a value of 0 indicates information that needs to be discarded, that is, the information retained in the Hadamard product is the embedded secret information, as shown in Fig. 2(b). For third parties, the information of digital Cardan Grille is strictly confidential.

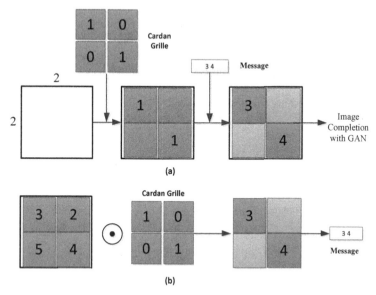

Fig. 2. The principles of digital Cardan Grille

3.2 Automatic Generation of Digital Cardan Grille

Although the traditional Cardan Grille can generate a stego cover, it needs to artificially construct a meaningful cover, such as Tibetan head poems, which is time-consuming and labor-intensive. The automatic generation of digital Cardan Grille not only meets the needs of the covert communication in reality, but also improves the security of the information, which greatly increases the obstacles for third-party attacks and deciphering.

According to the design principle of digital Cardan Grille, the message digest algorithm in cryptography is used for automatic generation. The message digest algorithm is mainly applied in the field of digital signature. Its main characteristic is that the encryption process does not require a key, and it has uniqueness and irreversibility. Only by entering the same plain text data can be obtained through the same message digest algorithm. Well-known digest algorithms include MD5 algorithm, SHA-1 algorithm, SHA-256 algorithm and a large number of variants. In this paper, the SHA-1 algorithm is used as an example for the automatic generation and interpretation of digital Cardan Grille, as shown in Fig. 3. SHA-1 algorithm can output a 20-byte hash value, which is usually represented in the form of 40 hexadecimal digits. There are four steps to automatically generating digital Cardan Grille:

1. Obtain the agreed public information as an input signal. The public information obtained at different times is different, such as the headlines on the front page of the Wall Street journal, which encode the information in binary;
2. Fill the key. The sender and receiver share the key in advance and fill the encoded public information according to the agreed filling rules. it is filled at the end and only serves as the principle explanation in Fig. 3;
3. Encryption. The filled information is input into the SHA-1 generator, and a 20-byte message digest hash value (160 bits) is output.
4. The message digests in step 3 are arranged in sequence to form a matrix, which is called as digital Cardan Grille. The extra digits can be discarded according to the amount of embedded secret information. If the number of matrix bits cannot fill the entire vector, the rest are filled with zeros.

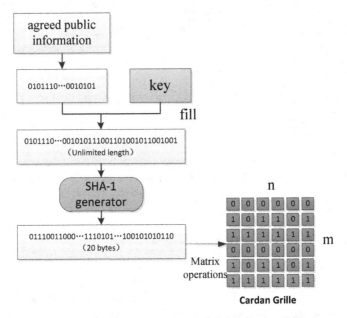

Fig. 3. The automatic generation of digital Cardan Grille

3.3 Completion Stego Image

In recent years, there are various techniques for image completion, but the image damage rate that needs to be completed in this article occupies more than 95%, which means that the completion here is not a small amount of repair, but more inclined to generate images on a large scale. As mentioned above, inspired by [19], this paper still uses the a image inpainting method which proposed by Yeh [20] based on a Deep Convolutional Generative Adversarial Network (DCGAN), and some parameters are adjusted to ensure that the generated image meets the needs of semantics.

Assuming that we need to complete the broken y, we still use the Hadamard product for pre-processing. First we choose a binary mask M that has values 0 or 1. According to the digital Cardan Grille's principle and Hadamard product, we can know that $M \odot y$ can represent the original part of the stego image, and $(1 - M) \odot y$ also means the other part that needs completion. Suppose we generate a reasonable $G(z')$ to complete the missing part, the original part and the completed part constitute the reconstruction of the image:

$$x_{reconstructed} = M \odot y + (1 - M) \odot G(z') \tag{2}$$

In the process of image completion, two loss functions need to be defined. The specific content is similar to the reference [20], where the embedded information part is equivalent to the unbroken area of the image.

Contextual Loss: To keep the same context as the input image, make sure the known pixel locations in the input image y are similar to the pixels in $G(z)$. We need to penalize $G(z)$ for not creating a similar image for the pixels that we know about. Formally, we do this by element-wise subtracting the pixels in y from $G(z)$ and looking at how much they differ:

$$L_{contextual(z)} = |M \odot G(z) - M \odot y| \tag{3}$$

In the ideal case, all of the pixels at known locations are the same between y and $G(z)$. Then $G(z)_i - y_i = 0$ for the known pixels i and thus $L_{contextual}(z) = 0$.

Perceptual Loss: To recover an image that looks real, let's make sure the discriminator is properly convinced that the image looks real. We'll do this with the same criterion used in training the DCGAN:

$$L_{perceptual(z)} = \log(1 - D(G(z))) \tag{4}$$

Contextual Loss and Perceptual Loss successfully predict semantic information in the missing region and achieve pixel-level photorealism. We're finally ready to find z' with a combination of context loss and perception loss:

$$L(z) \equiv L_{contextual}(z) + \lambda L_{perceptual}(z) \tag{5}$$

$$z' \equiv arg \, minL(z) \tag{6}$$

where λ is a hyper-parameter that controls how import the contextual loss is relative to the perceptual loss.

Compared with [20], the proposed scheme needs more information to complete, and we have now adjusted the parameters. In experiments, the effect of generating images is relatively good when $= 0.7$.

4 Experiment and Analysis

In order to verify the performance of the scheme, in the experiments we used the CelebA data set (Ziwei Liu and Tang 2015) and the LSUN dataset to train the network model separately. The former is openly provided by the Chinese University of Hong Kong and is widely used for face-related computer vision training tasks, including more than 200,000 images. The latter is a large-scale image data set constructed by humans performing deep learning in a loop, which contains 10 scene categories and 20 object categories, each category has about one million labeled images.

The experimental environment is shown in Table 1. We used the DCGAN model architecture from Yeh et al. [20] in this work. The optimizer in DCGAN uses an Adam-based optimization method with a learning rate of 0.0002. At each training, the weight of the discriminator D is updated once, the weight of the generator G is updated twice [21].

Table 1. Experimental environment

Software platform	Tensorflow v0.12	
Hardware environment	CPU	i7-8250U 3.2 GHz
	RAM	16 GB DDR3 1600 MHz
	GPU	NVIDIA 1080

In the experiment, we use the alignment tool to preprocess the image to 128 × 128. Based on the work of Brandon Amos's bamos/dcgan-completion.tensorflow [22], I modified the parameters. Digital Cardengo is automatically generated. Assuming that the embedded information is randomly distributed over 16 pixels, Fig. 4(a) is an example of generating a model complement CelebA image after training for 11 epochs. Figure 4(b) is an example of generating a model complement LSUN image after training for 7 epochs.

In order to better evaluate the quality characteristics of the generated images, we introduced the No Reference Image Quality Assessment (NR-IQA) method [23]. Because this method does not need to be compared with the original image, and the image steganography based on the generated model does not have the original carrier, the characteristics of the two are completely consistent, which makes up for the shortcomings of the current traditional steganographic evaluation system.

If there is no abnormality in visual observation, Magnitude spectrum, frequency histogram, and DCT coefficient histogram are commonly used as evaluation methods. 1,000 images randomly selected from the algorithm generated in this paper are analyzed, and specific examples are shown in Fig. 5. Through simulation experiments, compared with natural images, the generated samples have no statistical abnormalities in frequency characteristics, abnormal DCT coefficients, etc., which can basically meet the needs of realistic communication, and can also effectively resist detection based on statistical steganalysis.

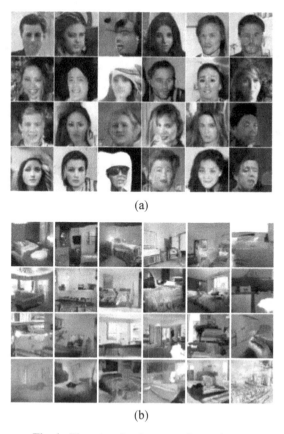

(a)

(b)

Fig. 4. The example of generated stego image

Next, we verify the detection resistance of the generated image, that is, use the current mainstream detection algorithms to verify the imperceptibility of the algorithm in this paper. This paper randomly chooses 6000 real images (from CelebA data set and the LSUN dataset) and 4000 images generated by this algorithm as the test set. The four detection algorithms selected are steganalysis in DCT domain [24], RS detection [25], nonlinear SVM detection [26], and S-CNN detection [27]. The DCT domain steganalysis mainly focuses on the statistical characteristics of DCT coefficients and their impact on spatial pixels. The RS detection method (regular groups and singular groups) is evaluated based on the gray value of the image. The essence of the nonlinear SVM detection method is to extract the feature data of the sample, which is a binary classification model. S-CNN detection is the latest method for image detection using convolutional neural networks. In the case of random group independent testing, the comparison results of steganography detection are shown in Table 2.

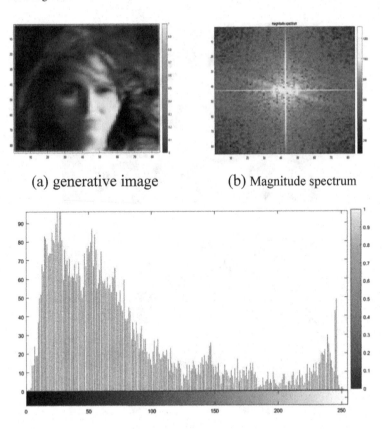

(a) generative image   (b) Magnitude spectrum

(c) Frequency histogram of corresponding image

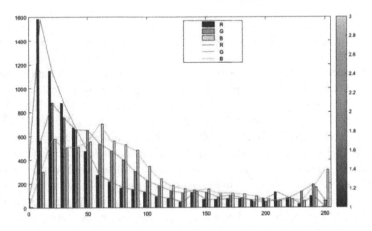

(d) DCT coefficient histogram of corresponding image

Fig. 5. The example of generated encrypted image

Except for the S-CNN detection method, the detection accuracy of the other three detection algorithms approaches 0.5, and the experimental results show that the algorithm in this paper has great advantages in terms of detection resistance. For the S-CNN detection method, the detection accuracy rate is close to 0.65, which needs to be further improved. In the actual communication process, imperceptibility can be better satisfied by reducing the amount of embedded information, so as to ensure the universality of steganography.

Table 2. Accuracy of the steganalysis test

Detection algorithm	Detection accuracy	
	Group accuracy	Average accuracy
Steganalysis in DCT domain	0.534	0.556
	0.579	
RS detection	0.510	0.516
	0.523	
Nonlinear SVM detection	0.573	0.562
	0.551	
S-CNN detection	0.672	0.658
	0.644	

Based on the above experiments, we theoretically analyze the security of the algorithm in this paper. The security of the scheme in this article is based on two aspects: First, it is based on the confidentiality of digital Cardan Grille. The security of digital Cardan Grille cannot be separated from the key and digital digest algorithm. In the case that the key is not public, the message digest algorithms such as SHA-1 and MD5 are irreversible And uniqueness, which guarantees that brute force cracking is not feasible. In other words, the security of the system depends on the confidentiality of the selected key, which fully complies with the Kerckhoffs criterion. At the same time, a message digest algorithm with different key lengths can be selected according to different secret levels. The algorithm is easy to implement and difficult to decipher under the premise of regularly changing keys. The second is that the transmitted dense image is directly generated by the generator, which can meet the statistical characteristics of the current specific steganography detection, and greatly increases the ability to resist steganalysis. But compared with the traditional steganography method based on cover modification, it is less versatile.

Assuming that the attacker suspects that the passed image contains secret information, it is also difficult to obtain the same cardengo to extract secret information. Even if the algorithm of information steganography is leaked, under the premise of no key, only meaningless results will be obtained, thereby ensuring the security of covert communication.

5 Conclusion and Future Work

In this paper, a generative image steganography algorithm based on digital Cardan Grille is proposed. Combining the ideas of traditional Cardan Grille and deep learning, and using the deep generation model to construct a normal semantic image, which is fully in line with the research direction of coverless information hiding. The program can be extended to other media such as text, video and other fields. We use the CelebA dataset and the LUSN dataset to evaluate the performance of the proposed scheme. Theoretical analysis and experimental results show the feasibility and safety of the method. However, the generality of the algorithm in this paper is still poor and needs to be improved.

In future work, we hope to pay more attention to the embedded information capacity. On the basis of ensuring the security of the generated image, the embedded capacity is increased to meet the needs of realistic covert communication.

Acknowledgment. This work was supported by National Key R&D Program of China (Grant No. 2017YFB0802000), National Natural Science Foundation of China (Grant Nos. 61379152, 61403417).

References

1. Utepbergenov, I., Mussin, T., Kuandykova, J.: Creating a program and research a cryptosystem on the basis of Cardan grille. In: 2013 Second International Conference on Informatics and Applications (ICIA). IEEE (2013)
2. Wang, H., Wang, S.: Cyber warfare: steganography vs. steganalysis. Commun. ACM **47**(10), 76–82 (2004)
3. Goodfellow, I.J., Pouget-Abadie, J., Mirza, M., et al.: Generative adversarial networks. Adv. Neural. Inf. Process. Syst. **3**, 2672–2680 (2014)
4. Katzenbeisser, S., Petitcolas, F.A.P.: Information Hiding Techniques for Steganography and Digital Watermarking (2000)
5. Fridrich, J.: Steganography in Digital Media: Principles, Algorithms, and Applications. Cambridge University Press, Cambridge (2010)
6. Zhou, Z.L., Cao, Y., Sun, X.M.: Coverless information hiding based on bag-of-words model of image. J. Appl. Sci. **34**(5), 527–536 (2016)
7. Otori, H., Kuriyama, S.: Data-embeddable texture synthesis. In: Butz, A., Fisher, B., Krüger, A., Olivier, P., Owada, S. (eds.) SG 2007. LNCS, vol. 4569, pp. 146–157. Springer, Heidelberg (2007). https://doi.org/10.1007/978-3-540-73214-3_13
8. Otori, H., Kuriyama, S.: Texture synthesis for mobile data communications. IEEE Comput. Graph. Appl. **29**(6), 74–81 (2009)
9. Volkhonskiy, D., Nazarov, I., Borisenko, B., et al.: Steganographic generative adversarial networks [EB/OL]. ArXiv e-prints, 2017, 1703, 16 March 2017. http://arxiv.org/abs/1703.05502.pdf
10. Radford, A., Metz, L., Chintala, S.: Unsupervised representation learning with deep convolutional generative adversarial networks. Comput. Sci. (2015)
11. Shi, H., Dong, J., Wang, W., et al.: SSGAN: secure steganography based on generative adversarial networks [EB/OL]. ArXiv e-prints, 2018, 1707, 06 November 2018. https://arxiv.org/abs/1707.01613v3.pdf

12. Wang, Y.J., Niu, K., Yang, X.Y.: Information hiding scheme based on generative adversarial network. J. Comput. Appl. **38**(10), 2923–2928 (2018). https://doi.org/10.11772/j.issn.1001-9081

13. Hu, D., Wang, L., Jiang, W., et al.: A novel image steganography method via deep convolutional generative. IEEE Access **6**, 38303–38314 (2018). https://doi.org/10.1109/ACCESS.2018.2852771

14. Hayes, J., Danezis, G.: Generating steganographic images via adversarial training [EB/OL]. ArXiv e-prints, 2017, 1703, 01 March 2017. https://arxiv.org/abs/1703.00371.pdf

15. Liu, J., Ke, Y., Lei, Y., et al.: The reincarnation of grille cipher: a generative approach (2018)

16. Liu, M.M., Zhang, M.Q., Liu, J., et al.: Coverless information hiding based on generative adversarial networks. J. Appl. Sci. **36**(2), 371–382 (2018). https://doi.org/10.3969/j.issn.0255-8297.2018.02.015

17. Ke, Y., Liu, J., Zhang, M.-Q., et al.: Steganography security: principle and practice. IEEE Access **6**, 73009–73022 (2018)

18. Novotny, M.A.: Matrix products with applications to classical statistical mechanics, 1. Reflectivity of one-dimensional solids, 2. **56**(3), 452–458 (1978)

19. Liu, J., Zhou, T., Zhang, Z., et al.: Digital cardan grille: a modern approach for information hiding. In: The 2018 2nd International Conference (2018)

20. Yeh, R., Chen, C., Lim, T. Y., Hasegawajohnson, M., Do, M.N.: Semantic image inpainting with perceptual and contextual losses (2016)

21. Dumoulin, V., Belghazi, I., Poole, B., et al.: Adversarially learned inference (2016)

22. Amos, B.: Image completion with deep learning in TensorFlow. https://github.com/bamos/dcgan-completion.tensorflow

23. Mandgaonkar, V.S., Kulkarni, C.V.: No reference image quality assessment. In: 2014 Annual IEEE India Conference (INDICON). IEEE (2015)

24. Suryawanshi, G.R., Mali, S.N.: Study of effect of DCT domain steganography techniques in spatial domain for JPEG images steganalysis. Int. J. Comput. Appl. **127**(6), 16–20 (2015)

25. Zhen, Y.U., Chen, K.: Analysis and improvement on RS detection algorithm. Comput. Eng. **34**(8), 170–178 (2008)

26. Yan, H., Qin, J.: Trojan Horse detection method based on nonlinear SVM model. Comput. Eng. **37**, 121–123 (2011)

27. Chen, T., Lu, S., Fan, J.: S-CNN: Subcategory-aware convolutional networks for object detection. IEEE Trans. Pattern Anal. Mach. Intell. 1

Pixel Grouping Based Image Hashing for DIBR 3D Image

Chen Cui[1], Xujun Wu[1], Jun Yang[2], and Juyan Li[1(✉)]

[1] School of Information Science and Technology, Heilongjiang University,
Harbin, Heilongjiang, China
2018012@hlju.edu.cn, 1149347287@qq.com, lijuyan587@163.com
[2] College of Mathematics Physics and Information Engineering, Jiaxing University,
Jiaxing 314000, China
yangj95@mail2.sysu.edu.cn

Abstract. Most of the traditional 2D image hashing schemes do not take into account the change of viewpoint to construct the hash vector, resulting in the classification accuracy rate is unsatisfactory when applied in identification for Depth-image-based rendering (DBIR) 3D image. In this work, pixel grouping according to histogram shape and Nonnegative matrix factorization (NMF) is applied to design DIBR 3D image hashing with better robustness resist to geometric distortions and higher classification accuracy rate for virtual images identification. Experiments show that the proposed hashing is robust to common signal and geometric distortion attacks, such as additive noise, blurring, JPEG compression, scaling and rotation. When compared with the state-of-art schemes for traditional 2D image hashing, the proposed hashing provides better performances under above distortion attacks when considering the virtual images identification.

Keywords: DIBR 3D image identification · Image hashing · Histogram · Nonnegative matrix factorization (NMF)

1 Introduction

Depth-image-based Rendering (DIBR) [1] is a kind of 3D representation technology, by which virtual right image and right image are generated from the center image according to the depth information represented with the depth image. Then, viewers can easily get stereo perception with the virtual image pair. In the digital communication model of DIBR 3D image, receiver performs

Supported by the National Natural Science Foundation of China (Grant Number: 61702224). The Special Funds of Heilongjiang University of the Fundamental Research Funds for the Heilongjiang Province (RCCXYJ201811, RCCXYJ201812). The Open Fund of the State Key Laboratory of Information Security (2019-ZD-05). The Natural Science Foundation of Zhejiang Province (No. LY18F020020).

D. Wang et al. (Eds.): SPNCE 2020, LNICST 344, pp. 356–369, 2021.
https://doi.org/10.1007/978-3-030-66922-5_25

depth-image-based Rendering operation to generated virtual image pair for 3D video. As a matter of fact, either of the center image, the virtual left image and virtual right image may suffer from illegal or unauthorized re-distributions. In order to resolve this problem, robust perceptual hashing has been widely used for digital multimedia protection. As variety of copies for the center image and virtual images existing, image hashing can also help us to find the similar one and detect the tempered. In this paper, we focus on designing a robust image hashing scheme for DIBR 3D image identification.

In DIBR system, virtual right image and left image are generated from the corresponding center image with pixels mapping. In a sense, virtual images have similar visual content with their corresponding center image, which demands the hashing method to be designed should identify the virtual images with the same content as the center image as shown in Fig. 1.

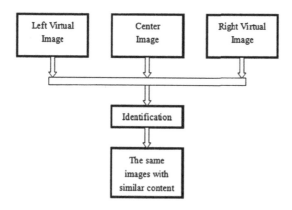

Fig. 1. The character of image hashing for DIBR 3D images

2   Related Work

Robust images hashing has been extensively studied for content-based identification of traditional 2D images. As feature extraction affects the identification performance for image hashing, many existing methods focus on extracting robust features resist to content-preserving operations [2,3]. In addition, some matrix analysis methods have been adopted to extract the robust features for hash generation, such as singular value decomposition (SVD) [4] and non-negative matrix factorization (NMF) [5], which are robust to most kinds of signal distortion attacks but sensitive to geometric distortions such as rotation.

In order to make the hashing scheme robust to geometric distortion attacks such as rotation, some robust image hashing schemes are proposed to extract the geometric-invariant features for generating the final hash vector [6]. Lv et al. [7] proposed a shape contexts and local feature points based image hashing scheme. Compressing the descriptors of stable SIFT feature points in each hash bin to form the final hash vector, their hashing scheme is robust to geometric distortion attacks, such as rotation. Tang et al. [8,9] proposed a kind of robust image hashing scheme based on ring partition. Using the pixels in every ring to form a secondary image resist to rotation, they extract the final hash vector from the secondary image. The experimental results show that their hash schemes are robust to rotation with good discriminative capability. This kind of method considers that the viewpoint dose not change, when the digital image is attacked by most of the content-preserving manipulations. In other words, the image center of original image and their copies would not change. In fact, the center of center image and virtual images are different, which is caused by DIBR 3D process. As a result, this kind of traditional 2D hashing scheme would not show good performance when applied for DIBR 3D image identification.

Some of the state-of-art traditional 2D image hashing schemes which resist to geometric distortions do not take into account the situation about viewpoint changing [7,8]. Dividing the image into several rings or constructing rotation invariant secondary image according to the unchanged image center is the key step to construct hash vector robust to rotation manipulation. However, the image center changes when generating virtual images in the DIBR system.

In this work, a pixel grouping and nonnegative matrix factorization based hashing scheme is designed for DIBR 3D images identification. The key contribution is using the approximate invariance of histogram shape to eliminate as much difference between center image and virtual image as possible. The rest of this paper is organized as follows: Sect. 2 briefly reviews the DIBR operations. Section 3 introduces the pixel grouping according to approximate invariance of histogram shape and nonnegative matrix factorization based image hashing. Section 4 shows the experimental results and performance comparisons. Section 5 gives the final conclusions.

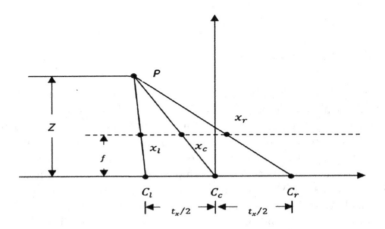

Fig. 2. The relationship of pixel in the left image, center image and right image

3 Review of Depth-Image-Based Rendering Process

Figure 2 illustrates the relationship between the center image and virtual images generated by DIBR operations [10]. Suppose P is a point in the space, C_c, C_l and C_r represent the center viewpoint, left viewpoint and the right viewpoint, respectively, f represents the focal length of the center viewpoint, Z represents the depth of P. x_c, x_l and x_r represent the x-coordinate of pixel in the center image, the virtual left image and the virtual right image, respectively. t_x represents the baseline distance, value of which is equal to the distance between the left and right viewpoints. As geometric relations shown in Fig. 2, x-coordinate of pixel in virtual images is computed as

$$x_l = x_c + \frac{t_x}{2}\frac{f}{Z},$$
$$x_r = x_c - \frac{t_x}{2}\frac{f}{Z}, \qquad \cdot \tag{1}$$

$$Z(v) = Z_{far} + v \times \frac{Z_{near}-Z_{far}}{255}, v \in [0,255] \tag{2}$$

Where x_l, x_c and x_r represent the x–coordinate of pixels in the left virtual image, center image and virtual right image, respectively. In fact, the gray values of pixels in gray image are not the real depth value. Pixel with gray value close to 255 indicates that P is close to the near clipping plane Z_{near}. On the other hand, pixel with gray value close to 0 indicates that P is close to the far clipping plane Z_{far}. According to formula 2, the depth value $Z(v)$ of P is computed, where v represents the gray value.

4 Proposed Image Hashing

Our DIBR 3D image hashing algorithm includes the following steps: the original center image is filtered with a Gaussian kernel low-pass filter to get the low frequency, and we standardize the low frequency of center image for hash generation. Then, pixels of normalized low frequency image are divided into different groups according to the histogram shape. Then these pixel groups are used to construct an secondary image, which is almost unchangeable under geometric distortions and slightly changes after DIBR operations. Lastly, the secondary image is discomposed by nonnegative matrix factorization to get the coefficient matrix, and the final hash is constructed with these coefficients.

4.1 Pre-processing

Low-pass filtering is adopted to extract the low-frequency component of original center image, aiming to enhance robustness of proposed hashing scheme to some common content-preserving manipulations [13]. The low-frequency component IC_{low} of original center image IC is obtained as

$$IC_{low}(x, y) = G(x, y, \sigma) * I(x, y) \tag{3}$$

$*$ represents the convolution operation, and the low-pass filter Gaussian function $G(x, y, \sigma)$ is represented as

$$G(x, y, \sigma) = \frac{1}{2\pi\sigma^2} e^{-\frac{x^2+y^2}{2\sigma^2}} \tag{4}$$

σ is the standard difference. According to parameters setting in [13], σ is set to 1.

4.2   Pixel Grouping

The gray levels of filtered image I_{low} also range from 0 to 255. In this paper, only pixels with M different gray levels are randomly selected to construct the secondary, aiming to ensure the security of proposed hashing algorithm. With a key-based sequence $P(M) = \{p_i | i = 1...M, 0 \leq p_i \leq 255\}$, M gray levels $h_1, h_2, ..., h_M$ are selected for pixels grouping, where $h_i = p_i$. The set of selected gray level is represented as

$$H_M = \{h_i | i = 1, ..., M\} \tag{5}$$

After resizing I_{low} to $m \times m$, pixels with L_B neighboring gray levels in H_M are selected to form one pixel group. In total, $n = \lfloor M/L_B \rfloor$ groups are formed, where $\lfloor \bullet \rfloor$ is a floor function.

Suppose g_i be one of the pixel groups. In order to form the i-th column of the secondary image, we sort and resize g_i to a new vector v_i sized $k \times 1$. Then the secondary image is modeled as

$$V = [v_1, v_2, v_3, ..., v_n] \tag{6}$$

It is clear that the histogram shape of V is the same as resized I_{low}, the secondary image V is robust to geometric distortions such as rotation. In this paper, M is set to 240, $m = 256$, $L_B = 6$, $k = 4m$.

4.3   Hash Generation

Since the histogram shape is almost unchangeable under geometric distortions and slightly changes after DIBR operations, features extracted from the secondary image V also have this property. NMF is used to get the base matrix W and coefficient matrix H, respectively. Concatenate the coefficient matrix H to obtain a the final hash vector, the length L of hash vector is $n \times r$, where n is the number of pixel groups and r is the rank for NMF. In this paper, r is set to 2.

In this paper, correlation coefficient is taken as the metric to measure similarity between two image hash vectors $Hash1$ and $Hash2$. The correlation coefficient $S(Hash1, Hash2)$ is defined as

$$S(Hash1, Hash2) = \frac{cov(Hash1, Hash2)}{\sqrt{D(Hahs1)}\sqrt{D(Hahs1)}} \tag{7}$$

According to formula 7, S ranges from -1 to 1, a bigger S value indicates that the input image is more similar with the original corresponding center image. If the correlation coefficient S is higher than the threshold predefined, the input image is viewed as perceptual content unchanged. If the correlation coefficient S is lower than the threshold predefined, the input image is viewed as a different image or a maliciously tempered version of the original corresponding center image. For DIBR 3D images, the virtual images should have much bigger S value when computing the perceptual distance from their corresponding original center image. According to experiment results listed in Table 1 and Table 2, some virtual images are viewed different from the original center image when the hashing method proposed in [9] is adopted. It is clearly that our DIBR 3D image hashing scheme can identify the virtual images with visual contents the same as the original center one.

Table 1. Perceptual distance between center image and left virtual image computed by different hashing method.

Image	Proposed method	Method in [9]
Breakdancers	0.9984	-0.3817
Dolls	0.9952	-0.7150
Books	0.9982	-0.7499
ballet	0.9984	0.8183

Table 2. Perceptual distance between center image and right virtual image computed by different hashing method.

Image	Proposed method	Method in [9]
Breakdancers	0.9990	0.5647
Dolls	0.9909	0.7812
Books	0.9982	0.8849
ballet	0.9980	0.9093

5 Experimental Results

120 different color images are collected from the Ground Truth Database [14] in order to test discriminative capability of proposed hashing. The hash vectors are generated for these 120 images, then 7140 correlation coefficients S are computed between each pair of different hash vectors. The maximum value of these correlation coefficients is 0.9785, and the minimum value is -0.5101. If the

threshold T is set as 0.92, 0.32% pairs of different images are identified with the similar content. 0.09% pairs of different images are identified with the similar content with T is set as 0.94. No pair of different images is identified with the similar content when T is set as 0.98.

Table 3. Content-preserving operations and the parameters setting.

Manipulation	Parameters setting	Copies
Additive noise		
Gaussian noise	$variance \in (0.0005{\sim}0.005)$	10
Salt & Paper noise	$variance \in (0.001{\sim}0.01)$	10
Speckle Noise	$variance \in (0.001{\sim}0.01)$	10
Blurring		
Gaussian blurring	filter size:3 $\sigma \in (0.5{\sim}5)$	10
Circular blurring	radius $\in (0.2{\sim}2)$	10
Motion blurring	len $= 1, 2, 3$ $\theta = 0°, 45°, 90°$	9
Geometric attacks		
Rotation	$\theta \in (-10°{\sim}+10°)$	12
Cropping & Rotation	$\theta \in (-10°{\sim}+10°)$	12
Scaling	$factor \in (0.5{\sim}2.0)$	6
JPEG compression	$QF \in (10{\sim}100)$	10

Database with 2727 images is constructed to evaluate the identification performance for DIBR 3D image. In this database, we select 9 pairs of center and depth images from Middlebury Stereo Datasets [11] and Microsoft Research 3D Video Datasets [12], and sizes of these images are ranging from 450×375 to 1390×1110. Hashes extracted from center image, virtual left image, virtual right image and their distorted versions are generated with our hashing scheme in order to calculate the identification accuracy rate. The distorted versions are generated by attacking the center and virtual images according to 10 classes of common content-preserving operations. In this paper, Matlab is exploited to implemented these 10 classes operations with different parameters. These operations include common signal and geometric distortion attacks such as JPEG compression, blurring, additive noise, scaling, rotation and cropping after rotation. The operations and their parameters are listed in Table 3.

Firstly, four pairs of the center and depth image are selected from above dataset. They are Breakdancers, Books, Dolls, and ballet as listed in Table 1. each virtual images pair and the center image are attacked by the content-preserving operations listed in Table 3. As shown in Fig. 3, no pair of visually identical images (including the distorted center and virtual images) is identified as different content when the threshold T is set to 0.96.

In order to show the identification performance of our DIBR 3D image hashing scheme is better than some other existing traditional 2D hashing schemes,

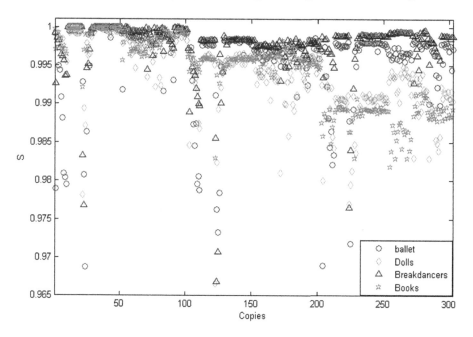

Fig. 3. Robustness test based on four test images.

two kinds of the current state-of-the-art 2D image hashing schemes are selected to for experimental comparisons. One is the Shape Contexts and Local Feature Points based hashing algorithm (RSCH) proposed in paper [7], and the other is the Ring Partition and NMF based hashing algorithm proposed in [9].

Suppose $IC = \{IC_i, 1 \leq i \leq S\}$ be the set of original center images. Then we generate the compact hash $H(IC_i)$ from each of the center images, $H(IC_i) = (h_1, h_2, ...h_L)$ is the hash vector with length L for center image IC_i.

In this paper, we use correlation coefficient as the performance metric to evaluate the distance between two different hash vectors $H(I_1)$ and $H(I_2)$. Suppose $H(IC_i)$ is the hash vector of one of the center image set, and $H(I_Q)$ is the query hash vector of distorted vision for either of the center image or their corresponding virtual images. Then, we calculate the correlation coefficient S between $H(I_Q)$ and $H(IC_i)$, and the query image is identified as the i-th original center image as

$$i = argmax_i \left\{S(H(I_Q), H(IC_i))\right\} \tag{8}$$

Where $S(H(I_Q), H(IC_i))$ is calculated as the correlation coefficient between $H(I_Q)$ and $H(IC_i)$.

Higher identification accuracy rate means that the images attacked by common content-preserving operations can still be identified having similar perceptual content with the original one, no matter attacked by common content-preserving operations. When considering the DIBR 3D image identification, high identification performance means that the virtual images should be identified

Table 4. Identification accuracy performances for center and virtual images by different methods.

Manipulation	Our method	Method in [9]	Method in [7]
Additive Noise			
Gaussian Noise	100%	78.89%	69.26%
Salt & Paper Noise	100%	81.48%	78.52%
Speckle Noise	100%	80.00%	77.78%
Blurring			
Gaussian Blurring	100%	81.11%	71.48%
Circular Blurring	100%	80.74%	78.14%
Motion Blurring	100%	79.84%	77.41%
Geometric Attacks			
Rotation	100%	73.46%	61.72%
Cropping & Rotation	100%	74.07%	71.91%
Scaling	100%	79.32%	64.20%
JPEG Compression	100%	80.25%	84.88%

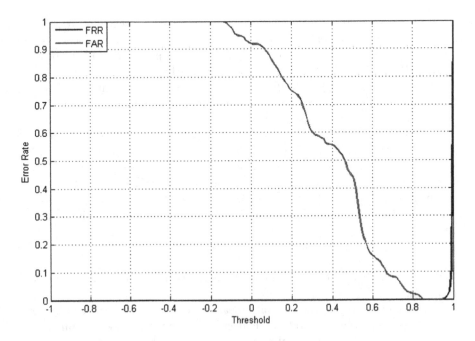

Fig. 4. Robustness test based on four test images.

having similar perceptual content with their corresponding center image even though the virtual images are attacked by common content-preserving operations. In this paper, FRR(false reject rate) and FAR(false accept rate) are used to evaluate the robustness of proposed DIBR 3D image hashing scheme. FRR describes the error identification probability, the smaller FRR is, the better robustness of hash algorithm. FAR reflects the discrimination of hashing algorithm, the smaller FAR is, the better the discrimination. It is clearly that an excellent hashing algorithm should have the minimum FRR and minimum FAR with a certain threshold. As shown in Fig. 4, the FRR and FAR are zero when the threshold is set from 0.86 to 0.93. This experiment shows that the proposed hashing scheme is robust to common signal and geometric distortion attacks, such as additive noise, blurring, JPEG compression, scaling and rotation.

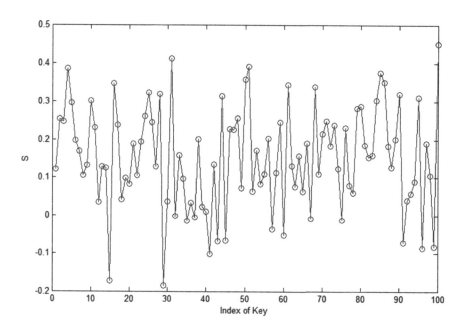

Fig. 5. Correlation coefficients between hashes of "Breakdancers" generated by different keys.

According to the experiment results listed in Table 4, it is clearly that our DIBR 3D hashing scheme outperforms Ring partition hashing scheme and RSCH hashing scheme under content-preserving operations listed in above section. The underlying reason is that these kinds of traditional 2D image hashing method consider that all perceptually insignificant distortions and malicious manipulations on a digital image would not lead to viewpoint changes, and the center of an image is generally preserved and thus relatively stable under geometric

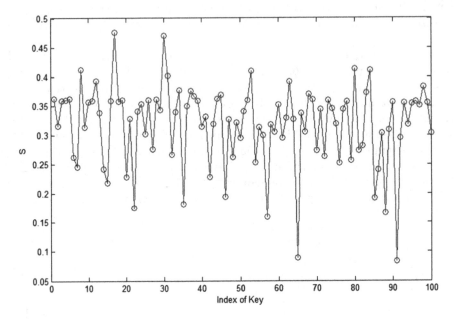

Fig. 6. Correlation coefficients between hashes of "Ballet" generated by different keys.

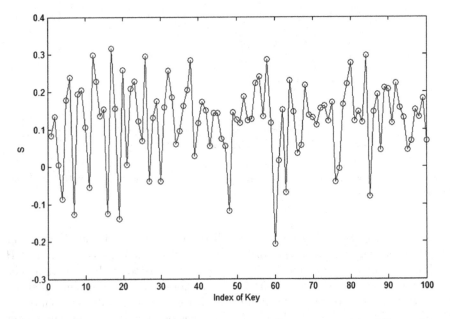

Fig. 7. Correlation coefficients between hashes of "Dolls" generated by different keys.

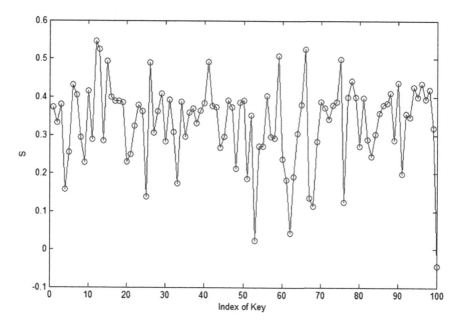

Fig. 8. Correlation coefficients between hashes of "Books" generated by different keys.

attacks such as rotation. In fact, virtual images are generated from center image by pixels shifting in the DIBR process.

In paper [7], they divide the image into several rings with the center of the image as the center, compressing the descriptors of selected SIFT feature points in every ring to generate the final hash. In fact, the center of virtual images and center image are different result in the hash vectors between the center image and virtual images are different. In paper [9], they also divide the image into several rings with the center of the image as the center. Using the pixels in every ring to form a secondary image, they extract the final hash from the secondary image. In the same way mentioned above, the different centers lead to form different secondary images, and the final hash vector of the center image is different from the hash vector of either virtual image.

To enhance the security of hashing scheme, a secret key is usually used in the processes of feature extraction and feature compression to generate the final hash. As a result, the key-based hashing scheme is key-dependent, making the hash unpredictable to prevent unauthorized access.

In the proposed hashing scheme, only pixels with M different gray levels are used to construct the secondary image. Using a key-based sequence P(M) to select pixel groups, the security of proposed hashing scheme is enhanced. To validate key dependence of proposed hashing scheme, four images "Breakdancers", "Ballet", "Dolls" and "Books" are adopted.

For each image, hashes are generated with 100 different keys. Then, we calculate the correlation coefficients between the original key-based hash and hashes with different keys, it can be found that all correlation coefficients between different hashes of the four images are smaller. It should be noted that the parameters of hash generation are kept unchanged except the key-based sequence P(M) for selecting pixel groups in this experiment. Then, the correlation coefficients between the original key-based hash and other 100 hashes with different keys are computed for the four images mentioned above, and the obtained results are illustrated in Fig. 5, Fig. 6, Fig. 7 and Fig. 8, where the x-axis is the index of key and the y-axis is the correlation coefficient S, which represents the hash distance. For the image of "Breakdancers", the maximum, the minimum and the average distances are 0.4507, −0.1849 and 0.1525, respectively. For the image of "Ballet", the maximum, the minimum and the average distances are 0.4754, 0.0838 and 0.3185, respectively. For the image of "Dolls", the maximum, the minimum and the average distances are 0.3162, −0.2067 and 0.1226, respectively. For the image of "Books", the maximum, the minimum and the average distances are 0.5470, −0.0440 and 0.3319, respectively. It is clearly that the maximum distances between the original key-based hash and other 400 hashes with different keys are lower than 0.96 as shown in Fig. 3. This experimental result shows that the security of proposed hashing scheme is enhanced with a key-based sequence P(M) to select pixel groups.

6   Conclusions

In this paper, we propose a pixel grouping and NMF based DIBR 3D image hashing scheme, which can be used for virtual images identification, authentication and retrieval. Low-pass filtering and histogram shape based pixel grouping are the key steps to make proposed hashing scheme robust to common content-preserving manipulations, and the approximate invariance of histogram shape to cropping and DIBR operations ensure that our DIBR 3D image hashing scheme also have better performance for virtual images identification. The experiment results have shown that the proposed DIBR 3D image hashing resists common content-preserving manipulations including signal distortion attacks and geometric distortion attacks. However, the proposed hashing method may identify an input image with different content to be visually identical, when the input image has the same histogram shape. We will solve this problem in the future work.

Acknowledgment. We would like to thank anonymous reviewers for their helpful comments and suggestions, and their comments and suggestions help us to improve this paper's quality. This work is supported by the National Natural Science Foundation of China (Grant Number: 61702224). The Special Funds of Heilongjiang University of the Fundamental Research Funds for the Heilongjiang Province (RCCXYJ201811, RCCXYJ201812). The Open Fund of the State Key Laboratory of Information Security (2019-ZD-05). The Natural Science Foundation of Zhejiang Province (No. LY18F020020).

References

1. Fehn, C.: Depth-image-based rendering (DIBR) compression and transmission for a new approach on 3D-TV. In: Proceedings of the SPIE Stereoscopic Displays and Virtual Reality Systems XI, pp. 93–104 (2004)
2. Ahmed, F., Siyal, M.Y., Abbas, V.U.: A secure and robust hash-based scheme for image authentication. Sig. Process. **90**(5), 1456–1470 (2010)
3. Monga, V., Evans, B.L.: Perceptual image hashing via feature points: performance evaluation and tradeoffs. IEEE Trans. Image Process. **15**(11), 3452–3465 (2006)
4. Kozat, S., Venkatesan, R., Mihcak, M.: Robust perceptual image hashing via matrix invariants. In: 2004 International Conference on Image Processing, pp. 3443–3446. IEEE, Singapore (2004)
5. Monga, V., Mhcak, M.K.: Robust and secure image hashing via non-negative matrix factorizations. IEEE Trans. Inf. Forensics Secur. **2**(3), 376–390 (2007)
6. Roy, S., Sun, Q.: Robust hash for detecting and localizing image tampering. In: 2007 IEEE International Conference on Image Processing, pp. 117–120. IEEE, San Antonio (2007)
7. Lv, X., Wang, Z.J.: Perceptual image hashing based on shape contexts and local feature points. IEEE Trans. Inf. Forensics Secur. **7**(3), 1081–1093 (2012)
8. Tang, Z.J., Zhang, X.Q., Li, X.X., Chao, S.C.: Robust image hashing with ring partition and invariant vector distance. IEEE Trans. Inf. Forensics Secur. **11**(1), 200–214 (2016)
9. Tang, Z.J., Zhang, X.Q., Chao, S.C.: Robust perceptual image hashing based on ring partition and NMF. IEEE Trans. Knowl. Data Eng. **26**(3), 711–724 (2014)
10. Zhang, L., Tam, W.: Stereoscopic image generation based on depth images for 3D TV. IEEE Trans. Broadcast. **51**(2), 191–199 (2005)
11. Scharstein, D., Pal, C.: Learning conditional random fields for stereo. In: 2007 IEEE Conference on Computer Vision and Pattern Recognition, pp. 1–8. IEEE, Minneapolis (2007)
12. Wang, Z., Bovik, A.C., Sheikh, H.R., Simoncelli, E.P.: Image quality assessment: from error visibility to structural similarity. IEEE Trans. Image Process. **13**(4), 600–612 (2007)
13. Xiang, S.J., Kim, H.J., Huang, J.W.: Invariant image watermarking based on statistical features in the low-frequency domain. IEEE Trans. Circuits Syst. Video Technol. **18**(6), 777–790 (2008)
14. Ground Truth Database. http://www.cs.washington.edu/research/imagedatabase/groundtruth/. Accessed 8 May 2008

Improved Conditional Differential Analysis on NLFSR Based Block Cipher KATAN32 with MILP

Zhaohui Xing[1,2]📖, Wenying Zhang[1(✉)]📖, and Guoyong Han[3]📖

[1] School of Information Science and Engineering, Shandong Normal University,
Jinan 250014, China
3613452@qq.com, zhangwenying@sdnu.edu.cn
[2] School of Science, Shandong Jiaotong University, Jinan 250357, China
[3] School of Management Engineering, Shandong Jianzhu University,
Jinan 250101, China
hgy_126@126.com

Abstract. This paper describes constructing a Mixed Integer Linear Programming (MILP) model for conditional differential cryptanalysis on nonlinear feedback shift register (NLFSR)-based block ciphers, and proposes an approach for detecting the bit with a strongly-biased difference. The model is successfully applied to the block cipher KATAN32 in the single-key scenario, resulting in practical key-recovery attacks covering more rounds than the previous. In particular, we present two distinguishers for 79 and 81 out of 254 rounds of KATAN32. Based on the 81-round distinguisher we recover 11 equivalent key bits of 98-round KATAN32 with the time complexity being less than 2^{31} encryptions of 98-round KATAN32 and recover 13 equivalent key bits of 99-round KATAN32 with the time complexity being less than 2^{33} encryptions of 99-round KATAN32. Thus far, our results are the best known practical key-recovery attacks for the round-reduced variants of KATAN32 as far as the number of rounds and the time complexity. All the results are verified experimentally.

Keywords: KATAN block cipher · Conditional differential cryptanalysis · Mixed Integer Linear Programming (MILP)

1 Introduction

Radio frequency identification (RFID) technology has been used in many aspects of life such as access control, parking management, identification, goods tracking, and wireless sensor networks (WSNs) have been used for various critical industrial applications, such as heart beat monitoring, temperature monitoring for precision agriculture, and power usage monitoring for smart grid [22]. In this new cryptography environment, applications of RFID technology and sensor networks have similar features such as weak computation ability, small

© ICST Institute for Computer Sciences, Social Informatics and Telecommunications Engineering 2021
Published by Springer Nature Switzerland AG 2021. All Rights Reserved
D. Wang et al. (Eds.): SPNCE 2020, LNICST 344, pp. 370–393, 2021.
https://doi.org/10.1007/978-3-030-66922-5_26

storage space, and strict power constraints, and such very constrained environments require new cryptographic primitives such as tiny and efficient ciphers. This tends to make traditional block ciphers such as AES not suitable for these constrained environments. To meet these needs, a number of lightweight ciphers, including PRESENT [5], KATAN and KTANTAN family [6], have been proposed.

The KATAN and KTANTAN block ciphers were proposed by Christophe DeCannière, Orr Dunkelman and Miroslav Knezevic at CHES 2009 [6]. To increase its speed, KATAN uses nonlinear feedback shift registers (NLFSRs) as well as linear key schedule. Both KATAN and KTANTAN have three variants with 32-bit, 48-bit and 64-bit block sizes, each requiring an 80-bit user key. In addition, KATAN and KTANTAN share the same data path specification, including round transformation and round constants. The only difference between KATAN and KTANTAN is the generation of subkeys. For KTANTAN, two bits of the 80-bit $K = k_{79}k_{78}...k_1k_0$ are selected each round. But the key schedule of the KATAN32 cipher (and the other two variants KATAN48 and KATAN64) loads the 80-bit key into an LFSR (the least significant bit of the key is loaded to position 0 of the LFSR), and for each round positions 0 and 1 of the LFSR are generated as the round subkey k_{2i} and k_{2i+1}, and the LFSR is clocked twice. Because of the simple key schedule, KTANTAN was broken by Wei et al. [23], and while a more complex key schedule makes KATAN secure and stronger, the key schedule is also linear.

1.1 Related Work

KATAN family ciphers have been analyzed by extensive cryptanalysis. At ASIACRYPT 2010, Knellwolf et al. analyzed KATAN&KTANTAN [14] using conditional differential cryptanalysis [3] and recovered four equivalent key bits for 78 of 254 rounds of KATAN32 in the single-key scenario. They subsequently analyzed KATAN32 in the related-key scenario with an improved technique using automatic tools, then obtained key-recovery attacks for 120 of 254 rounds of KATAN32 [15]. Finding the non-uniformity of the difference distribution after 91 rounds, Martin R. Albrecht and Gregor Leander proposed an 91-round distinguisher with the time complexity being 2^{32} encryptions [2]. These results on KATAN32 are listed in Table 1.

Other types of attacks formally published on this cipher, such as all subkeys recovery (ASR) which is a variant of the meet-in-the-middle (MITM) attack [12], Match Box MITM attack [9], Dynamic Cube attack [1], Multidimensional MITM attack [18,28], are also listed in Table 1. As can be seen from the details in Table 1, each time complexity is too high to present a practical attack.

As stated in [11], related-key attacks are arguable in a practical sense, because a related-key attack is under the assumption that the attacker had known and even controlled the relation between multiple unknown keys. Because of this assumption, the related-key attack is arguable from the aspect of practical security, though it is meaningful during the design and certification of a cipher. In particular, the key of an ultra lightweight block cipher in low-end devices such

Table 1. Cryptanalytic results on KATAN32

Type	Scenario	Rounds	Time complexity	Reference
Conditional differential	Single-key	78	2^{22}	[14]
	Related-key	120	2^{31}	[15]
Improved conditional differential	Single-key	97	2^{30}	This paper
	Single-key	98	2^{31}	This paper
	Single-key	99	2^{33}	This paper
Differential	Single-key	91	$2^{32}*$	[2]
	Single-key	115	2^{78}	
MIMT ASR	Single-key	119	$2^{79.1}$	[12]
Match box MITM	Single-key	153	$2^{78.5}$	[9]
Dynamic cube	Single-key	155	$2^{78.3}$	[1]
Multidimensional MITM	Single-key	175	$2^{79.3}$	[28]
	Single-key	206	2^{79}	[18]

*A 91-round differential distinguisher

as a passive RFID tag may not be changed during its life cycle so, in a practical sense, the security of a lightweight cipher under the single-key scenario is the most important. As shown in [13], even though the result of an attack in the related-key scenario is better, it is still meaningful to explore an attack in the single-key scenario.

Conditional differential cryptanalysis was first introduced by Biham and Ben-Aroya at Crypto 1993 [3], since this technique, which is a very popular technique in hash functions cryptanalysis, allows increasing the probability of a differential characteristic being satisfied under some conditions, it also can be useful for block ciphers. Mixed integer linear programming (MILP) is a general mathematical tool for optimization that takes as inputs a linear objective function and a system of linear inequalities and finds solutions that optimize the objective function under the constraints of all inequalities. It was first applied by Mouha et al. in [17] and Wu in et al. [25] to count the active Sboxes of word-based block ciphers, and Sun et al. in [21] used it to search for differential characteristics and linear approximations. It has been also applied to search for integral distinguishers and division trails [26] and impossible differentials [19]. In particular, it has been applied to key-recovery attacks of keyed Keccak MAC, where attackers implemented conditional cube attacks on Keccak with the propagation of cube variables controlled under conditions in the first several rounds and attacked keyed Keccak [10,16,20].

1.2 Our Contributions

In this paper, we improved conditional differential attacks from two aspects. On the one hand, we propose a method of automatic conditional differential cryptanalysis using MILP. This method helps us minimize the number of conditions under which the differential characteristic can hold, because of the fewer the conditions, the higher the probability of the differential path. On the other hand, we propose a method to calculate the bias of every bit quickly and detect the bit which has a strongly biased difference. Finally, using the standard differential attack, we extended the conditional differential attack to more rounds. The details are described in the following paragraphs.

We first propose a novel method using MILP to automatically search an initial difference and conditions for conditional differential cryptanalysis. In [14], Knellwolf chose initial differences manually and it is difficult to find the optimal choice, a crucial element in this attack. In this paper, we solve this problem using MILP. We analyze how to identify conditions on internal state variables, and then, by modeling relations between differences in state bits and conditions, we construct a linear inequality system. The object function of this MILP problem is the minimum number of conditions in a certain number rounds. Based on the method using MILP, we automatically obtain the initial difference and conditions.

Second, we present an approach to detect the bias in the difference of the update bit. In [14], Knellwolf detected the bias experimentally by observing certain non-randomness of a difference of the update bit. We find that the probability of a difference in the update bit is determined by the probabilities of differences in bits that generate the update bit. After the analysis we present a formula for evaluating the probability of the difference in the update bit, helping us detect which bit has a strongly biased difference.

Given the initial difference, the conditions, and the position of the bit with a bias, we are able to mount a key-recovery attack.

We apply conditional differential cryptanalysis with these two improvements to analyze the security of KATAN32. It is shown that we can retrieve ten equivalent key bits for the variant of KATAN32 with 79 initialization rounds and four equivalent key bits with 81 initialization rounds.

Using standard differential attacks, we extend the 81-round conditional differential key-recovery attacks to 97-round, 98-round and 99-round with time complexity being 2^{30}, 2^{31}, 2^{33} encryptions respectively. Extended key-recovery attacks can recover 10,11 and 13 equivalent key bits respectively. This is the best known practical cryptanalytic result on KATAN32 so far.

All of our attacks succeed experimentally. All of our source codes and experiments results are available at https://www.dropbox.com/sh/028s4f06f363b2h/AADItFkz-N1KaAMZR7nIPTawa?dl=0.

1.3   Organization

The paper is organized as follows. In Sect. 2, some preliminaries are introduced. Section 3 describes the two improvements in conditional differential attacks. In Sect. 4, with these improvements, the attacks mounted on 79 and 81 of 254 rounds of KATAN32 are presented in detail. In Sect. 5 we extend the attacks to 97, 98 and 99 of 254 rounds of KATAN32 combined with standard differential attacks. Finally, we conclude the paper in Sect. 6.

2   Preliminaries

We present our notations in Table 2.

Table 2. The notations used throughout the paper

Symbol	Definition
F_2	The finite field of two elements
F_2^n	The n-dimensional vector space over F_2
S_t	The state of the 13-bit NLFSR at round t
L_t	The state of the 19-bit NLFSR at round t
s_{t+i}	The i-th bit of the 13-bit NLFSR at round t
l_{t+i}	The i-th bit of the 19-bit NLFSR at round t
Δs_{t+i}	The difference in s_{t+i}
Δl_{t+i}	The difference in l_{t+i}
X	A 32-bit plaintext block
x_i	The i-th bit of the plaintext
K	An 80-bit key
k_i	The i-th bit of the key

2.1   Description of KATAN

KATAN and KTANTAN are composed of three block ciphers with 32, 48, 64-bit block sizes, respectively, denoted by KATANn and KTANTANn for $n = 32, 48, 64$. These 6 ciphers all have 80-bit keys and the only difference between KATAN and KTANTAN is the key schedule.

Here we will briefly describe KATAN32, which is analyzed in this paper.

Key Schedule. The master key $K = (k_0, \cdots, k_{79})$ is loaded into an 80-bit linear feedback register and new round keys are generated by the linear feedback relation:

$$k_{i+80} = k_i \oplus k_{i+19} \oplus k_{i+30} \oplus k_{i+67}, 0 \leq i \leq 427.$$

In the reminder of this paper, for any $i \geq 80$, we call k_i one equivalent key bit.

Round Function. In initialization, a 32-bit plaintext block $X=(x_{31}, \ldots, x_0)$ is loaded into two NLFSRs with lengths 13 and 19 bits, respectively. Denote states of the 13-bit NLFSR and the 19-bit NLFSR at round t as $S_t = (s_t, s_{t+1}, \ldots, s_{t+12})$ and $L_t = (l_t, l_{t+1}, \ldots, l_{t+18})$.

When $t = 0$, the plaintext is loaded as $l_{t+i} = x_{18-i}$ for $0 \leq i \leq 18$ and $s_{t+i} = x_{31-i}$ for $0 \leq i \leq 12$.

At round t, for $0 \leq t \leq 253$, two new bits s_{t+13} and l_{t+19} are produced according to following equations:

$$s_{t+13} = l_t \oplus l_{t+11} \oplus l_{t+6}l_{t+8} \oplus l_{t+10}l_{t+15} \oplus k_{2t+1}, \tag{1}$$

$$l_{t+19} = s_t \oplus s_{t+5} \oplus s_{t+4}s_{t+7} \oplus s_{t+9}a_t \oplus k_{2t}, \tag{2}$$

where a_t is a round constant generated by the 8-bit LFSR using the recursive relation $a_t = a_{t-3} \oplus a_{t-5} \oplus a_{t-7} \oplus a_{t-8}(t \geq 8)$ with the seed value $(a_0, a_1, a_2, a_3, a_4, a_5, a_6, a_7) = (1, 1, 1, 1, 1, 1, 1, 0)$. After 254 rounds, the state is outputted as the ciphertext. The round function is depicted in Fig. 1.

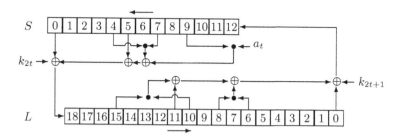

Fig. 1. The round function of KATAN32 cipher

2.2 Conditional Differential Analysis

Knellwolf et al. applied conditional differential cryptanalysis to NLFSR-based cryptosystems at ASIACRYPT 2010 [14]. This technique is based on differential cryptanalysis used to analyze initialization mechanisms of stream ciphers in [4,7,24]. After choosing an initial difference, it studies the propagation of an input difference through NLFSR-based cryptosystems and identifies conditions on internal state bits to prevent propagation whenever possible. By taking the plaintext pairs conforming with these conditions as input, biases can be detected in differences of update bits at some rounds. Once a bias is detected, the key is considered to obey the expected conditions and we obtain information for secret key bits. In some cases where there are single key bits and relations of key bits in the conditions, we call each of them one equivalent key bit, leading to a key-recovery attack.

3   Improved Conditional Differential Cryptanalysis

In [14], authors traced differences through cryptosystems and prevented their propagation whenever possible by identifying conditions on internal state variables. They gave suggestions on how to manually choose an initial difference rather than a specific method for acquiring it. These suggestions were that the difference propagation should be controllable for as many rounds as possible with a small number of conditions and that there should not be too many conditions involving bits of K during initial rounds. While the initial difference is of crucial importance with respect to the number of rounds attacked, it is not easy to manually choose a suitable initial difference. In this paper, we propose a novel method using MILP to search for an initial difference, deriving as few conditions as possible and the differential characteristic that covers as many rounds as possible. We also present a method for evaluating the probability of the difference in the update bit, by which we can detect the bit with an obvious bias.

Using these two improvements, we apply the improved conditional differential cryptanalysis to NLFSR based block cipher KATAN32. The framework of the analysis is divided into the following four steps.

Search for an Initial Difference with MILP. With the method described in Subsect. 3.1, one can formulate an MILP model of difference propagation during each round, search for a differential characteristic with minimum conditions, and obtain the initial difference of the characteristic and conditions.

Choose Conditions. We trace the propagation of the initial difference and identify conditions that prevent propagation of differences until the number of key bits and plaintext bits involved in conditions becomes too great(exceed the enumeration capability) to mount an attack.

Calculate the Bias. Given the initial difference and conditions chosen in the previous steps, the probability of the difference in each bit of the two NLFSR at the round when the conditions cease being applied can be easily derived. Taking this probability as input of the method described in Subsect. 3.2, we can calculate the probability of the difference in update bit at each subsequent round. According to these probabilities, we can locate the bit whose difference has an obvious bias and number of rounds is largest.

Mount the Key-Recovery Attack. Since the conditions include some equivalent key bits, if plaintexts are selected with the conditions consisting of correct equivalent key bits, the difference in the update bit located during the previous step would show the bias. Utilizing conditions and the bit with the bias, equivalent key bits involved in the conditions can be recovered. The attack is involved in Algorithm 1.

3.1 Modeling the Difference Propagation of the Round Function

By modeling the propagation of differences under the control of conditions that can prevent the generation of a difference in the update bit, we obtain a conditional differential characteristic with the fewest conditions, and we can also obtain the initial difference and conditions applied from this conditional differential characteristic. The steps are as follows:

Algorithm 1: The framework of the conditional differential attack

---

Input: Equation (1) and Equation (2) ;

Output: g: correct equivalent key bits;

Obtain an initial difference ΔX and a conditions set κ by MILP technique;

$\kappa \leftarrow$ {conditions chosen from κ in the previous rounds to make sure that the number of key bits and plaintext bits involved in conditions should not exceed the enumeration capability} ;

$\lambda \leftarrow$ {the probability of the difference of each bit at round r from which conditions just cease being applied. It is derived from ΔX and κ};

$P \leftarrow$ {the probabilities of the differences of each subsequent update bits after round r calculated by λ using the method described in subsection 3.2 };

$t \leftarrow$ the bit derived from P having the non-zero bias and at the highest possible number of rounds;

for $g \in$ {*enumerate equivalent key bits involved in* κ} **do**

    $count1 = 0$;

    $count0 = 0$;

    for $x \in$ {*enumerate plaintext bits involved in* κ} **do**

        if x, g satisfy κ **then**

            calculate Δt from x and ΔX;

            if $\Delta t = 1$ **then**

                | $count1 + +$;

            else

                | $count0 + +$;

            end

        end

    end

    $P\{\Delta t = 1\} = count1/(count1 + count0)$;

    $A \leftarrow A \cup \{(g, |P\{\Delta t = 1\} - \frac{1}{2}|)\}$;

end

searching in A for the max $|P\{\Delta t = 1\} - \frac{1}{2}|$;

return g in accordance with the max $|P\{\Delta t = 1\} - \frac{1}{2}|$;

---

1. Finding all Modes of Difference Propagation Under the Control of Conditions. For KATAN32 based on NLFSR, at each round only two bits are generated by some bits from the previous round, so the differences in these two update bits are caused only by the bits involved in Eq. (1) and Eq. (2).

There are linear and non-linear terms in Eq. (1) and Eq. (2). If there are differences in non-linear terms, the difference in the update bit can be canceled by imposing conditions even if there are at the same time differences in linear terms. If differences appear only in linear terms, there are no possible conditions that could be applied to cancel the differences; they only can be canceled by one another or they propagate to the next round.

For example, for Eq. (1): $s_{t+13} = l_t \oplus l_{t+11} \oplus l_{t+6}l_{t+8} \oplus l_{t+10}l_{t+15} \oplus k_{2t+1}$, if $\Delta l_{t+6} = 1$, with the other bits having no differences, we add the condition $l_{t+8} = 0$ to ensure that $\Delta s_{t+13} = 0$. The number of conditions is 1 and the difference of the update bit is 0. If $\Delta l_{t+11} = 1$, with the other bits have no differences, no conditions could cancel the difference which appears in the update bit and propagates to the next round. In this case, the number of conditions is 0 and the difference of the update bit is 1.

This shows that we can apply conditions to prevent the propagation of differences when the difference state (we call the difference of the internal state the difference state) is at some particular value, while for some other values there are no conditions that could prevent the propagation of the differences.

For each exact difference state, it can be confirmed whether conditions could be applied and whether there would be a difference in the update bit according to the previous strategy that aimed at preventing the propagation of differences.

With respect to Eq. (1), s_{t+13} is generated by 6 bits in the 19-bit NLFSR of round t, so that the difference of s_{t+13} depends on the values and the differences of these 6 bits. Let c (the flag of adding a condition) denote whether a condition is applied to cancel the difference of the update bit, and let us search all values of the vector $(\Delta l_t, \Delta l_{t+11}, \Delta l_{t+6}, \Delta l_{t+8}, \Delta l_{t+10}, \Delta l_{t+15}, \Delta s_{t+13}, c)$ following the following strategies.

If $\Delta s_{t+13} = 0$ according to Eq. (1), c takes value 0.

Example 1. If $(\Delta l_t, \Delta l_{t+11}, \Delta l_{t+6}, \Delta l_{t+8}, \Delta l_{t+10}, \Delta l_{t+15}) = (1, 1, 0, 0, 0, 0)$, according to Eq. (1), $\Delta s_{t+13} = 0$. Since no conditions need to be added, $(\Delta l_t, \Delta l_{t+11}, \Delta l_{t+6}, \Delta l_{t+8}, \Delta l_{t+10}, \Delta l_{t+15}, \Delta s_{t+13}, c) = (1, 1, 0, 0, 0, 0, 0, 0)$ is the vector we hunt.

Assuming that Δs_{t+13} may be 1 or 0 according to Eq. (1). If a condition could be applied to ensure that $\Delta s_{t+13} = 0$, Δs_{t+13} takes value 0 and c takes value 1.

Example 2. Suppose that $(\Delta l_t, \Delta l_{t+11}, \Delta l_{t+6}, \Delta l_{t+8}, \Delta l_{t+10}, \Delta l_{t+15}) = (0, 0, 0, 0, 0, 1)$, according to Eq. (1), Δs_{t+13} could be either 1 or 0. But if we impose the condition $l_{t+10} = 0$, Δs_{t+13} must be 0, and c takes value 1, so $(0,0,0,0,0,1,0,1)$ is the vector we hunt.

If there must be a difference in s_{t+13} and no conditions can cancel it, Δs_{t+13} takes value 1 and c takes value 0.

Example 3. Suppose that $(\Delta l_t, \Delta l_{t+11}, \Delta l_{t+6}, \Delta l_{t+8}, \Delta l_{t+10}, \Delta l_{t+15}) = (1, 0, 0, 0, 0, 0)$, according to Eq. (1) $\Delta s_{t+13} = 1$, and it cannot be canceled by any conditions. Then Δs_{t+13} takes value 1 and c takes value 0. So we obtain $(1,0,0,0,0,0,1,0)$.

The difference state $(\Delta l_t, \Delta l_{t+11}, \Delta l_{t+6}, \Delta l_{t+8}, \Delta l_{t+10}, \Delta l_{t+15})$ can take on one of $2^6 = 64$ values. We derive the exact values of c and Δs_{t+13} from each value of the difference state in accordance with the above strategies. Then, with respect to Eq. (1), we get all 64 values of the 8-dimensional vector $(\Delta l_t, \Delta l_{t+11}, \Delta l_{t+6}, \Delta l_{t+8}, \Delta l_{t+10}, \Delta l_{t+15}, \Delta s_{t+13}, c)$, presented in Table 3. Meanwhile, with respect to Eq. (2), we can also find all the difference state values $(\Delta s_t, \Delta s_{t+5}, \Delta s_{t+4}, \Delta s_{t+7}, \Delta s_{t+9}, \Delta l_{t+19}, c)$. It should be noted that in Eq. (2) there is a constant a_t that achieves an exact value at each round. To simplify constraints of the MILP, we model two cases corresponding to the values of a_t.

Table 3. 64 vectors $(\Delta l_t, \Delta l_{t+11}, \Delta l_{t+6}, \Delta l_{t+8}, \Delta l_{t+10}, \Delta l_{t+15}, \Delta s_{t+13}, c)$

$(0,0,0,0,0,0,0,0)$	$(0,1,0,1,1,0,0,1)$	$(1,0,1,1,0,0,0,1)$
$(0,0,0,0,0,1,0,1)$	$(0,1,0,1,1,1,0,1)$	$(1,0,1,1,0,1,0,1)$
$(0,0,0,0,1,0,0,1)$	$(0,1,1,0,0,0,0,1)$	$(1,0,1,1,1,0,0,1)$
$(0,0,0,0,1,1,0,1)$	$(0,1,1,0,0,1,0,1)$	$(1,0,1,1,1,1,0,1)$
$(0,0,0,1,0,0,0,1)$	$(0,1,1,0,1,0,0,1)$	$(1,1,0,0,0,0,0,0)$
$(0,0,0,1,0,1,0,1)$	$(0,1,1,0,1,1,0,1)$	$(1,1,0,0,0,1,0,1)$
$(0,0,0,1,1,0,0,1)$	$(0,1,1,1,0,0,0,1)$	$(1,1,0,0,1,0,0,1)$
$(0,0,0,1,1,1,0,1)$	$(0,1,1,1,0,1,0,1)$	$(1,1,0,0,1,1,0,1)$
$(0,0,1,0,0,0,0,1)$	$(0,1,1,1,1,0,0,1)$	$(1,1,0,1,0,0,0,1)$
$(0,0,1,0,0,1,0,1)$	$(0,1,1,1,1,1,0,1)$	$(1,1,0,1,0,1,0,1)$
$(0,0,1,0,1,0,0,1)$	$(1,0,0,0,0,0,1,0)$	$(1,1,0,1,1,0,0,1)$
$(0,0,1,0,1,1,0,1)$	$(1,0,0,0,0,1,0,1)$	$(1,1,0,1,1,1,0,1)$
$(0,0,1,1,0,0,0,1)$	$(1,0,0,0,1,0,0,1)$	$(1,1,1,0,0,0,0,1)$
$(0,0,1,1,0,1,0,1)$	$(1,0,0,0,1,1,0,1)$	$(1,1,1,0,0,1,0,1)$
$(0,0,1,1,1,0,0,1)$	$(1,0,0,1,0,0,0,1)$	$(1,1,1,0,1,0,0,1)$
$(0,0,1,1,1,1,0,1)$	$(1,0,0,1,0,1,0,1)$	$(1,1,1,0,1,1,0,1)$
$(0,1,0,0,0,0,1,0)$	$(1,0,0,1,1,0,0,1)$	$(1,1,1,1,0,0,0,1)$
$(0,1,0,0,0,1,0,1)$	$(1,0,0,1,1,1,0,1)$	$(1,1,1,1,0,1,0,1)$
$(0,1,0,0,1,0,0,1)$	$(1,0,1,0,0,0,0,1)$	$(1,1,1,1,1,0,0,1)$
$(0,1,0,0,1,1,0,1)$	$(1,0,1,0,0,1,0,1)$	$(1,1,1,1,1,1,0,1)$
$(0,1,0,1,0,0,0,1)$	$(1,0,1,0,1,0,0,1)$	
$(0,1,0,1,0,1,0,1)$	$(1,0,1,0,1,1,0,1)$	

When $a_t = 1$, Eq. (2) contains five boolean variables s_t, s_{t+5}, s_{t+4}, s_{t+7}, s_{t+9}, so the difference state $(\Delta s_t, \Delta s_{t+5}, \Delta s_{t+4}, \Delta s_{t+7}, \Delta s_{t+9})$ can take on one of $2^5 = 32$ different values deriving the 32 values of the 7-dimensional vector $(\Delta s_t, \Delta s_{t+5}, \Delta s_{t+4}, \Delta s_{t+7}, \Delta s_{t+9}, \Delta l_{t+19}, c)$ shown in Table 4.

When $a_t = 0$, Eq. (2) contains four boolean variables s_t, s_{t+5}, s_{t+4}, s_{t+7}, so the difference state $(\Delta s_t, \Delta s_{t+5}, \Delta s_{t+4}, \Delta s_{t+7})$ can take on one of $2^4 = 16$ different values that lead to the 16 values of the 6-dimensional vector $(\Delta s_t, \Delta s_{t+5}, \Delta s_{t+4}, \Delta s_{t+7}, \Delta l_{t+19}, c)$ shown in Table 5.

2. Modeling the Vector Sets Using Linear Inequalities. Via SageMath at http://www.sagemath.org, we obtain 19 linear inequalities that accurately describe the set of the 64 8-dimensional vectors in Table 3, that is, this set of linear inequalities characterize the difference propagation of Eq. (1) under the control of conditions. There are 10 inequalities remaining after a simple reduction. L_1 shows the 10 inequalities.

Table 4. 32 vectors $(\Delta s_t, \Delta s_{t+5}, \Delta s_{t+4}, \Delta s_{t+7}, \Delta s_{t+9}, \Delta l_{t+19}, c)$

$(0,0,0,0,0,0,0)$	$(0,1,0,1,1,0,1)$	$(1,0,1,1,0,0,1)$
$(0,0,0,0,1,1,0)$	$(0,1,1,0,0,0,1)$	$(1,0,1,1,1,0,1)$
$(0,0,0,1,0,0,1)$	$(0,1,1,0,1,0,1)$	$(1,1,0,0,0,0,0)$
$(0,0,0,1,1,0,1)$	$(0,1,1,1,0,0,1)$	$(1,1,0,0,1,1,0)$
$(0,0,1,0,0,0,1)$	$(0,1,1,1,1,0,1)$	$(1,1,0,1,0,0,1)$
$(0,0,1,0,1,0,1)$	$(1,0,0,0,0,1,0)$	$(1,1,0,1,1,0,1)$
$(0,0,1,1,0,0,1)$	$(1,0,0,0,1,0,0)$	$(1,1,1,0,0,0,1)$
$(0,0,1,1,1,0,1)$	$(1,0,0,1,0,0,1)$	$(1,1,1,0,1,0,1)$
$(0,1,0,0,0,1,0)$	$(1,0,0,1,1,0,1)$	$(1,1,1,1,0,0,1)$
$(0,1,0,0,1,0,0)$	$(1,0,1,0,0,0,1)$	$(1,1,1,1,1,0,1)$
$(0,1,0,1,0,0,1)$	$(1,0,1,0,1,0,1)$	

Table 5. 16 vectors $(\Delta s_t, \Delta s_{t+5}, \Delta s_{t+4}, \Delta s_{t+7}, \Delta l_{t+19}, c)$

$(0,0,0,0,0,0)$	$(0,1,1,0,0,1)$	$(1,1,0,0,0,0)$
$(0,0,0,1,0,1)$	$(0,1,1,1,0,1)$	$(1,1,0,1,0,1)$
$(0,0,1,0,0,1)$	$(1,0,0,0,1,0)$	$(1,1,1,0,0,1)$
$(0,0,1,1,0,1)$	$(1,0,0,1,0,1)$	$(1,1,1,1,0,1)$
$(0,1,0,0,1,0)$	$(1,0,1,0,0,1)$	
$(0,1,0,1,0,1)$	$(1,0,1,1,0,1)$	

$$
L_1 = \begin{cases}
-\Delta s_{t+13} - c + 1 \geq 0 \\
-\Delta l_{t+6} + c \geq 0 \\
-\Delta l_t - \Delta l_{t+11} - \Delta s_{t+13} + 2 \geq 0 \\
-\Delta l_{t+8} + c \geq 0 \\
-\Delta l_{t+10} + c \geq 0 \\
-\Delta l_{t+15} + c \geq 0 \\
\Delta l_t + \Delta l_{t+11} - \Delta s_{t+13} \geq 0 \\
-\Delta l_t + \Delta l_{t+11} + \Delta s_{t+13} + c \geq 0 \\
\Delta l_t - \Delta l_{t+11} + \Delta s_{t+13} + c \geq 0 \\
\Delta l_{t+6} + \Delta l_{t+8} + \Delta l_{t+10} + \Delta l_{t+15} - c \geq 0
\end{cases}
\tag{3}
$$

Using the same method, we obtain two sets of linear inequalities L_2 and L_3 that accurately describe the 32 7-dimensional vectors given in Table 4 and the 16 6-dimensional vectors given in Table 5. The two sets are shown below:

$$L_2 = \begin{cases} -\Delta s_t - c + 1 \geq 0 \\ -\Delta s_{t+7} + c \geq 0 \\ -\Delta s_{t+4} + c \geq 0 \\ \Delta s_{t+4} + \Delta s_{t+7} - c \geq 0 \\ \Delta s_t - \Delta s_{t+5} - \Delta s_{t+9} - \Delta s_{t+2} \geq 0 \\ -\Delta s_t + \Delta s_{t+5} - \Delta s_{t+9} - \Delta s_{t+2} \geq 0 \\ -\Delta s_t - \Delta s_{t+5} + \Delta s_{t+9} - \Delta s_{t+2} \geq 0 \\ \Delta s_t + \Delta s_{t+5} + \Delta s_{t+9} - \Delta s_t \geq 0 \\ -\Delta s_t - \Delta s_{t+5} - \Delta s_{t+9} + \Delta s_t + c + 2 \geq 0 \\ \Delta s_t + \Delta s_{t+5} - \Delta s_{t+9} + \Delta s_t + c \geq 0 \\ \Delta s_t - \Delta s_{t+5} + \Delta s_{t+9} + \Delta s_t + c \geq 0 \\ -\Delta s_t + \Delta s_{t+5} + \Delta s_{t+9} + \Delta s_t + c \geq 0 \end{cases} \qquad (4)$$

$$L_3 = \begin{cases} -\Delta s_t - c + 1 \geq 0 \\ -\Delta s_{t+4} + c \geq 0 \\ -\Delta s_{t+7} + c \geq 0 \\ \Delta s_{t+4} + \Delta s_{t+7} - c \geq 0 \\ \Delta s_t - \Delta s_{t+5} + \Delta s_t + c \geq 0 \\ -\Delta s_t + \Delta s_{t+5} + \Delta s_t + c \geq 0 \\ \Delta s_t + \Delta s_{t+5} - \Delta s_t \geq 0 \\ -\Delta s_t - \Delta s_{t+5} - \Delta s_{t+2} \geq 0 \end{cases} \qquad (5)$$

3. Formulating the MILP Model to Determine an Initial Difference and Minimum Conditions.

With these linear inequalities, we can obtain the relationships among the differences of bits that generate the update bit, the flag of adding a condition and the difference of the update bit in one round. We then expand the linear inequalities to n rounds, where n is a selected number, to obtain constraints of the MILP model. The objective function to be minimized is $\sum_{i=0}^{n} c_i$. The constraint of the initial difference is $\sum_{i=0}^{31} \Delta x_i \geq 1$. In our work, the MILP problem is solved by Cplex. With this solution, we can obtain both an initial difference and minimum conditions.

There are too many plaintext bits and key bits in the conditions applied in the later rounds, so we prefer applying the conditions in earlier rounds rather than all of them. No more conditions have been applied since a particular round, which leads to uncontrollable difference propagation in subsequent rounds. After several rounds, the probability of the difference in the update bit would always be $1/2$. In Subsect. 3.2, we propose a method to evaluate the difference probability of the update bit, which helps us to find the bit whose difference probability deviates significantly from $1/2$ and has the largest number of rounds, and utilize this bit to mount key-recovery attacks.

3.2   Detecting the Bias of the Difference

In [14,15], a bias was detected by experimentally observing certain non-randomness, and we now present a method for automatically detecting the bias by programming. The method produces a formula for calculating the probability of the update bit difference, enabling us to find the bit whose probability of the difference has a bias from $1/2$. The greater the bias, the higher probability of a successful attack.

The properties below show that we can evaluate the probability of difference in the update bit, given all the probabilities of difference in the bits that generate the update bit. Thus, using the probability of difference in each bit of the two NLFSR at the round where the conditions cease being applied, we can calculate probabilities of update bits in each and every subsequent round.

Property 1. Let a, b be two independent random boolean variables, then the probability $P\{\Delta(a \oplus b) = 1\} = P\{\Delta a = 1\} + P\{\Delta b = 1\} - P\{\Delta a = 1\}P\{\Delta b = 1\}$.

With Property 1, if the probabilities of the differences in a and b were known, we could evaluate the probability of the difference in $a \oplus b$. It can be extended to the sum of four boolean variables:

Property 2. Let x, y, z, w be independent random boolean variables, then the probability

$$P\{\Delta(x \oplus y \oplus z \oplus w) = 1\}$$
$$= \frac{1}{2} - \frac{1}{2}(1 - 2P\{\Delta x = 1\})(1 - 2P\{\Delta y = 1\})$$
$$(1 - 2P\{\Delta z = 1\})(1 - 2P\{\Delta w = 1\}).$$

In the following, we consider the case when the boolean variable where the difference is in is the product of the other two boolean variables. Property 3 shows me how to evaluate the probability.

Property 3. Let a, b be the same as defined in Property 1, then the probability
$$P\{\Delta(a \cdot b) = 1\} = (P\{\Delta a = 1\} + P\{\Delta b = 1\} - P\{\Delta a = 1\}P\{\Delta b = 1\}) \cdot \frac{1}{2}.$$

In Eq. (1) and Eq. (2), since there is no difference in the key and const, k_{2t+1}, k_{2t} and a_t do not influence the probability of the difference. Accordingly, we can derive the results as follows:

From Eq. (1), we can obtain the formula to calculate the probability of $\Delta s_{t+13} = 1$:

$$P\{\Delta s_{t+13} = 1\}$$
$$= P\{\Delta(l_t \oplus l_{t+11} \oplus l_{t+6}l_{t+8} \oplus l_{t+10}l_{t+15}) = 1\}$$
$$= \frac{1}{2} - \frac{1}{2} \cdot (1 - 2P\{\Delta l_t = 1\})(1 - 2P\{\Delta l_{t+11} = 1\}) \tag{6}$$
$$(1 - 2P\{\Delta(l_{t+6}l_{t+8}) = 1\})(1 - 2P\{\Delta(l_{t+10}l_{t+15}) = 1\}),$$

where

$$P\{\Delta(l_{t+6}l_{t+8}) = 1\} = (P\{\Delta l_{t+6} = 1\} + P\{\Delta l_{t+8}$$

$$= 1\} - P\{\Delta l_{t+6} = 1\}P\{\Delta l_{t+8} = 1\}) \cdot \frac{1}{2},$$

$$P\{\Delta(l_{t+10}l_{t+15}) = 1\} = (P\{\Delta l_{t+10} = 1\} + P\{\Delta l_{t+15}$$

$$= 1\} - P\{\Delta l_{t+10} = 1\}P\{\Delta l_{t+15} = 1\}) \cdot \frac{1}{2}.$$

From Eq. (2), we can obtain the formula to calculate the probability of $\Delta l_{t+19} = 1$:

$$
\begin{aligned}
&P\{\Delta l_{t+19} = 1\} \\
&= P\{\Delta(s_t \oplus s_{t+5} \oplus s_{t+4}s_{t+7} \oplus s_{t+9}a_t) = 1\} \\
&= \frac{1}{2} - \frac{1}{2} \cdot (1 - 2P\{\Delta s_t = 1\})(1 - 2P\{\Delta s_{t+5} = 1\}) \\
&\quad (1 - 2P\{\Delta(s_{t+4}s_{t+7}) = 1\})(1 - 2P\{\Delta(s_{t+9}a_t) = 1\}),
\end{aligned}
\tag{7}
$$

where

$$P\{\Delta(s_{t+4}s_{t+7}) = 1\} = (P\{\Delta s_{t+4} = 1\} + P\{\Delta s_{t+7}$$

$$= 1\} - P\{\Delta s_{t+4} = 1\}P\{\Delta s_{t+7} = 1\}) \cdot \frac{1}{2},$$

$$P\{\Delta(s_{t+9}a_t) = 1\} = a_t P\{\Delta s_{t+9} = 1\}.$$

Using the two formulas, we can calculate the probabilities of the differences in the update bits in Algorithm 2 at every subsequent round after the conditions stop being applied. After a certain round the probability forever becomes 1/2. Before that, we can find the biased bit corresponding to the longest conditional differential characteristic.

4   Application to KATAN32

We have applied MILP method to KATAN32 for different rounds to obtain different differential characteristics and minimum conditions. We chose two results with fewer conditions in the previous rounds.

For 64-round KATAN32 (we have modeled 64-round KATAN32 together), the minimum number of conditions is 27. We cannot, however, apply all these conditions, since there are too many key bits and plaintext bits involved in them, resulting in attack failure. We only choose 11 conditions from the first 23 rounds to impose in this analysis. Since other conditions from round 24 have not been applied, difference propagation becomes out of control, with more and more probabilities of differences in update bits tending to be 1/2. We calculate the probabilities of $\Delta s_{t+13} = 1$ and $\Delta l_{t+19} = 1$ after round 23, and we find that finally the probability of $\Delta s_{t+13} = 1$ would always be 1/2 starting from s_{79} and the probability of $\Delta l_{t+19} = 1$ would always be 1/2 starting from l_{82}. Before l_{82}, we detect an obvious bias in Δl_{79}. l_{79} is generated at round 60 and is the

---

Algorithm 2: Calculating the probabilities of the differences in the update bits from round t to round u

---

Input: $\{P\{\Delta s_t = 1\}, P\{\Delta s_{t+1} = 1\}, \ldots P\{\Delta s_{t+12} = 1\}\}$: the set of probabilities of the difference for each bit of the 13-bit NLFSR at round t ; $\{P\{\Delta l_t = 1\}, P\{\Delta l_{t+1} = 1\}, \ldots P\{\Delta l_{t+18} = 1\}\}$: the set of probabilities of the difference for each bit of the 19-bit NLFSR at round t.

Output: A : the set of the probabilities of the differences for update bits from round t to round u, there are two update bits at each round.

$S := \{P\{\Delta s_t = 1\}, P\{\Delta s_{t+1} = 1\}, \ldots P\{\Delta s_{t+12} = 1\}\}$;
$L := \{P\{\Delta l_t = 1\}, P\{\Delta l_{t+2} = 1\}, \ldots P\{\Delta s_{t+18} = 1\}\}$;
$A := \emptyset$;

for $i \in \{t, t+1, \ldots, u\}$ **do**

$\quad$ $P\{\Delta s_{i+13} = 1\} :=$ The probability calculated from L according to formula 6
$\quad$;
$\quad$ $P\{\Delta l_{i+19} = 1\} :=$ The probability calculated from S according to formula 7
$\quad$;
$\quad$ $S := \{P\{\Delta s_{i+1} = 1\}, P\{\Delta s_{i+2} = 1\}, \ldots P\{\Delta s_{i+13} = 1\}\}$;
$\quad$ $L := \{P\{\Delta l_{i+1} = 1\}, P\{\Delta l_{i+2} = 1\}, \ldots P\{\Delta l_{i+19} = 1\}\}$;
$\quad$ $A := A \cup \{(P\{\Delta s_{i+13} = 1\}, P\{\Delta l_{i+19} = 1\})\}$

end
return A;

---

rightmost bit of the 19-bit NLFSR at round 79. Utilizing the bias of Δl_{79}, we can recover 10 equivalent key bits of the 79-round KATAN32.

For 77-round KATAN32, the minimum number of conditions is 34. We only impose seven conditions from the first 16 rounds and recover four equivalent key bits of the 81-round KATAN32 with a bias in Δl_{81}. l_{81} is generated at round 62 and is the rightmost bit of the 19-bit NLFSR at round 81.

In this section, we present the details of our analysis and attacks on these two results.

4.1   Key-Recovery Attack on 79-Round KATAN32

The differential characteristic of 64-round KATAN32 has the initial difference of weight six at the positions $0, 11, 21, 26, 30, 31$ of the plaintext block, $\Delta X = 0xc4200801$. We only apply 11 conditions in the first 23 rounds.

At round 1, we have $\Delta s_{14} = x_9, \Delta l_{20} = x_{23}$, and we impose conditions $x_9 = 0, x_{23} = 0$. At round 3, we have $\Delta s_{16} = x_5, \Delta l_{22} = x_{24}$, and we impose conditions $x_5 = 0, x_{24} = 0$. At round 6, we have $\Delta l_{25} = s_{13}$, and we impose the condition

$$s_{13} = x_{18} \oplus x_7 \oplus x_{12}x_{10} \oplus x_8x_3 \oplus k_1 = 0. \tag{8}$$

At round 8, we have $\Delta s_{21} = l_{23}$, and we impose the condition

$$l_{23} = x_{27} \oplus x_{22} \oplus k_8 = 0. \tag{9}$$

At round 10, we have $\Delta s_{23} = x_2$, and we impose the condition $x_2 = 0$. At round 12, we have $\Delta s_{25} = l_{20}$, and we impose the condition

$$l_{20} = x_{30} \oplus x_{25} \oplus x_{21} \oplus k_2 = 0. \tag{10}$$

At round 14, we have $\Delta s_{27} = l_{24}$, and we impose the condition

$$l_{24} = x_{26} \oplus x_{21} \oplus x_{22}x_{19} \oplus x_{17} \oplus x_6 \oplus k_3 \oplus k_{10} = 0. \tag{11}$$

At round 19, we have $\Delta s_{32} = l_{34}$. If we try to impose the condition $l_{34} = 0$, it has too many variables, which would make the attack unavailable because of the significantly high computing complexity. So we skip this condition, and assume $P\{\Delta s_{32} = 1\} = 1/2$. At round 21, we have $\Delta s_{34} = l_{27}$, and we impose the condition

$$l_{27} = x_{19}(x_{16} \oplus x_{10}x_8 \oplus x_6x_1 \oplus k_5) \oplus k_{16} = 0. \tag{12}$$

At round 23, we have $\Delta s_{36} = l_{31}$, and we impose the condition

$$l_{31} = x_{19} \oplus x_{14} \oplus x_3 \oplus x_8x_6 \oplus x_4(x_{31} \oplus x_{26} \oplus x_{22} \oplus k_0) \oplus \\ (x_{15} \oplus x_4 \oplus k_7)(x_{12} \oplus x_1 \oplus x_6x_4 \oplus k_{13}) \oplus k_9 \oplus k_{24} = 0. \tag{13}$$

The difference propagation and the conditions applied are presented in Table 8.

After imposing these conditions, we obtain the probability of difference in each bit at round 24 as follows: $(0, 0, 0, 0, 0, 0, 0, \frac{1}{2}, 0, 0, 0, 0, 0, 0, 0, 0, 0, 0, 1, 0, 0, 0, 0, 0, 0, 0, 0, 0, 0, 0, 0, 0)$.

According to Algorithm 2, we can compute the bias of the difference in the update bit for each round after round 24, and find that starting from l_{82} the probability of $\Delta l_{t+19} = 1$ would always be $1/2$, and among the bits whose positions are very close to l_{82}, l_{79} has the maximum biased difference, shown as follows:

$$P\{\Delta l_{79} = 1 | \text{all the conditions satisfied}\} \approx 0.5 - 0.00001.$$

We confirmed the strongly biased difference in bit l_{79} experimentally. Let us consider the conditions applied. There are 10 equivalent key bits k_0, k_1, k_2, $k_3 \oplus k_{10}$, k_5, k_7, k_8, $k_9 \oplus k_{24}$, k_{13}, k_{16} and 21 bits of plaintext x_1, x_3, x_4, x_6, x_7, x_8, x_{10}, x_{12}, x_{14}, x_{15}, x_{16}, x_{17}, x_{18}, x_{19}, x_{21}, x_{22}, x_{25}, x_{26}, x_{27}, x_{30}, x_{31} involved in the conditions. We choose 2^8 key in which bits $k_0, k_1, k_2, k_3, k_5, k_7, k_8, k_9$ are free and others fixed. For each key, we enumerate 2^{21} plaintexts of which the 21 bits involved in the conditions are free and other bits are zero. We then can use the conditions (8)–(13) to filter the 2^{21} plaintexts, and if the plaintext satisfied the conditions, we calculate Δl_{79} with the initial difference $\Delta X = 0xc4200801$ and count $P\{\Delta l_{79} = 1\}$ at last. The complexity of each experiment is less than $2^{21+1} = 2^{22}$ evaluations of the 60-round KATAN32 encryption while not every plaintext can be filtered. The experimental results verify the strongly biased difference in bit l_{79}. All the results of these 256 experiments are that $P\{\Delta l_{79} = 1\}$ is lower than $0.5 - 0.00001$.

Furthermore, we can mount a key-recovery attack. Looking at the conditions (8)–(13), we consider k_0, k_1, k_2, $k_3 \oplus k_{10}$, k_5, k_7, k_8, $k_9 \oplus k_{24}$, k_{13}, k_{16} , the 10 equivalent key bits, as ten variables. In a key-recovery attack, since the key is unknown to the attacker, we enumerate 2^{10} guesses of these 10 equivalent key bits. For each guess, similar to the verification, we use conditions (8)–(13) to filter 2^{21} plaintexts of which the 21 bits involved in the conditions (8)–(13) are free and other 11 bits are fixed to zero, then calculate Δl_{79} with initial difference $\Delta X = 0xc4200801$ and finally count $P\{\Delta l_{79} = 1\}$.

When the guess is correct, plaintexts are filtered by the conditions corresponding to the correct guessed equivalent key bits, then $P\{\Delta l_{79} = 1\}$ shows the obvious bias. In the 1024 statistical results from 1024 guesses of 10 equivalent key bits, the maximum bias in the results corresponds to the correct values of the ten equivalent key bits. This allows us to recover k_0, k_1, k_2, $k_3 \oplus k_{10}$, k_5, k_7, k_8, $k_9 \oplus k_{24}$, k_{13}, k_{16}, with experimental complexity less than $2^{10+21+1} = 2^{32}$ evaluations of the 60-round KATAN32 encryption. We randomly chose four 80-bit keys and mounted four key-recovery attack experiments, and each time the 10 equivalent key bits were recovered correctly, as shown by the results listed in Table 6.

Table 6. Four key-recovery attack experiments on 79-round KATAN32

No.	80-bit key	Equivalent key bits with the maximum bias									
-----	------------	k_0	k_1	k_2	$k_3 \oplus k_{10}$	k_5	k_7	k_8	$k_9 \oplus k_{24}$	k_{13}	k_{16}
1	$0x68b1644ead28b1644e8e$	0	1	1	1	0	0	1	0	0	0
2	$0xf8b164cead28b1644e8e$	1	1	1	0	0	0	1	1	0	0
3	$0x38116486a99ab3664d9e$	0	0	1	1	0	0	0	1	0	0
4	$0x2d11e4062b92bb2e4d9f$	0	0	1	0	1	1	0	0	0	1

4.2 Key-Recovery Attack on 81-Round KATAN32

The initial difference of the differential characteristic of 77-round KATAN32 weights three at position 7, 18 and 28 of the plaintext block, $\Delta X = 0x10040080$.

At round 1, we have $\Delta s_{14} = x_2$, then impose the condition $x_2 = 0$ to prevent difference propagation.

Similarly, at round 3, 5, 7, we have $\Delta s_{16} = x_9, \Delta s_{18} = x_5, \Delta s_{20} = x_1$, so we require bits x_9, x_5, x_1 to be zero.

At round 12, we have $\Delta s_{25} = l_{27}$, and we impose the condition

$$l_{27} = x_{23} \oplus x_{18} \oplus x_7 \oplus x_{12}x_{10} \oplus x_8x_3 \oplus x_{19}(x_{16} \oplus x_{10}x_8 \oplus k_5) \oplus k_{16} \oplus k_1 = 0. \tag{14}$$

At round 14, we have $\Delta s_{27} = l_{20}$ and we impose the condition

$$l_{20} = x_{30} \oplus x_{25} \oplus x_{26}x_{23} \oplus x_{21} \oplus k_2 = 0. \tag{15}$$

At round 16, we have $\Delta s_{29} = l_{24}$ and we impose the condition

$$l_{24} = x_{26} \oplus x_{21} \oplus x_{22}x_{19} \oplus x_{17} \oplus x_6 \oplus k_3 \oplus k_{10} = 0. \tag{16}$$

The differences in propagation and conditions applied are presented in Table 9. After imposing these conditions, we obtain the probability of difference in each bit at round 17 as follows: $(0,0,0,0,0,0,0,0,0,0,0,0,0,0,0,0,\ 0,0,0,1,0,\ 0,\ 0,\ 0,\ 0,\ 0,\ 0,\ 0,\ 0,\ 0,\ 0,\ 0)$. We compute the bias of the update bit of each round from the 17th round and find that starting from l_{84} the probability of $\Delta l_{t+19} = 1$ would always be $1/2$. Among the bits whose positions are very close to l_{84}, l_{81} has the maximum biased difference, shown as follows:

$$P\{\Delta l_{81} = 1 | \text{all the conditions satisfied}\} \approx 0.5 + 0.000226.$$

We experimentally verified the strongly-biased difference in bit l_{81}. There are four equivalent key bits $k_5, k_{16} \oplus k_1, k_2, k_3 \oplus k_{10}$ and 16 bits of plaintext $x_{30}, x_{26},$ $x_{25}, x_{23}, x_{22}, x_{21}, x_{19}, x_{18}, x_{17}, x_{16}, x_{12}, x_{10}, x_8, x_7, x_6, x_3$ in conditions (14)–(16). We choose 2^4 keys of which k_1, k_2, k_3, k_5 are free and the others fixed. For each key, we enumerate 2^{16} plaintexts of which the 16 bits involved in conditions are free and other bits are fixed to 0. We then use conditions (14)–(16) to filter the 2^{16} plaintexts, and if a plaintext satisfies the conditions, we calculate Δl_{81} with the initial difference $\Delta x = 0x10040080$ and count $P\{\Delta l_{81} = 1\}$. The complexity of each experiment is less than $2^{16+1} = 2^{17}$ evaluations of the 62-round KATAN32 encryption. In all the results of these 16 experiments $P\{\Delta l_{81} = 1\}$ is greater than $0.5 + 0.000226$.

We now will describe mounting the key-recovery attack. Looking at conditions (14)–(16), we consider $k_1 \oplus k_{16}, k_2, k_3 \oplus k_{10}, k_5$ these four equivalent key bits as four boolean variables. There are 16 bits of plaintext involved in conditions (14)–(16). To enlarge the space of plaintexts after filtering, we choose other three bits of plaintext not included in any condition as free bits in addition to the 16 bits of plaintext involved in conditions (14)–(16). For each of the 2^4 guesses of these four variables, we use conditions (14)–(16) to filter the 2^{19} plaintexts enumerated by the 19 bits we just choose with the remaining 13 bits fixed to 0. We then calculate Δl_{81} with initial difference $\Delta x = 0x10040080$ and count $P\{\Delta l_{81} = 1\}$. In the 16 statistical results obtained from 16 guesses of four equivalent key bits, the maximum bias in the results corresponds to the correct value of the four equivalent key bits, allowing us to recover $k_1 \oplus k_{16}, k_2, k_3 \oplus k_{10}, k_5$. The complexity of the experiment is less than $2^{4+19+1} = 2^{24}$ evaluations of the 62-round KATAN32 encryption. We chose five 80-bit keys randomly and mounted five key-recovery attack experiments, and each time the four equivalent key bits can be correctly recovered. The results of these five key-recovery attack experiments are listed in Table 7.

Table 7. Five key-recovery attack experiments on 81-round KATAN32

No.	80-bit key	Equivalent key bits with the maximum bias			
		$k_1 \oplus k_{16}$	k_2	$k_3 \oplus k_{10}$	k_5
1	0x68b1644ead28b1644e8e	1	1	1	0
2	0x48b1644ead28b1644e8e	1	0	1	0
3	0xcda964ceb98cb7644e8e	1	0	1	1
4	0xc9a1448cb886b7644f86	1	0	1	0
5	0x99a1448cb88680644f86	0	0	0	0

5 Extension with the Standard Differential Attack

Combined with the standard differential attack, the conditional differential attack on 81-round KATAN32 can be extended to 97-round, 98-round and 99-round key-recovery attacks.

5.1 Key-Recovery Attack on 97-Round KATAN32

Inspired by the technique representing the dependence of the intermediate state on the output by an algebraic representation in [27], we give the algebraic representation of the intermediate state using the ciphertext and round keys.

Using Eq. (1) and Eq. (2), we can get the expression of l_t, s_t in decryption direction:

$$l_t = s_{t+13} \oplus l_{t+11} \oplus l_{t+6}l_{t+8} \oplus l_{t+10}l_{t+15} \oplus k_{2t+1}, \tag{17}$$

$$s_t = l_{t+19} \oplus s_{t+5} \oplus s_{t+4}s_{t+7} \oplus s_{t+9}a_t \oplus k_{2t}, \tag{18}$$

Suppose the output bits of 97-round KATAN32 corresponding to plaintext X are $S_{97} = (s_{97}, s_{98}, \ldots, s_{110})$ and $L_{97} = (l_{97}, l_{98}, \ldots, l_{115})$, and the output bits of 97-round KATAN32 corresponding to plaintext $X + \Delta X$ are $S'_{97} = (s'_{97}, s'_{98}, \ldots, s'_{110})$ and $L'_{97} = (l'_{97}, l'_{98}, \ldots, l'_{115})$. For decryption direction, Δl_{81} can be expressed by round keys and the ciphertext of 97-round KATAN32 by using Eq. (17) and Eq. (18) iteratively.

$\Delta l_{81} = l_{113} \oplus s_{99} \oplus s_{98}s_{101} \oplus s_{105} \oplus l_{103} \oplus l_{98}l_{100} \oplus l_{102}l_{107} \oplus (s_{100} \oplus l_{98} \oplus (s_{106} \oplus l_{104} \oplus l_{99}l_{101} \oplus l_{103}l_{108} \oplus k_{187})(s_{108} \oplus l_{106} \oplus l_{101}l_{103} \oplus l_{105}l_{110} \oplus k_{191}) \oplus l_{97}l_{102} \oplus k_{175})(s_{102} \oplus l_{100} \oplus (s_{108} \oplus l_{106} \oplus l_{101}l_{103} \oplus l_{105}l_{110} \oplus k_{191})l_{97} \oplus l_{99}l_{104} \oplus k_{179}) \oplus (s_{104} \oplus l_{102} \oplus l_{97}l_{99} \oplus l_{101}l_{106} \oplus k_{183})(s_{109} \oplus l_{107} \oplus l_{102}l_{104} \oplus l_{106}l_{111} \oplus k_{193}) \oplus l'_{113} \oplus s'_{99} \oplus s'_{98}s'_{101} \oplus s'_{105} \oplus l'_{103} \oplus l'_{98}l'_{100} \oplus l'_{102}l'_{107} \oplus (s'_{100} \oplus l'_{98} \oplus (s'_{106} \oplus l'_{104} \oplus l'_{99}l'_{101} \oplus l'_{103}l'_{108} \oplus k_{187})(s'_{108} \oplus l'_{106} \oplus l'_{101}l'_{103} \oplus l'_{105}l'_{110} \oplus k_{191}) \oplus l'_{97}l'_{102} \oplus k_{175})(s'_{102} \oplus l'_{100} \oplus (s'_{108} \oplus l'_{106} \oplus l'_{101}l'_{103} \oplus l'_{105}l'_{110} \oplus k_{191})l'_{97} \oplus l'_{99}l'_{104} \oplus k_{179}) \oplus (s'_{104} \oplus l'_{102} \oplus l'_{97}l'_{99} \oplus l'_{101}l'_{106} \oplus k_{183})(s'_{109} \oplus l'_{107} \oplus l'_{102}l'_{104} \oplus l'_{106}l'_{111} \oplus k_{193})$.

According to this expression, one can calculate Δl_{81} by using the ciphertexts of 97-round KATAN32 and 6 equivalent key bits $k_{175}, k_{179}, k_{183}, k_{187}, k_{191}, k_{193}$. We extend the attack described in Subsect. 4.2 to 97-round as follows. Plaintexts being filtered by the conditions are encrypted to get ciphertexts by 97-round

Table 8. Differential characteristic and conditions for $\Delta X = 0xc4200801$

Round	Difference state	conditions
0	1100010000100 0000000100000000001	
1	1000100001000 0000001000000000010	$x_9 = 0, x_{23} = 0$
2	0001000010000 0000010000000000100	
3	0010000100000 0000100000000001000	$x_5 = 0, x_{24} = 0$
4	0100001000000 0001000000000010000	
5	1000010000000 0010000000000100000	
6	0000100000000 0100000000001000000	condition (8)
7	0001000000000 1000000000010000000	
8	0010000000000 0000000000100000000	condition (9)
9	0100000000000 0000000001000000000	
10	1000000000000 0000000010000000000	$x_2 = 0$
11	0000000000000 0000000100000000001	
12	0000000000000 0000001000000000010	condition (10)
13	0000000000000 0000010000000000100	
14	0000000000000 0000100000000001000	condition (11)
15	0000000000000 0001000000000010000	
16	0000000000000 0010000000000100000	
17	0000000000000 0100000000001000000	
18	0000000000000 1000000000010000000	
19	0000000000000 0000000000100000000	
20	000000000000* 0000000001000000000	
21	00000000000*0 0000000010000000000	condition(12)
22	0000000000*00 0000000100000000000	
23	000000000*000 0000001000000000000	condition(13)
24	00000000*0000 0000010000000000000	

The red bits denote the update bits.
The blue bits denote the bits that generate the update bits.
The differential probability of the bit $*$ is $\frac{1}{2}$

KATAN32. Δl_{81} could be computed from ciphertexts of 97-round KATAN32 and the guess of these 6 equivalent key bits $k_{175}, k_{179}, k_{183}, k_{187}, k_{191}, k_{193}$. Given every guess of 10 equivalent key bits ($k_1 \oplus k_{16}$, k_2, $k_3 \oplus k_{10}$, k_5, k_{175}, k_{179}, k_{183}, k_{187}, k_{191}, k_{193}), we can calculate and count Δl_{81} with respect to a set of filtered plaintexts. If the guess is right, the $P\{\Delta l_{81} = 1\}$ shows a obvious bias. The computational cost of the experiment is less than 2^{24+6} encryptions of 97-round KATAN32. We mounted five key-recovery attack experiments with the same key as the experiments in Subsect. 4.2 and each time the 10 equivalent key bits can be correctly recovered.

Table 9. Differential characteristic and conditions for $\Delta X = 0x10040080$

Round	Difference state	conditions
0	0001000000000 1000000000010000000	
1	0010000000000 0000000000100000000	$x_2 = 0$
2	0100000000000 0000000001000000000	
3	1000000000000 0000000010000000000	$x_9 = 0$
4	0000000000000 0000000100000000001	
5	0000000000000 0000001000000000010	$x_5 = 0$
6	0000000000000 0000010000000000100	
7	0000000000000 0000100000000001000	$x_1 = 0$
8	0000000000000 0001000000000010000	
9	0000000000000 0010000000000100000	
10	0000000000000 0100000000001000000	
11	0000000000000 1000000000010000000	
12	0000000000000 0000000000100000000	condition(14)
13	0000000000000 0000000001000000000	
14	0000000000000 0000000010000000000	condition(15)
15	0000000000000 0000000100000000000	
16	0000000000000 0000001000000000000	condition(16)
17	0000000000000 0000010000000000000	

The red bits denote the update bits.
The blue bits denote the bits that generate the update bits.

5.2 Key-Recovery Attack on 98-Round KATAN32

If Δl_{81} is expressed by the ciphertext of 98-round KATAN32 and round keys, there are 7 equivalent key bits involved in the expression. So the computational cost of the key-recovery attack is less than 2^{24+7} encryptions of 98-round KATAN32. In this attack, 11 equivalent key bits $k_1 \oplus k_{16}$, k_2, $k_3 \oplus k_{10}$, k_5, k_{175}, k_{179}, k_{183}, k_{187}, k_{191}, k_{193}, k_{195} can be correctly recovered. Every experiment requires about 2.4 hours on a 2.5 Ghz PC with our implementation.

5.3 Key-Recovery Attack on 99-Round KATAN32

If Δl_{81} is expressed by the ciphertext of 99-round KATAN32 and round keys, there are 9 equivalent key bits involved in the expression. So the computational cost of the key-recovery attack is less than 2^{24+9} encryptions of 99-round KATAN32. In this attack, 13 equivalent key bits $k_1 \oplus k_{16}$, k_2, $k_3 \oplus k_{10}$, k_5, k_{175}, k_{179}, k_{183}, k_{187}, k_{191}, k_{193}, k_{195}, k_{196}, k_{197} can be correctly recovered. Every experiment requires about 9.64 hours on a 2.5 Ghz PC with our implementation.

It is thus possible to extend the conditional differential attack on 81-round KATAN32 to 114-round with a time complexity of 2^{63} encryptions.

6 Conclusion

In this paper, we propose two strategies for improving conditional differential analysis on the NLFSR based block cipher KATAN32. We first apply the MILP model to automatically search for the conditional differential characteristic of NLFSR based block ciphers, helping us efficiently obtain the initial difference and conditions of the conditional differential analysis. We propose a new method to calculate the probability of the difference to help quickly detect the bit with a bias. We apply the improved conditional differential analysis to KATAN32 and obtain two results, recovering 10 equivalent key bits of 79-round KATAN32 and four equivalent key bits of 81-round KATAN32, respectively.

Combined with standard differential attack, we extend the 81-round conditional key-recovery attack to 99-round with the time complexity being 2^{33} encryptions of 99-round KATAN32 and recover 13 equivalent key bits. Compared with previously best practical distinguisher on KATAN32, our results are extended more than 7 rounds with less cost of computation time and memory. We believe both strategies to be general to NLFSR based ciphers. Applying these two strategies on other NLFSR based ciphers will be one topic of interest in our future works.

Acknowledgements. We are very grateful to the anonymous reviewers. This work was supported by the National Natural Science Foundation of China under Grant 61672330, and Grant 11771256.

References

1. Ahmadian, Z., Rasoolzadeh, S., Salmasizadeh, M., Aref, M.R.: Automated dynamic cube attack on block ciphers: cryptanalysis of SIMON and KATAN. IACR Cryptology ePrint Archive 2015, 40 (2015). http://eprint.iacr.org/2015/040
2. Albrecht, M.R., Leander, G.: An all-in-one approach to differential cryptanalysis for small block ciphers. IACR Cryptology ePrint Archive 2012, 401 (2012). http://eprint.iacr.org/2012/401
3. Ben-Aroya, I., Biham, E.: Differential cryptanalysis of lucifer. In: Stinson, D.R. (ed.) CRYPTO 1993. LNCS, vol. 773, pp. 187–199. Springer, Heidelberg (1994). https://doi.org/10.1007/3-540-48329-2_17
4. Biham, E., Dunkelman, O.: Differential cryptanalysis in stream ciphers. IACR Cryptology ePrint Archive 2007, 218 (2007). http://eprint.iacr.org/2007/218
5. Bogdanov, A., et al.: PRESENT: an ultra-lightweight block cipher. In: Paillier, P., Verbauwhede, I. (eds.) CHES 2007. LNCS, vol. 4727, pp. 450–466. Springer, Heidelberg (2007). https://doi.org/10.1007/978-3-540-74735-2_31
6. De Cannière, C., Dunkelman, O., Knežević, M.: KATAN and KTANTAN — a family of small and efficient hardware-oriented block ciphers. In: Clavier, C., Gaj, K. (eds.) CHES 2009. LNCS, vol. 5747, pp. 272–288. Springer, Heidelberg (2009). https://doi.org/10.1007/978-3-642-04138-9_20
7. De Cannière, C., Küçük, Ö., Preneel, B.: Analysis of grain's initialization algorithm. In: Vaudenay, S. (ed.) AFRICACRYPT 2008. LNCS, vol. 5023, pp. 276–289. Springer, Heidelberg (2008). https://doi.org/10.1007/978-3-540-68164-9_19

8. Abed, F., et al.: Pipelineable on-line encryption. In: Cid, C., Rechberger, C. (eds.) FSE 2014. LNCS, vol. 8540, pp. 205–223. Springer, Heidelberg (2015). https://doi.org/10.1007/978-3-662-46706-0_11

9. Fuhr, T., Minaud, B.: Match box meet-in-the-middle attack against KATAN. In: Cid and Rechberger [8], pp. 61–81. https://doi.org/10.1007/978-3-662-46706-0_4

10. Huang, S., Wang, X., Xu, G., Wang, M., Zhao, J.: Conditional cube attack on reduced-round Keccak sponge function. In: Coron, J.-S., Nielsen, J.B. (eds.) EUROCRYPT 2017. LNCS, vol. 10211, pp. 259–288. Springer, Cham (2017). https://doi.org/10.1007/978-3-319-56614-6_9

11. Isobe, T.: A single-key attack on the full GOST block cipher. J. Cryptol. **26**(1), 172–189 (2013). https://doi.org/10.1007/s00145-012-9118-5

12. Isobe, T., Shibutani, K.: Improved all-subkeys recovery attacks on fox, KATAN and SHACAL-2 block ciphers. In: Cid and Rechberger [8], pp. 104–126. https://doi.org/10.1007/978-3-662-46706-0_6

13. Jiang, Z., Jin, C.: Impossible differential cryptanalysis of 8-round Deoxys-BC-256. IEEE Access **6**, 8890–8895 (2018). https://doi.org/10.1109/ACCESS.2018.2808484

14. Knellwolf, S., Meier, W., Naya-Plasencia, M.: Conditional differential cryptanalysis of NLFSR-based cryptosystems. In: Abe, M. (ed.) ASIACRYPT 2010. LNCS, vol. 6477, pp. 130–145. Springer, Heidelberg (2010). https://doi.org/10.1007/978-3-642-17373-8_8

15. Knellwolf, S., Meier, W., Naya-Plasencia, M.: Conditional differential cryptanalysis of trivium and KATAN. In: Miri, A., Vaudenay, S. (eds.) SAC 2011. LNCS, vol. 7118, pp. 200–212. Springer, Heidelberg (2012). https://doi.org/10.1007/978-3-642-28496-0_12

16. Li, Z., Bi, W., Dong, X., Wang, X.: Improved conditional cube attacks on Keccak keyed modes with MILP method. In: Takagi, T., Peyrin, T. (eds.) ASIACRYPT 2017. LNCS, vol. 10624, pp. 99–127. Springer, Cham (2017). https://doi.org/10.1007/978-3-319-70694-8_4

17. Mouha, N., Wang, Q., Gu, D., Preneel, B.: Differential and linear cryptanalysis using mixed-integer linear programming. In: Wu, C.-K., Yung, M., Lin, D. (eds.) Inscrypt 2011. LNCS, vol. 7537, pp. 57–76. Springer, Heidelberg (2012). https://doi.org/10.1007/978-3-642-34704-7_5

18. Rasoolzadeh, S., Raddum, H.: Multidimensional meet in the middle cryptanalysis of KATAN. IACR Cryptology ePrint Archive 2016, 77 (2016). http://eprint.iacr.org/2016/077

19. Sasaki, Yu., Todo, Y.: New impossible differential search tool from design and cryptanalysis aspects. In: Coron, J.-S., Nielsen, J.B. (eds.) EUROCRYPT 2017. LNCS, vol. 10212, pp. 185–215. Springer, Cham (2017). https://doi.org/10.1007/978-3-319-56617-7_7

20. Song, L., Guo, J., Shi, D.: New MILP modeling: improved conditional cube attacks to Keccak-based constructions. IACR Cryptology ePrint Archive 2017, 1030 (2017). http://eprint.iacr.org/2017/1030

21. Sun, S., Hu, L., Wang, P., Qiao, K., Ma, X., Song, L.: Automatic security evaluation and (related-key) differential characteristic search: application to SIMON, PRESENT, LBlock, DES(L) and other bit-oriented block ciphers. In: Sarkar, P., Iwata, T. (eds.) ASIACRYPT 2014. LNCS, vol. 8873, pp. 158–178. Springer, Heidelberg (2014). https://doi.org/10.1007/978-3-662-45611-8_9

22. Wang, D., Li, W., Wang, P.: Measuring two-factor authentication schemes for real-time data access in industrial wireless sensor networks. IEEE Trans. Ind. Inform. **14**(9), 4081–4092 (2018). https://doi.org/10.1109/TII.2018.2834351

23. Wei, L., Rechberger, C., Guo, J., Wu, H., Wang, H., Ling, S.: Improved meet-in-the-middle cryptanalysis of KTANTAN. IACR cryptology eprint archive 2011, 201 (2011)
24. Wu, H., Preneel, B.: Resynchronization attacks on WG and LEX. In: Robshaw, M. (ed.) FSE 2006. LNCS, vol. 4047, pp. 422–432. Springer, Heidelberg (2006). https://doi.org/10.1007/11799313_27
25. Wu, S., Wang, M.: Security evaluation against differential cryptanalysis for block cipher structures. IACR Cryptology ePrint Archive 2011, 551 (2011). http://eprint.iacr.org/2011/551
26. Xiang, Z., Zhang, W., Bao, Z., Lin, D.: Applying MILP method to searching integral distinguishers based on division property for 6 lightweight block ciphers. In: Cheon, J.H., Takagi, T. (eds.) ASIACRYPT 2016. LNCS, vol. 10031, pp. 648–678. Springer, Heidelberg (2016). https://doi.org/10.1007/978-3-662-53887-6_24
27. Zhang, W., Cao, M., Guo, J., Pasalic, E.: Improved security evaluation of SPN block ciphers and its applications in the single-key attack on SKINNY. IACR Trans. Symmetric Cryptol. **2019**(4), 171–191 (2019). https://doi.org/10.13154/tosc.v2019.i4.171-191
28. Zhu, B., Gong, G.: Multidimensional meet-in-the-middle attack and its applications to KATAN32/48/64. Cryptog. Commun. **6**(4), 313–333 (2014). https://doi.org/10.1007/s12095-014-0102-9

Applied Cryptography

A Verifiable Combinatorial Auction with Bidder's Privacy Protection

Mingwu Zhang[1,2,3(✉)] and Bingruolan Zhou[1]

[1] School of Computer Science, Hubei University of Technology, Wuhan 430000, China
csmwzhang@gmail.com, brlzhou@163.com
[2] State Key Laboratory of Cryptology, P. O. Box 5159, Beijing 100878, China
[3] School of Computer Science and Information Security,
Gulin University of Electronic Technology, Guilin, China

Abstract. Combinatorial auctions are employed in many fields such as spectrum auction and energy auction. However, data concerning bidders' bid and bundle might reveal sensitive information, such as personal preference and competitive relation. In order to solve this problem, this paper proposes a privacy-preserving and verifiable combinatorial auction scheme to protect bidders' privacy and ensure the correctness of the result. In our scheme, we employ a one-way and monotonically increasing function to protect each bidder's bid, so that the auctioneer is able to pick out the largest bid without disclosing any information about bids. Moreover, we convert the question of judging whether a bidder is a winner to the question of judging whether the vector product is 0. In our scheme, crypto service provider (CSP) is responsible for key distribution and blind signature to verify the authenticity and correctness of the result. Besides, we put forward a privacy-preserving and verifiable payment determination model to compute the payment the winner should pay.

Keywords: Privacy-preserving · Combinatorial auction

1 Introduction

With the rapid development and wide application of Internet, the number of online e-commerce activities is increasing. The auction is gradually changing from traditional auction to electronic auction and becoming an important part of e-commerce. For example, spectrum [2] and energy [4] can be auctioned on the Internet. The electronic auction system generally consists of auctioneer, sellers and bidders. The seller entrusts the auctioneer to arrange the auction, accept the bids, and declare the winner [1]. In a single auctioneer combinatorial auction, the auctioneer sells multiple heterogeneous goods simultaneously, and bidders bid on any combination of the goods (called bundle or set) instead of just one [5]. Such

© ICST Institute for Computer Sciences, Social Informatics and Telecommunications Engineering 2021
Published by Springer Nature Switzerland AG 2021. All Rights Reserved
D. Wang et al. (Eds.): SPNCE 2020, LNICST 344, pp. 397–405, 2021.
https://doi.org/10.1007/978-3-030-66922-5_27

auctions have been researched extensively recently, in part due to the generality of it, and in part due to growing application scenarios where combinatorial auction is necessary [14].

In privacy-preserving combinatorial auction protocols, bidders protect their private information using cryptographic technique. After the execution of the auction, only the auction outcomes, i.e., who are winners and the corresponding payments, are revealed. The losers' bids and bundles are kept private in the auction because the auctioneers may use losers' bids to maximize their revenues in future auctions [1]. For example, the average of losers' bids can motivate auctioneers to increase the starting price in future auction of similar goods. In addition, private information of bidders, such as bundle and bids, can be used to disclose personal preference and how much bidders want to pay. In auctions where there is serious competition between bidders, these information are vital and need to be protected.

In private-preserving combinatorial auction, an important problem to be solved is how to determine the winner, i.e., how to pick out a set of disjoint goods, the value of which is the maximized. [12] use dynamic programming approach to solve the problem of winner determination in privacy-preserving combinatorial auction because dynamic programming can well solve the problem of finding the shortest path of directed graph.

Shamir's threshold secret sharing scheme can also be used to solve the privacy-preserving problem in combinatorial auctions. For example, [6] employs secret sharing scheme to share bids between the evaluators, which could resist the passive adversary model. All evaluators come together to find out the optimal solution through secure dynamic programming. Considering the communication cost of the protocols, [3] proposed an authentic property without increasing the communications cost in combinatorial auctions. Homomorphic encryption provides an available approach to protect each bidder's bidding values with a vector of cipher texts, and ensure the auctioneer to figure out the maximum value securely [7–9,11,13].

Various approaches are proposed to achieve the privacy-preserving combinatorial auction, such as dynamic programming, Shamir's threshold secret sharing scheme, homomorphic encryption and secure multi-party computation, etc.

[12] employed dynamic programming to solve the problem in the combinatorial auction. However, with the increase of the number of bidders and goods, dynamic programming will lead to non-polynomial time computation time. [6] implemented the privacy-preserving combinatorial auction through Shamir's threshold secret sharing scheme, and through further improvements, [3] reduced the communications cost in designing the secure auction protocol. [7–9,11,13] gave combinatorial auction protocols that are based on homomorphic encryption technique in ciphertext fields, however, these protocols need a high computational cost. [10] employed the technique of secure multi-party computation to implement privacy-preserving combinatorial auction, where the protocols are not scalable since the inputs of combinatorial auction can not be pre-determined.

2   Privacy Preserving Combinatorial Auction Model

2.1   System Model

As shown in Fig. 1, auctioneer has a series of goods $G = \{g_1, ..., g_m\}$, which will be auctioned to N bidders $B = \{B_1, ..., B_n\}$. Each bidder B_i gives his own bundle $S_i \in G$ that he expects to obtain and his bid $b_i(S_i)$, i.e. the price B_i is willing to pay on his bundle S_i. Crypto service provider is responsible for key distribution and collaborative computation. Besides, CSP will generate blind signature for bidders' bid and bundle, which will be used to verify the correctness of the result later.

The winners are chosen by the auctioneer as follows:

$$W = \operatorname{argmax}_{B_i} \sum b_i(S_i) \quad s.t. \quad \cap_{B_i} S_i = \emptyset \tag{1}$$

i.e., a set of conflict-free bidders whose total bid is maximized, and $A = \cup_{B_i \in W} S_i$ is the set of winners' bundle. After that, the auctioneer will determine the price that the winner should pay according to some mechanism.

We assume that each bidder has only one sequence of goods expected to buy. That is, if at least one good in the bundle that a bidder expects to get has been auctioned, the bidder will not get the remaining goods. This assumption is equivalent to the restriction that each bidder is limited to one bid only. We simplify bid $b_i(S_i)$ as b_i and denote the auctioneer as E.

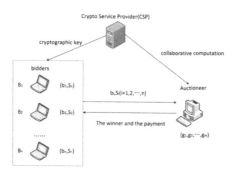

Fig. 1. System model

2.2   Adversary Model

When the allocation terminates, the auctioneer is supposed to only know the winners, their bundles and their bids. Each bidder only knows whether he is a winner. The bidder will also be informed the price he should pay, if he is the winner. Each bidder does not know anything about others' bundle or bid. CSP is responsible for key distribution and signature generation. In addition, CSP will help auctioneer to decrypt but will know nothing about auction results.

The auctioneer is assumed to be curious, malicious and ignorant. He is interested in bidders' bundles and bids to improve his business (i.e., "*curious*"). For example, the auctioneer may try to infer bidders' preferences and competitive relationship based on the bundles and bids. The auctioneer may also report a fake price to the winners (i.e., "*malicious*"), but he is not aware of bidders' side information such as distribution of bid or bidders' preference on goods (i.e., "*ignorant*").

Bidders are assumed to be curious, and non-cooperative. They are interested in others' bundles and bids to help them make decision (i.e., "*curious*"). However, they will not collude with each other or the auctioneer (i.e., "*non-cooperative*").

CSP is assumed to be honest, curious and non-cooperative. CSP follows the protocol steps honestly but try to learn bidders' bundles and bids (i.e., "*curious*"). But CSP will not collude with the auctioneer (i.e., "*non-cooperative*").

3   Our Proposed Scheme

Before auction, all bidders blind sign their bundle S_i and average value φ_i through the crypto service provider (CSP). Since we use the blind signature scheme, CSP will not get any relevant information. These signatures will be used for verification later.

3.1   Privacy-Preserving Winner Determination Model

Algorithm 1. Greedy Winner Determination

1: Mark the set of auctioned goods as A, the set of winners as W. During the initial phase, $A = \emptyset, W = \emptyset$. Each $B_i (i = 1, \cdots, n)$ computes average value $\varphi_i = \frac{b_i}{|S_i|}$.

2: Sort B_i in a non-increasing good according to the value of the φ_i, that is, the bigger the φ_i, the former the B_i. The sorted sequence is called L.

3: Check the B_i in L from front to back to see whether $A \cap S_i = \emptyset$. If true, update sets A and W, $A = A \cup S_i$, $W = W \cup B_i$.

4: After auction, W is the set of winners, and A is the set of goods that have been auctioned.

Algorithm 1 proposes a greedy winner determination model. Because the comparison and sorting will reveal the private information S_i and b_i of the bidders, we cannot directly compare φ_i and sort B_i on the plaintext (step 2) or directly select the winner (step 3). We use the monotonically increasing and one-way function to protect the bidder's b_i, which enables the auctioneer to pick out the largest b_i without knowing anything about b_i. Besides, the auctioneer needs to check whether B_i's bundle S_i contains the good that has already been auctioned. We use m-dimensional binary vector A to represent the auction status of m goods, where the k-th bit $a_k = 1$ if the k-th good g_k has already been auctioned and $a_k = 0$ if the k-th good g_k has not been auctioned. Similarly, we use another m-dimensional binary vector $\boldsymbol{S}_i$ to represent B_i's bundle S_i, where k-th bit $s_{i,k} = 1$ if the k-th good $g_k \in S_i$ and $s_{i,k} = 0$ if the k-th good $g_k \notin S_i$.

If B_i's bundle S_i does not contain the good that has already been auctioned, then

$$\boldsymbol{A}_i \cdot \boldsymbol{S}_i = 0 \Leftrightarrow \Sigma_{k=1}^{m} a_k \cdot s_{i,k} = 0$$

If vector product is θ, that means B_i's bundle S_i includes θ already-auctioned goods.

Thus, we can propose a privacy-preserving winner determination model (Algorithm 2), which can be regarded as a black-box algorithm and only outputs the winner and the corresponding bundle.

Algorithm 2. Privacy-preserving Winner Determination

1: Mark the set of auctioned goods as A, the set of winners as W. During the initial phase, $A = \emptyset, W = \emptyset$. Each $B_i (i = 1, \ldots, n)$ computes average value $\varphi_i = \frac{b_i}{|S_i|}$.

2: CSP picks a pair of Elgamal algorithm key: $mpk = (h_1 = g^{s_1}, h_2 = g^{s_2}, \cdots, h_m = g^{s_m})$, $msk = S = (s_1, s_2, \cdots, s_m)$, and publishes mpk.

3: Each B_i picks a random number r_i and encrypts $\boldsymbol{S}_i = (s_{i,1}, s_{i,2}, \cdots, s_{i,m})$

$$c_{i,1} = h_1^{r_i} \cdot g^{s_{i,1}}, c_{i,2} = h_2^{r_i} \cdot g^{s_{i,2}}, \cdots, c_{i,m} = h_m^{r_i} \cdot g^{s_{i,m}}, c_{i,m+1} = g^{r_i}$$

4: Each B_i sends a request to CSP.

5: CSP selects a large number U, calculates $\Delta = l \cdot U^2$ and selects $a_1, a_2, \cdots, a_n$ that satisfy $a_i > \Delta^i$ for $i = 1, 2, \cdots, n$. And then, CSP randomly choose noise e from $(\Delta, a_1 + a_2 + \cdots + a_n)$. Finally, CSP sends $a_1, a_2, \cdots, a_n$, e and Δ to B_i

6: After receiving $a_1, a_2, \cdots, a_n$, e and Δ, B_i computes $f(\varphi_i) = a_1(\varphi_i(\text{mod } \Delta)) + a_2(\varphi_i(\text{mod } \Delta))^2 + \cdots + a_n(\varphi_i(\text{mod } \Delta))^n + e$ and sends $f(\varphi_i)$ to E.

7: E picks out the B_i with the largest average value φ_i, and checks whether the bundle of B_i contains the good that has already been auctioned through the step 8 - step 12.

8: E sends $A = (a_1, a_2, \cdots, a_m)$ to CSP.

9: CSP computes $sk_y = S \cdot A = (s_1 a_1 + s_2 a_2 + \cdots + s_m a_m)$ and sends sk_y to E.

10: E picks the largest $f(\varphi_i)$ and asks the corresponding B_i to send $c_{i,1}, c_{i,2}, \cdots, c_{i,m}, c_{i,m+1}$ to the auctioneer E.

11: Upon receiving the ciphertext $c_{i,1}, c_{i,2}, \cdots, c_{i,m}, c_{i,m+1}$, the auctioneer E computes

$$\Pi_{j=1}^{m} \frac{c_{i,j}^{a_j}}{c_{i,m+1}^{sk_y}} = g^{\langle S_i \cdot A \rangle} \text{ and check whether it is equal to 1.}$$

12: If the result is 1 after decryption, B_i is the winner. E puts the B_i into the winner set W and marks its corresponding bundle as auctioned in set A. Otherwise, B_i is not the winner. E will remove B_i from bidders. Then repeat step 8 - step 12 until no set can be updated.

In Algorithm 2, each B_i calculates $f(\varphi_i)$ and sends $f(\varphi_i)$ to the auctioneer E. Because $f(\varphi_i)$ is a one-way increasing function, the auctioneer E picks the largest $f(\varphi_i)$ by comparing the value of $f(\varphi_i)$, which is equivalent to picking the largest φ_i. Besides, E verifies whether the bidder's bundle contains the good that has already been auctioned through judging whether $g^{S_i \cdot A}$ is equal to 1. If $g^{S_i \cdot A} = 1$, that means compared with other bidders, the average value of B_i is the largest, and the corresponding bundle is also available, which means B_i

is the winner of this round. The auctioneer will update A and W to continue the search for the next winner. If $g^{S_i \cdot A} \neq 0$, the final output is indistinguishable from a random number in $\mathbb{Z}_n$ from the auctioneer's perspective, which means the bundle of B_i contains at least one good that has been auctioned. E will remove B_i from bidders and re-select the bidder with the largest average value.

After the winner is selected, E will inform the winner to send the average value φ_i, bundle S_i, $Sig(\varphi_i)$, $Sig(S_i)$ to E, and the signatures $Sig(\varphi_i)$ and $Sig(S_i)$ can guarantee the integrity of φ_i and S_i.

3.2 Privacy-Preserving Verifiable Payment Determination Model

We propose privacy-preserving verifiable payment determination model (Algorithm 3) as follow. Because E cannot know any information about B_j' bundle S_j from Algorithm 2, so E cannot know any information about b_j from $\frac{b_j}{|S_j|}$. Similarly, the winner B_i can't get any information about B_j's bundle S_j and b_j, and B_i even does not who is B_j. B_j does not get any information in this process. Since E and B_i have the signature $Sig(\varphi_j)$ generated by CSP, E can believe that B_j gives him the correct φ_j, and B_i can verify that E does not send the wrong p_i.

Algorithm 3. Privacy-preserving and Verifiable Payment Determination

1: E removes the winner B_i from bidders, and modifies A to $(A - S_i)$, where A is the set of auctioned goods and S_i is the bundle of B_i. Then thorough Algorithm 3, E chooses a new winner B_j, who is the candidate of B_i. E notifies B_j to send average value $\varphi_j = \frac{b_j}{|S_j|}$ and $Sig(\varphi_j)$ to E.

2: If the candidate of B_i can be successfully found, E computes $p_i = \frac{b_j}{|S_j|}|S_i|$ and sends p_i and $Sig(\varphi_j)$ to B_i. If no candidate is found, E sets p_i as the agreed default value and notifies B_j that p_i is the default value.

3: If p_i is not the default value, B_i can recover φ_j from $\frac{p_i}{|S_i|}$ and verify whether φ_j is correct through $Sig(\varphi_j)$. If they are not equal to each other, B_i knows that the payment is not correct.

---

4 Security Analysis

Theorem 1. *An adversarial auctioneer E's advantage abv_{S_i} is negligible.*

Proof. Every winner's bundle S_i is given to E, therefore we have:

$$adv_{S_i} = Pr[S_i|\mathcal{S}, Output \leftarrow \mathcal{A}_{our}(1^k)] - [S_i|Output \leftarrow \mathcal{A}_{black}] = 1 - 1 = 0$$

if B_i is a winner. Further, because the ElGamal encryption algorithm is semantically secure, during the privacy-preserving Winner Determination (Algorithm 2), all that an adversarial E learns is whether there exists a feasible bundle. This reveals nothing about losers' S_j, therefore any adversary's view on losers' bundle in our model is the same as the one in an ideal black-box algorithm. Therefore,

$$adv_{S_j} = Pr[S_j|\mathcal{S}, Output \leftarrow \mathcal{A}_{our}(1^k)] - [S_j|Output \leftarrow \mathcal{A}_{black}] < negl(\kappa)$$

if B_j is a loser, where $negl(\cdot)$ is a negligible function.

Theorem 2. *An adversarial auctioneer E's advantage adv_{b_j} is negligible for all loser.*

Proof. In the payment determination model of the winner B_i, the candidate B_j's average value φ_j is disclosed to E. Because of the privacy-preserving Winner Determination (Algorithm 2), E knows nothing about S_j, and he does not learn b_j from $\varphi_j = \frac{b_j}{|S_j|}$. Therefore,

$$adv_{b_j} = Pr[b_j|\mathcal{S}, Output \leftarrow \mathcal{A}_{our}(1^k)] - [b_j|Output \leftarrow \mathcal{A}_{black}] < negl(\kappa)$$

5 Performance Analysis

In our combinatorial auction scheme, each bidder needs to transfer $(m + 1)$ ciphertext, so N bidders need to transfer a total of $N \cdot (m + 1)$ ciphertext, and the auctioneer needs to return the result. The security parameter used in our scheme is τ, and the length of the ciphertext of Elgamal is 2τ. Because the length of the result is relatively small compared to τ, so it can be ignored. Therefore, in our combinatorial auction scheme, the communication overhead is $N \cdot (m + 1) \cdot (2\tau) = 2N(m + 1)\tau$.

To evaluate the computation overhead, we conducted a simulation experiment. The experimental environment was Windows 8 64-bit operating system, memory 4G, Intel(R) Core(TM) i5-4210U CPU @ 1.70 GHz. In order to exclude the communication I/O during the simulation, we generated all strings in the communication and conducted the computation in the local instance. Security parameter κ is 128-bit and every operation is run 1000 *times* to get the average run time.

As can be seen from Fig. 2, the auctioneer's computation overhead increases linearly with the increase of the amount of total bidders. But Fig. 2 shows that in our protocol, the auctioneer's computation overhead grows with a small constant factors linearly. Figure 2 shows that the amount of total bidders do not have a big impact on bidder's computation overhead, because each bidder calculates the average values φ_i, $f(\varphi_i)$ and encrypts the bundle S_i locally.

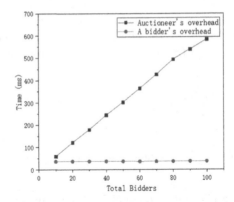

Fig. 2. Auctioneer's and bidder's overhead in winner determination.

6 Conclusion

In this paper, we proposed a privacy-preserving combinatorial auction scheme to protect bidder's privacy and ensure the security and verifiability of the result. We employed a one-way and monotonically increasing function to ensure the auctioneer to pick out the largest bid without discloseing the bid. We designed a privacy-preserving winner determination model to guarantee the correctness of the result. In our scheme, crypto service provider (CSP) is responsible for key distribution and blind signature. Besides, we put forward a privacy-preserving verifiable payment determination model to compute the payment the winner should pay. We extensively analyzed the security of our scheme to show that any adversary's view is the same as the one in a black-box algorithm. However, our work is not necessarily suitable for any context, we firmly believe we can continue to improve our work in the future.

Acknowledgment. This work is supported by the National Natural Science Foundation of China under grants 61672010 and 61702168, the Open Research Project of State Key Laboratory of Cryptology of China, and the Key projects of Guangxi Natural Science Foundation under grant 2019JJD170020.

References

1. Alvarez, R., Nojoumian, M.: Comprehensive survey on privacy-preserving protocols for sealed-bid auctions. Comput. Secur. **88**, 101502 (2019)
2. Chen, Y., Ma, Z., Wang, Q., Huang, J., Zhang, Q.: Privacy-preserving spectrum auction design: challenges, solutions and research directions. IEEE Wirel. Commun. **PP**(99), 1–9 (2019)
3. Hu, C., Li, R., Mei, B., Li, W., Alrawais, A., Bie, R.: Privacy-preserving combinatorial auction without an auctioneer. Eurasip J. Wirel. Commun. Netw. **2018**(1), 38 (2018). https://doi.org/10.1186/s13638-018-1047-z

4. Lin, J., Pipattanasomporn, M., Rahman, S.: Comparative analysis of auction mechanisms and bidding strategies for P2P solar transactive energy markets. Appl. Energy **255**, 113687 (2019)
5. Jung, T., Li, X.Y.: Enabling privacy-preserving auctions in big data (2013)
6. Kikuchi, H.: $(M + 1)$st-price auction protocol. In: Syverson, P. (ed.) FC 2001. LNCS, vol. 2339, pp. 351–363. Springer, Heidelberg (2002). https://doi.org/10.1007/3-540-46088-8_27
7. Larson, M., Li, R., Hu, C., Li, W., Cheng, X., Bie, R.: A bidder-oriented privacy-preserving VCG auction scheme. In: Xu, K., Zhu, H. (eds.) WASA 2015. LNCS, vol. 9204, pp. 284–294. Springer, Cham (2015). https://doi.org/10.1007/978-3-319-21837-3_28
8. Larson, M., Li, W., Hu, C., Li, R., Cheng, X., Bie, R.: A secure multi-unit sealed first-price auction mechanism. In: Xu, K., Zhu, H. (eds.) WASA 2015. LNCS, vol. 9204, pp. 295–304. Springer, Cham (2015). https://doi.org/10.1007/978-3-319-21837-3_29
9. Miao, P., Zhu, X., Fang, Y.: Using homomorphic encryption to secure the combinatorial spectrum auction without the trustworthy auctioneer. Wirel. Netw. **18**(2), 113–128 (2012). https://doi.org/10.1007/s11276-011-0390-3
10. Palmer, B., Bubendorfer, K., Welch, I., Development and evaluation of a secure, privacy preserving combinatorial auction. In: Australasian Information Security Conference (2011)
11. Pan, M., Sun, J., Fang, Y.: Purging the back-room dealing: secure spectrum auction leveraging Paillier cryptosystem. IEEE J. Sel. Areas Commun. **29**(4), 866–876 (2011)
12. Parkes, D.C., Rabin, M.O., Thorpe, C.: Cryptographic combinatorial clock-proxy auctions (2009)
13. Xing, K., Hu, C., Yu, J., Cheng, X., Zhang, F.: Mutual privacy preserving k-means clustering in social participatory sensing. IEEE Trans. Ind. Inform. **13**, 2066–2076 (2017)
14. Zaman, S., Grosu, D.: Combinatorial auction-based allocation of virtual machine instances in clouds. J. Parallel Distrib. Comput. **73**(4), 495–508 (2013)

A Multi-user Shared Searchable Encryption Scheme Supporting SQL Query

Mingyue Li[1,2]([⊠]), Ruizhong Du[3], and Chunfu Jia[1,2]

[1] College of Cyber Science,
Nankai University, Tianjin 300350, China
15630424277@163.com
[2] Tianjin Key Laboratory of Network and Data Security Technology,
Nankai University, Tianjin 300350, China
[3] School of Cyber Security and Computer,
Hebei University, Baoding 071002, China

Abstract. Due to the tremendous benefits of cloud computing, organizations are highly motivated to store electronic records on clouds. However, outsourcing data to cloud servers separates it from physical control, resulting in data privacy disclosure. Although encryption enhances data confidentiality, it also complicates the execution of encrypted database operations. In this paper, we propose a multi-user shared searchable encryption scheme that supports multi-user selective authorization and secure access to encrypted databases. First, we apply the Diffie-Hellman protocol to a trapdoor generate algorithm to facilitate fine-grained search control without incremental conversions. Second, we utilize a private key to generate an encrypted index by bilinear mapping, which makes it impossible for an adversary to obtain trapdoor keywords by traversing the keyword space and to carry out keyword guessing attacks. Third, we use double-layered encryption to encrypt a symmetric decryption key. Only the proxies whose attributes are matched with access control list can obtain the key of decrypted data. Through theoretical security analysis and experimental verifications, we show that our scheme can provide secure and efficacious ciphertext retrieval without the support of a secure channel.

Keywords: Data privacy · Searchable encryption · Structured Query Language (SQL) · Multi-user shared

1 Introduction

With the rapid development of computer technology and internet applications, the demand for data access and the information storage capacity is increasing [1]. Cloud computing enables users to enjoy high-quality services and ubiquitous network access on demand. Outsourcing data to cloud servers separates it from

© ICST Institute for Computer Sciences, Social Informatics and Telecommunications Engineering 2021
Published by Springer Nature Switzerland AG 2021. All Rights Reserved
D. Wang et al. (Eds.): SPNCE 2020, LNICST 344, pp. 406–422, 2021.
https://doi.org/10.1007/978-3-030-66922-5_28

physical control, resulting in data privacy disclosure [2]. To protect user data privacy, plaintext is typically encrypted before being outsourced to clouds. While many organizations (such as governments, hospitals or companies) always store their data in relational databases, data encryption may lead to the fact that widely used structured queries language (SQL) of plaintext databases cannot be directly applied to encrypted data, e.g., the probabilistic encryption generates different ciphertext each time the same plaintext is encrypted and minimizes leaked information. However, it makes encrypted data inaccessible.

There are two ways to construct an SQL query scheme for encrypted data in a database. The first is to directly operate over encrypted database data, and the second is to use an index. The problem of performing calculation directly on encrypted data was put forward long before the application scenario of cloud computing appeared. Rivest et al. [3] presented the concept of full homomorphic encryption to solve this problem. Full homomorphism encryption supports any computation and does not disclose any information about the data itself. It can achieve ideal data confidentiality. However, the performance overhead is very large. Before the performance of full homomorphic encryption reaches practical level, we have to try other methods with lower performance overhead. Curtmola et al. [4] established an index for a whole database based on a pseudo-random function and a bloom filter to improve the query efficiency. Encryption index can quickly locate the records satisfying query conditions by making the encrypted data independent of query operations; thus, the subsequent searchable encryption schemes supporting SQL queries generally adopt this approach.

Unfortunately, existing solutions that support SQL queries have some limitations in terms of multi-user sharing, e.g., since different institutions encrypt indexes with different keys, the traditional schemes construct the token adjustment search scheme based on a key derivation algorithm, that is, for each query from user, data owner has to generate an encryption different of two tokens, which are respectively encrypted by the owner and user for keyword w, to realize the conversion between trapdoors. While the repeated encryption conversion lead to huge computations. Besides, some mechanisms that resist keyword guessing attacks require a data owner to encrypt a special keyword set by public key of authorized users. If the number of authorized users is large, such computation will leads to enormous overhead in terms of computing resources and storage space.

1.1   Our Contributions

In this paper, we propose a multi-user shared searchable encryption scheme supporting the SQL query (MSE-SQ) for secure and effective data sharing. MSE-SQ constructs a reversed index for each encrypted in its database to improve search efficiency. Moreover, the effectiveness of MSE-SQ is verified by theoretical analysis and experimentation. Our contributions are as follows:

- We propose a new trapdoor generation method based on the Diffie-Hellman key exchange protocol, which preserves the search functionality without frequent key exchanges when a user search different indexes encrypted by different keys. It can reduce vast communication overhead of trapdoor transformation.

- We mix the private key of a data owner into the encrypted index by bilinear mapping to resist the keyword guessing attacks from malicious servers, thus an adversary cannot obtain the trapdoor keywords by traversing the keyword space.
- We design a novel double-layered encryption algorithm based on user's attributes to ensure a secure transmission of symmetric decryption key without the support of a secure channel.

1.2   Related Work

Nowadays, data stored on a server in plaintext no longer meets the requirement of privacy protection; thus, data encryption technology has received increasingly greater attention. However, the encrypted storage of private data produces data query problems; thus, the searchable encryption scheme is proposed. In recent years, searchable encryption mechanisms have been widely studied [5,6,8–10]. However, these mechanisms focus on the retrieval of text files and cannot meet certain requirements, i.e., they do not support basic SQL queries for stored and queried ciphertext data in the database.

At present, there are many researches on searchable database encryption systems, which are mainly divided into two types: direct operations on encrypted database data and index methods.

Direct Operations on Encrypted Database Data. Raluca et al. [11] used onion encryption to combine different encryption methods, that is, they encrypt a data field by multiple nested encryption methods. The more secure the outer layer encryption method is, the weaker the function of this layer is. Mit et al. [12] used homomorphic encryption to support secure aggregation queries, wherein operations on aggregate fields are obtained by calculating the aggregation of aggregate fields on the server side and decrypting them on the client side. However, the encryption efficiency is very low when Mit et al.'s method is used to encrypt large-scaled data. Wong et al. [13] proposed a data interoperability scheme based on secure multi-party computing that uses the RSA encryption method to change the key re-encryption; thus, their scheme supports the operations such as addition, comparison and connection, etc. However, Wong et al.'s scheme does not consider the problem of multi-user key updates. Liu et al. [14] proposed a practical and secure homomorphic order-preserving encryption (FHOPE) scheme to solve efficiency problems associated with multiple encryption combinations. FHOPE allows cloud servers to perform complex SQL queries with different operators on encrypted data without repeated encryption. FHOPE directly operates on encrypted database data. Its efficiency is low. Subsequently, the encryption index method was established wherein encrypted data are independent of query operations. This enables the records satisfying query conditions to be quickly localed by an index.

Encrypted Index. Encrypted indexes enable records that meet the query conditions to be quickly located and improve query efficiency. Golle et al. [6] first proposed a keyword ciphertext query scheme that establish an index for each data table and define a security model called IND-CKA (indistinguishability against adaptive chosen keyword attack) for the secure index. This index scheme can quickly determine matched data table and does not reveal unnecessary matching information. Curtmola et al. [4] established an index for a whole database based on a pseudo-random function and a bloom filter which improved the query efficiency. However, it cannot achieve fine-grained access control. Li et al. [15] designed an encrypted index with a tree structure. They transform ciphertext sequence relations into data structure by a cryptography method and perform a strict security analysis. Since each node in the tree includes a bloom filter, the disadvantage of the tree index is its high miscalculation rate. To improve upon this database index, Karras et al. [16] offered a self-adaptive concept of database domain and put forward the self-adaptive encrypted index suitable for the range query of a column database. However, there is no strict security analysis of the self-balanced binary tree structure of the self-adaptive scheme. Subsequently, Monir et al. [17] extend the searchable encryption index from a text to an SQL database. This SE index supports the functions of range and boolean queries. The time complexity that judges whether a keyword is contained in a data table is $O(1)$. However, the range query of their scheme is a simple multi-keyword query, which needs to specify the range of values in advance. Thus the efficiency is relatively low for a non-standard query.

2 Preliminaries

In this section, we review some basic cryptographic notions and definitions used later in the paper, such as the notation of our encryption scheme, adversary modeling and security definitions.

2.1 Notations

In a cloud database system, the data owner outsources tremendous amount of two-dimensional table set $T = \{T_1, T_2, \ldots, T_m\}$ to a database server in encrypted form $C = \{C_1, C_2, \ldots, C_m\}$, where m is the number of two-dimensional tables. $R_{i,j}$ can represent a record in the two-dimensional table T_i, where $1 \le j \le \#T_i$. The keyword set is extracted from T_i and denoted by $W = \{w_1, w_2, \ldots, w_n\}$, where n is the number of keywords. For each $w_i \in W$, the $A(w_i)$ represents a collection of records containing the keyword w_i. Besides, the proxy of the data owner generates a secure index set I. We use h to denote the number of users and χ to denote the number of attributes of users.

2.2 Adversary Model

To standardize the scope of the study, we present three hypotheses as follows:

- The cloud server faithfully stores the encrypted data of the data owner and responds to the data requests, but it will try to analyze the encrypted data and index of the data owner to obtain privacy information.
- The authorized users are trusted. He/She neither actively disclose the plaintext of the data owner to an unauthorized entity nor actively disclose the obtained decryption information.
- The proxy server is fully trusted, which executes all the user-side functions of our protocol.

2.3   Complexity Assumptions

Assumption 1 *(DDH, Decisional Diffie-Hellman [18]). Let G_1 is multiplicative cyclic groups of a large prime order p and g is the generators of G_1. Given $(g, g^a, g^b, g^c) \in G_1$ as the challenge input, the DDH problem is to distinguish between the distributions (g, g^a, g^b, g^{ab}) and (g, g^a, g^b, g^c), where a, b, c are random in Z_p. If (ε, k) – DDH problem assumption stands on group G_1 and a polynomial time adversary A can solve the problem with a probability $Adv_{A,G_1}^{DDH}(k)$, we say that there is no such adversary A can solve DDH problem with a non-negligible probability.*

$$Adv_{A,G_1}^{DDH}(k) = \left| \begin{matrix} Pr(A(g, g^a, g^b, g^{ab})) - \\ Pr(A(g, g^a, g^b, g^c)) \end{matrix} \right| \leq \varepsilon \tag{1}$$

2.4   Security Definition

Let A be a polynomial bounded adversary and B be a challenger. After A has issued a table and a subset $L \subseteq \{1, 2, ..., n\}$, B responds with two encrypted tables associated with L, then A cannot but distinguish the encrypted tables created by L. Therefore, this security game can indicate a secure goal which it requires that A is unable to deduce the plaintext from other tables. Moreover, the scheme should resist keyword guessing attacks. We define an indistinguishability of ciphertext from limited random (ICLR) game on the basis of traditional security model in [20].

Game 1: The ICLR game is defined between a polynomially bounded adversary A and a challenger B, where adversary A can request the encryption $EncIndex(r, T_i, W)$ of any two-dimensional table T and any search capabilities (trapdoors). The ICLR game process is as follows.

Setup. Adversary A generates the public/private key pairs.

Phase 1. Adversary A requests the encryption $EncIndex(r, T_i, W)$ of any two-dimensional table T and any search capabilities (trapdoors).

Challenge. Phase 1 ends. A chooses a two-dimensional table T, and then sends it to the challenger B. B creates two tables $T_0 = Rand(T, L - \{t\})$ and $T_1 = Rand(T, L)$, where $L \subseteq \{1, 2, ..., n\}$ and a value $t \in L$, and chooses a random bit $b \in \{0, 1\}$, and then gives $EncIndex(r, T_\beta, W)$ to A.

Phase 2. A continues to ask for encrypted tables and capabilities, with the restriction that A may not ask for a capability that is distinguishing for T_0 and T_1. The challenger answers as in Phase 1.

Guess. A outputs $\beta' \in \{0,1\}$ and wins the game if $\beta' = \beta$. The advantage of A winning is defined as $|pr[\beta' = \beta] - 1/2|$.

Definition 1. *If polynomially bounded adversary A wins Game 1 with a negligible advantage, then the scheme is semantically secure in ICLR (Indistinguishability of Ciphertext from Limited Random) games.*

3   SQL Database

3.1   SQL Database

As shown in Table 1, an SQL database organizes data in a set of two-dimensional tables $(T_1, T_2, ..., T_m)$. Each row of a two-dimensional table is a record. Each column is an attribute. Each table item corresponds to a value.

Table 1. Example of an SQL database table

Per-id	Per-sex	Per-age	Per-marital
213781001	Male	20	Never-married
310034025	Female	18	Married-civ-spouse
209020132	Male	39	Widowed Mexico
243020122	Male	54	Married-civ-spouse
...	...	...	...

3.2   Dictionary Creation

There are two types of SQL data. One is insensitive plaintext data, the other is sensitive encrypted data. The former accounts for the majority of the database data. In view of the characteristics of the data, we only use algorithm 1 to create an appropriate keyword set for sensitive data.

We consider the following select operation for our data-base queries: SELECT attributes FROM table WHERE conditions. This query retrieves the records that match the conditions of a two-dimensional table (or view). Authorized users can perform the following three types of queries:

(1) Simple keyword query: Only contains one condition, $p_1 = $ <attr = val>.

(2) Conjunctive query: Contains multiple query conditions, such as SELECT attributes FROM table WHERE $p_1 \wedge p_2 \wedge \cdots \wedge p_x$, where $p_i = $ <attr = val> $(1 < i < n)$. If the SQL query statement contains two query criteria: $p_1 = $ <Per-sex = 'Male'> and $p_2 = $ <Per-age = '20'>, the proxy server generates $trapdoor_1 = $ trapdoor(tk, Per-sex = 'Male') and $trapdoor_2 = $ trapdoor(tk, Per-age = '20').

(3) Disjunctive query: The select operation is SELECT attributes FROM table WHERE $p_1 \vee p_2 \vee \cdots \vee p_x$. The SQL query is converted to an SE query by the same process as the connection query.

Algorithm 1. $\{W, A(w)_{w \in W}\} \leftarrow CreateDict(D)$.

Require: database , T;

Ensure: Keyword set, W; Record set ,$A(w_i)_{w_i \in W}$.

1: $W = \phi$;
2: **for** T_i in T **do**
3:     **for** $attr$ **do**
4:         $w \leftarrow < attr = val >$;
5:         **if** w not in W **then**
6:             $W = W \cup \{w\}$;
7:         **end if**
8:     **end for**
9: **end for**
10: **for** $w_i \in W$ **do**
11:     determine $A(w_i)$;
12: **end for**
13:
14: **return** W and $A(w_i)_{w_i \in W}$.

---

4   The Detailed Scheme

The system is divided into two modules encompassing multi-user ciphertext retrieval and the decryption of encrypted data.

4.1   System Model

As shown in Fig. 1, there are four entities in the system model: data user (User), proxy server (PS), database server (DBMS) and certification authority (CA). The User refers to the data owner, such as a hospital or medical institution user who is responsible for collecting, classifying data and searching encrypted data through SQL query statements. The PS is responsible for establishing the encrypted index, generating the trapdoor, sharing the secret key and decrypting the data. The DBMS is responsible for storing the encrypted database and executing specific queries according to the user's query requirements. The CA is a trusted certificate authority that provides trusted credentials for users during key sharing.

4.2   Multi-user Fine-Grained Access Control

Since different proxy servers generate different trapdoors and indexes for the same keywords. The query trapdoor is frequently converted when multi-user share data, which leads to vast computation overhead [17]. To this end, we propose a new trapdoor generation method based on the Diffie-Hellman key conversion protocol. It does not convert trapdoors through multiple encryption when multi user share encrypted data that reduces the computation overhead of

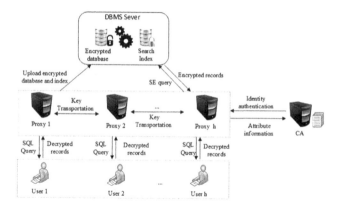

Fig. 1. System model

ciphertext retrieval. Moreover, only authorized users can search encrypted data to achieve fine-grained access control, which improves the security of the system. The specific algorithms of our scheme are as follows.

(1) $params \leftarrow Setup(1^k)$: Given a security parameter k and output system parameters $params = <q, g, \hat{e}, P, G, H_1, H_2>$, where q is a large prime number related to security parameter k, G and G_T are cyclic groups of order q, g is the generator of group G, $\hat{e} : G \times G \to G_T$ is an efficient non-degenerate bilinear map. H_1 and H_2 are hash functions, where $H_1 : \{0,1\}^* \to G_q$, $H_2 : G_{q^2} \to \{0,1\}^n$, $H_3 : \{0,1\}^* \to Z_q$.

(2) $(SK, PK) \leftarrow KeyGen(params)$:
  - The proxy server of the data owner uses system parameters to generate the private key $x \in Z_q$ and the public key g^x.
  - The proxy server of the data user uses system parameters to generate private key $y \in Z_q$ and public key g^y.

(3) $I \leftarrow EncIndex(r, T_i, W)$: The index generation algorithm specifies that the proxy server of data owner creates a keyword index for each table T_i in the database.
  - Randomly select $r \in Z_q$.
  - For each w_i in encrypted T_i, let $I_{w_i} = \hat{e}(H_1(w_i)^x, g^r)$ and the encrypted index $I = \{I_{w_1}, \ldots, I_{w_n}\}$.

(4) $C \leftarrow Enc(T, K)$: Proxy servers encrypt data uploaded to the cloud server. For each $T_i \in T$, the data owner runs $Enc_K(T_i)$ to generate ciphertext C_i and ciphertext set $C = \{C_1, C_2, \ldots, C_m\}$.

(5) $Td \leftarrow Trapdoor(g^x, y, Qc)$: It takes public key g^x of the data owner, the private key y of the data user and the query keyword set Qc as the input and executes the flowing steps:
  - The proxy server of the data user follows the Elliptic Curve Diffie-Hellman protocol and calculates $(g^x)^y$ using its private key y and public key g^x of the data owner, where $(g^x)^y$ is a point on an elliptic curve. The horizontal

coordinate a and vertical coordinate of this point b are concatenated and hashed to a point in G as $tk = H_3(a||b)$ and used as secret key for generating the trapdoor.

$$tk = H_3(a||b) \tag{2}$$

- The proxy server of the data user uses tk to generate trapdoor Td for $Qc = \{w_1', w_2', ..., w_t'\}$, where $1 \le t \le n$.

$$Td = \{H_1(w_1')^{tk}, H_1(w_2')^{tk}, ..., H_1(w_t')^{tk}\} \tag{3}$$

(6) $Acd \leftarrow Delegate(x, g^y, C)$: The proxy server of the data owner generates access control data Acd for the data user.
  - The proxy server of the data owner uses private key x and public key of user g^y to compute $(g^y)^x$ and obtain (g^{xy}). Then, the horizontal and vertical coordinates of point g^{xy} are joined to calculate $tk = H_3(a||b)$.
  - For each encrypted two-dimensional table C_j that the user is allowed access, entry (C_j, ζ_j) is created, where

$$\zeta_j = g^{rx} \tag{4}$$

$$\eta_j = g^{tk} \tag{5}$$

and list $Acd = \{(C_j, \zeta_j, \eta_j) | j \in (1, m)\}$

(7) $Dlist \leftarrow Search(Acd, Td)$: For the encrypted two-dimensional table C_j queried by the data user, the DBMS obtains the search result $Dlist$ by computing $\frac{\hat{e}(H_1(w_i')^{tk}, \zeta_j)}{\hat{e}(\eta_j, g)} \overset{?}{=} I_{w_i'}$, $1 \le i \le t$.

4.3 Secure Key Transmission with Dual Encryption

In the application of a public key searchable encryption scheme, considering the efficiency of plaintext encryption and decryption, another symmetric encryption scheme is typically used to encrypt and decrypt plaintext to protect data privacy. In our encryption scheme, we encrypt the tables in the database with symmetric encryption. As the security of symmetric algorithms depends on the encryption key used, we design a double-layered encryption algorithm based on attribute encryption to ensure the secure transmission of symmetric keys.

The key transfer process is shown in Fig. 2. After receiving the ciphertext encrypted by the proxy server of the data owner (PS_j) from the DBMS, the proxy server of the data user (PS_i) requests the attribute private key from the CA. The CA first confirms the data user's identity according to his/her attributes list and then generates the attribute private key ak for the proxy PS_i, where the identity list is shown in Table 2. Simultaneously, the CA sends the attribute information about the proxy PS_i to PS_j. PS_i requests the key from PS_j. The proxy PS_j uses double-layered encryption algorithm to encrypt the key requested by the PS_i and send it to PS_j. In addition, PS_j uses the private key sk and the attribute private key ak to decrypt the corresponding key. Finally, PS_i decrypts the encrypted two-dimensional tables with the decrypted keys to obtain the plaintext data. The main key sharing algorithms are as follows.

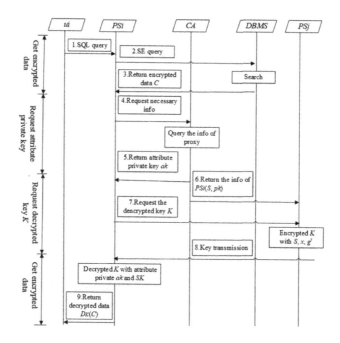

Fig. 2. The progress of decryption data

(1) $(MSK, P) \leftarrow Init(params)$: The CA randomly selects $MSK \in Z_q$ and generates $P = g^{MSK}$.

(2) $ak \leftarrow Skeygen(S, MSK, g^y)$: Take attribute set S and public key g^y of PS_j as input and complete the flowing steps:
   - Map the attributes of PS_j to point $Q = H_1(attr_1 \text{——} attr_2... \text{——} attr_\chi)$.
   - Generate attribute private key $ak = Q^{MSK}$.
   - Use public key g^y to encrypt ak and send it to PS_j, The proxy then uses the private key to decrypt ak.

(3) $c \leftarrow EncryTrans(S, x, g^y, K)$: The proxy of the data owner encrypts symmetric key $K \in \{0,1\}^N$ using the private key x, attributes and the public key g^y of the data user.
   - Map the attributes of the data user to point Q as done for the CA.
   - Randomly select $\vartheta \in Z_q$.
   - Use g^{xy} to encrypt ϑ to obtain $\vartheta' = g^{xy} + \vartheta$ and let $c = <\vartheta', K \oplus H_2(\hat{e}(Q, P))>$, where $\hat{e}(Q, P) \in G_T$.

(4) $K \leftarrow DecryKey(c, d, y, g^x)$: Decrypt K using private key y of the data user and the attribute private key ak given by the CA.
   - For $c = <\vartheta', K \oplus H_2(\hat{e}(Q, P))>$, decrypt ϑ' with the private key y and the public key g^x of the data owner and obtain $\vartheta = \vartheta' - g^{xy}$.
   - Let $V = K \oplus H_2(\hat{e}(Q, P))$, decrypt V with attribute private key ak: $V \oplus H_2(\hat{e}(Q, P)) = K$.

Table 2. Attributes list of proxies

Proxy ID	$attr_1$	$attr_2$	$attr_3$	...	$attr_\chi$
ID_1	$S_{1,1}$	$S_{2,1}$	$S_{3,1}$	...	$S_{\chi,1}$
ID_2	$S_{1,2}$	$S_{2,2}$	$S_{3,2}$	...	$S_{\chi,2}$
ID_3	$S_{1,3}$	$S_{2,3}$	$S_{3,3}$	...	$S_{\chi,3}$
...	...	...	...	...	...

5   Security Analysis

5.1   Confidentiality of Queries

Theorem 1. *Our proposed MSE-SQ scheme is $(1 - 1/2e^n q_T n)$ secure under the ICLR game without random oracle if DDH assumption is intractable.*

Proof. Let A be a polynomially bounded adversary with an advantage ε in Game 1, there is an emulator B that has advantage of $\varepsilon' = \varepsilon/e^n q_T n$ at minimum for solving the DDH problem, where q_T is the number of times the trapdoor is queried in $O(time(A))$.

Setup. A randomly selects element $x \in Z_q$ and set it as the private key, then he generates his public key g^x. Finally, he sents the private key $y \in Z_q$ and the public key g^y to B.

Phase 1. A forms the following queries:

- Index query. A sends an index generation request to B for the keyword set $W = \{w_1, w_2, ..., w_n\}$ in the two-dimensional table T_i. To simulate the EncIndex algorithm, B selects a random element γ_j in G for each key w_j. Furthermore, B selects random number u_i in Z_q and returns the searchable ciphertext $I = \{\hat{e}(\gamma_1^x, g^{u_i}), ..., \hat{e}(\gamma_\tau^x, g^{bu_i}), ..., \hat{e}(\gamma_n^x, g^{u_i})\}$, $\zeta_j = g^{xu_i}$ and $\eta_j = g^{tk}$, where $tk = H_3(a_x \| b_y)$, a_x and b_y are the horizontal coordinate and vertical coordinate, respectively, of g^{xy} on the elliptic curve.
- Trapdoor query. In order to test the search algorithm, A sends a trapdoor generation request $Qc = \{w_1', w_2', ..., w_t'\}$ to B. After receiving the trapdoor generation request, B asserts $\{w_1', w_2', ..., w_t'\} \subseteq W$ and computes $Td^* = \{H_1(w_1')^{tk}, H_1(w_2')^{tk}, ..., H_1(w_t')^{tk}\}$ for A.

Challenge. Once adversary A decides that Phase 1 is over, it sends a triple $\{T_i, t, L\}$ to emulator B, in which $L \subseteq \{1, ..., n\}$ and $t \in L$. Notice that the random element τ is independent of the location t selected. B responds as follows:

- If $j \neq t$ and $j \in L$, let $h_j = k_j$, where k_j is a random element in G.
- If $j \neq t$ and $j \notin L$, let $h_j = \hat{e}(\gamma_j^x, g^a)$.
- If $j = t$:
  * $t = \tau$, let $h_t = \hat{e}(\gamma_t^x, g^c)$.
  * $t \neq \tau$, let $h_t = k_t \in Z_q$.

– Send $\{h_1, ..., h_n, g^{ax}, g^{tk}\}$ to A.

Phase 2. Adversary A issues more queries, where A cannot query the keywords that have been queried. Emulator B responds as in Phase 1.

Guess. A outputs a guess $\beta' \in \{0,1\}$. If $\beta' = \beta$ and $\beta = 1$, then (g^a, g^b, g^c) is considered to be a DDH tuple. So when $\tau = t$, B can prove that (g^a, g^b, g^c) is a DDH tuple. The proof is as follows.

$$\frac{\hat{e}(\gamma_\tau^x, g^{bu_i})}{\hat{e}(g^{xu_i}, g^{tk})} = \frac{\hat{e}(\gamma_t^x, g^c)}{\hat{e}(g^{ax}, g^{tk})} \Leftrightarrow \frac{\hat{e}(\gamma_\tau, g)^{xbu_i}}{\hat{e}(g,g)^{xu_i tk}} = \frac{\hat{e}(\gamma_j, g)^{xc}}{\hat{e}(g,g)^{axtk}}$$

$$\Leftrightarrow \frac{\hat{e}(\gamma_\tau, g)^b}{\hat{e}(g,g)^{tk}} = \frac{\hat{e}(\gamma_j, g)^c}{\hat{e}(g,g)^{atk}} \Leftrightarrow \frac{\hat{e}(\gamma_\tau, g)^b}{\hat{e}(g,g)^{tk}} = \frac{\hat{e}(\gamma_j, g)^c}{\hat{e}(g,g)^a \hat{e}(g,g)^{tk}}$$

$$\Leftrightarrow \hat{e}(\gamma_\tau, g)^b \hat{e}(g,g)^a = \hat{e}(\gamma_t, g)^c \tag{6}$$

$$\Leftrightarrow \hat{e}(\gamma_\tau, g^{ab}) = \hat{e}(\gamma_t, g^c)$$

$$\Leftrightarrow g^{ab} = g^c$$

If $\beta = 0$, the ciphertext on position t is a random value. Thus, B cannot prove that (g^a, g^b, g^c) is a DDH tuple. To ensure a random encryption at position t, the challenge must not be a DDH-tuple. Whereas, the A's advantage in winning the ICLR game is the same as that of B settles the DDH challenge.

The above steps compose the simulation process of the security proof. We analyze the probability of B solving the DDH problem. Event E_1 indicates that B responds to the trapdoor query request for n keywords of adversary A. Event E_2 denotes that B gives up responding in the progress of the challenge. When q_T is large enough, $Pr(E_1) = 1/e^n$, $Pr(E_2) = 1/q_T n$. Thus, the probability for B solving the DDH problem is $\varepsilon' = \varepsilon \cdot Pr[E_1 \cup E_2] \geq \varepsilon/e^n q_T n$, where $\varepsilon/e^n q_T n \in [0, 1/2e^n q_T n]$ is negligible. Therefore, the probability that our encryption scheme is secure is at least $(1 - 1/2e^n q_T n)$ in Game 1.

5.2 Security of Secret Keys

Theorem 2. *Neither the certification authority, nor any other user can get the symmetric key of a document.*

Proof. Before uploading data to the DBMS, all the database tables are encrypted by standard symmetric cryptography. To enable only the authorized users to decrypt the encrypted records, the symmetric key for each table is encrypted to be transmitted by the following formula.

$$K \oplus H_2((Q, g^{MSK})^\vartheta) \tag{7}$$

Even if the certification authority CA has the attribute private key ak of a proxy, it cannot recover ϑ because it does not have the private key SK of the proxy. Since ϑ is random, the CA cannot get K and $H_2(\hat{e}(Q, P))$ must be obtained. Furthermore, if a malicious attacker steals the private key SK, it means the attacker can get ϑ'. However, it does not have the corresponding attribute private key ak and cannot compute $H_2(\hat{e}(Q, P))$. That is to say, the malicious attacker cannot restore K. Therefore, the proposed encryption scheme effectively ensures the secure transmission of secret keys and enables only the authorized users to decrypt the encrypted records.

6 Experimental Analysis

6.1 Experimental Environment

Implementation. Our SE scheme is implemented based on the architecture shown in Fig. 1 and tested with three identically configuration PCs. The system environment is based on a windows 7(64-bit) system, the hardware configuration is an Intel(R)Core(TM)i7-6700 (3.40 GHz) processor equipped with 8 GB of memory and a fast Ethernet network (1 Gbps). We use Alibaba cloud storage platform, a domestic cloud storage provider, to build a storage system. Multiple virtual machines son a single PC are used to simulate the proxies. All the functions are implemented in Java. The User and CA are deployed on separate PCs. Communication between the proxy and CA is mainly realized via socket transmission.

In the key sharing scenario, the OpenSSL library is used to generate 4096-bit PKE key pairs and an elliptic curve is used to implement IBE encryption. The PBC library is called to implement bilinear pairing. The search uses the PL/Java in Java, which is an additional feature of the server-side program. The code is packed into the JAR and the JAR is further loaded into the SQL server backstage for direct searches.

Data. The data set was extracted by Barry Becker from the census database in 1994 called Adult [19]. This database lists 16,281 records and comprises eight attributes, including identity information (Per-id), age (Per-age), gender (Per-sex), monthly wage (Per-wage) and other sensitive information composition.

6.2 Query Security

The experiment makes use of the privacy protection level of the scheme of the following formula, where $0 < p(T_i) < 1, \sum_{i=1}^{m} p(T_i) = 1$. The larger the $H(T)$, the lower the possibility of privacy leakage. The value of $H(T)$ is determined if there are no other external conditions.

$$H(T) = -\sum_{i=1}^{m} p(T_i) log_2 p(T_i) \tag{8}$$

Figure 3 shows a comparison of the privacy protection of the SE schemes, wherein our scheme provides the highest and SDSE [17] provides the lowest level of privacy protection. This is because our scheme can provide secure ciphertext retrieval without the support of a secure channel.

6.3 Multi-user Keyword Query Performance

Index Construction. As shown in Fig. 4, the time required by the three SE schemes to establish an encrypted index, which increases with the number of keywords, is compared experimentally. As the most basic scheme, SDSE adopts symmetric encryption index, which has the shortest time requirement. Our scheme

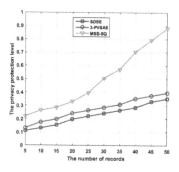

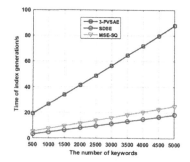

Fig. 3. Comparison of privacy protection level

Fig. 4. Encryption index time requirements

Table 3. Compare of single keyword query time

Single keyword query	$|A_W|$	Search time		
		3-PVSAE	SDSE	MSE-SQ
Per-sex = 'Male'	10860	8.231 s	1.005 s	1.543 s
Per-age = '25'	354	2.503 s	0.025 s	0.031 s
Per-id = '2016126911'	1	0.392 s	0.009 s	0.012 s

Table 4. Time of conjunctive query and disjunctive query

Conjunctive query	$|A_W|$	Search time
Per-sex = 'Male' and Per-age = '39'	808	0.564 s
Per-age = '25' and Per-marital = 'Never-married'	247	0.215 s
Per-sex = 'Male' and Native-country = 'Peru'	11	0.015 s
Disjunctive query	$|A_W|$	Search time
Per-age = '25' or Per-marital = 'Never-married'	5541	0.819 s
Native-country = 'Peru' or Per-age = '39'	405	0.158 s
Per-id = '310034025' or Native-country = 'Peru'	15	0.013 s

and 3-PVSAE [7] both are public key encryption schemes, but 3-PVSAE requires the longest time to generate its encryption index due to multiple modular exponentiation.

Trapdoor Generation. Figure 5 presents an experimental comparison of time required by the three SE schemes for trapdoor generation. The time required by the 3-PVSAE scheme is nearly 3.5 times greater than our scheme. The time cost of the SDSE scheme is the lowest.

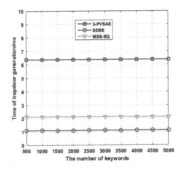

Fig. 5. Time of Trapdoor generation

Query Overhead. Table 3 shows an experimental time cost comparison of a single keyword query in each of the three SE schemes. The search time of the 3-PVSAE scheme is the longest and at nearly 7 times greater than our scheme using the same data set and index structure. The time of the SDSE scheme is the shortest. In addition, the query time overhead of the conjunctive and disjunctive queries are shown in Table 4.

7   Conclusion and Future Work

In this paper, we present a multi-user shared ciphertext retrieval scheme for SQL databases, which supports multi user using SQL statements to perform boolean query. To reduce the enormous computing burden of repeated encryption conversions of trapdoors, we use the new trapdoor generation method based on the Diffie-Hellman key exchange protocol to implement queries without converting the trapdoors and transmitting the trapdoor keys of different users. Furthermore, the double-layered encryption method provides secure retrieval without the support of a secure channel. Through theoretical security analysis, our SE scheme is statistically secure in ICLR game. We present a framework that embeds the search algorithm into SQL sever. This framework is evaluated in terms of practicality using a census database.

Due to the variety of cloud applications available nowadays, our future work consists in implementing automatic search matches for a greater number of protocols.

Acknowledgments. This work is supported by National Key R&D Program of China (2018YFA0704703); National Natural Science Foundation of China (61972215, 61702399, 61972073); Natural Science Foundation of TianJin (17JCZDJC30500).

References

1. Xu, L., Yuan, X., Wang, C., Wang, Q., Xu, C.: Hardening database padding for searchable encryption. In: INFOCOM, pp. 2503–2511 (2019)

2. Kasra Kermanshahi, S., Liu, J.K., Steinfeld, R., Nepal, S.: Generic multi-keyword ranked search on encrypted cloud data. In: Sako, K., Schneider, S., Ryan, P.Y.A. (eds.) ESORICS 2019. LNCS, vol. 11736, pp. 322–343. Springer, Cham (2019). https://doi.org/10.1007/978-3-030-29962-0_16

3. Ronald, R.L., Len, A., Michael, D.L.: On data banks and privacy homomorphisms. Found. Secure Comput. **4**, 169–180 (1978)

4. Curtmola, R., Garay, J.A., Kamara, S., Ostrovsky, R.: Searchable symmetric encryption improved definitions and efficient constructions. In: Proceedings of the 13th ACM Conference on Computer and Communications Security, pp. 79–88 (2006)

5. Song, D.X., Wagner, D., Perrig, A.: Practical techniques for searches on encrypted data. In: Proceedings of the 21st IEEE Symposium on Security and Privacy 2000, pp. 44–55 (2000)

6. Golle, P., Staddon, J., Waters, B.: Secure conjunctive keyword search over encrypted data. In: Jakobsson, M., Yung, M., Zhou, J. (eds.) ACNS 2004. LNCS, vol. 3089, pp. 31–45. Springer, Heidelberg (2004). https://doi.org/10.1007/978-3-540-24852-1_3

7. Zhang, R., Xue, R., Yu, T., Liu, L.: PVSAE: a public verifiable searchable encryption service framework for outsourced encrypted data. In: 2016 IEEE International Conference on Web Services (ICWS)

8. Xia, Z., Wang, X., Sun, X., Wang, Q.: A secure and dynamic multi-keyword ranked search scheme over encrypted cloud data. IEEE Trans. Parallel Distrib. Syst. **27**, 951–963 (2016)

9. Patel, S., Persiano, G., Yeo, K.: Symmetric searchable encryption with sharing and unsharing. In: Lopez, J., Zhou, J., Soriano, M. (eds.) ESORICS 2018. LNCS, vol. 11099, pp. 207–227. Springer, Cham (2018). https://doi.org/10.1007/978-3-319-98989-1_11

10. Zhang, Z., Wang, J., Wang, Y., Su, Y., Chen, X.: Towards efficient verifiable forward secure searchable symmetric encryption. In: Sako, K., Schneider, S., Ryan, P.Y.A. (eds.) ESORICS 2019. LNCS, vol. 11736, pp. 304–321. Springer, Cham (2019). https://doi.org/10.1007/978-3-030-29962-0_15

11. Popa, R.A., Redfield, C.M.S., Zeldovich, N., Balakrishnan, H.: CryptDB: protecting confidentiality with encrypted query processing. In: Proceedings of the 23rd ACM Symposium on Operating Systems Principles, pp. 85–100 (2011)

12. Catherine, M.S., Nickolai, Z.: CryptDB: protecting confidentiality with encrypted query processing. In: Proceedings of the 23rd ACM Symposium on Operating Systems Principles, pp. 85–100 (2011)

13. Wong, W.K., Kao, B., Cheung, D.W.-L., Li, R., Yiu, S.-M.: Secure query processing with data interoperability in a cloud database environment. In: ACM SIGMOD Conference 2014, pp. 1395–1406 (2014). https://doi.org/10.1145/2588555.2588572

14. Liu, G., Yang, G., Wang, H., Xiang, Y., Dai, H.: A novel secure scheme for supporting complex SQL queries over encrypted databases in cloud computing. Secur. Commun. Netw. **2018**, 7383514:1–7383514:15 (2018). https://doi.org/10.1155/2018/7383514

15. Li, R., Liu, A.X., Wang, A.L., Bruhadeshwar, B.: Fast and scalable range query processing with strong privacy protection for cloud computing. IEEE/ACM Trans. Netw. **24**, 2305–2318 (2016). https://doi.org/10.1109/TNET.2015.2457493

16. Karras, P., Nikitin, A., Saad, M., Bhatt, R., Antyukhov, D., Idreos, S.: Adaptive indexing over encrypted numeric data. In: Proceedings of the 2016 International Conference on Management of Data, pp. 171–183 (2016)

17. Azraoui, M., Önen, M., Molva, R.: Framework for searchable encryption with SQL databases. In: CLOSER, pp. 57–67 (2018)
18. Ning, J., Xu, J., Liang, K., Zhang, F., Chang, E.: Passive attacks against searchable encryption. IEEE Trans. Inf. Forensics Secur. **14**(3), 789–802 (2019)
19. Dua, D., Graff, C.: UCI Machine Learning Repository. http://archive.ics.uci.edu/ml
20. Jiang, S., Zhu, X., Guo, L., Liu, J.: Publicly verifiable Boolean query over outsourced encrypted data. IEEE Trans. Cloud Comput. **7**(3), 799–813 (2019)

Forward Secure Searchable Encryption with Conjunctive-Keyword Supporting Multi-user

Zhongyi Liu, Chungen Xu$^{(\boxtimes)}$, and Zhigang Yao

School of Science, Nanjing University of Science and Technology, Nanjing, China
ZhongyiLiu950217@outlook.com, xuchung@njust.edu.cn,
zhigangyaocrypto@outlook.com

Abstract. Searchable symmetric encryption (SSE) enables users to efficiently search ciphertext in the cloud and ensures the security of encrypted data. Recent works show that forward security is an important property in dynamic SSE. Many forward secure searchable symmetric encryption (FSSE) schemes supporting single-keyword search have been proposed. Only a few SSE schemes can satisfy the forward security and support conjunctive keyword search at the same time, which are realized by adopting inefficient or complicated cryptography tools. Very recently, Hu proposed a novel construction to achieve conjunctive-keyword search, that is, using inner-product encryption (IPE) to design a conjunctive-keyword FSSE scheme. However, IPE scheme is a conceptually complex and low efficient scheme. In this paper, we use a more efficient cryptographic tool, asymmetric scalar-product-preserving encryption (ASPE), to design an efficient and secure conjunctive-keyword FSSE scheme. To improve practicality, we design our scheme to support multi-user setting. Our scheme achieves sub-linear efficiency, and can easily be used in any single-keyword FSSE scheme to obtain a conjunctive-keyword FSSE scheme supporting multi-user. Compared with the current conjunctive-keyword FSSE scheme, our scheme has a better update and search efficiency.

Keywords: Forward security · Conjunctive-keyword search · Multi-user

1 Introduction

With the development of network technology, cloud computing technology has been widely used by companies and individuals. Using cloud storage service, users can outsource large amounts of data to the cloud. For data security, users need to encrypt their data before uploading. Searchable symmetric encryption

This work was supported in part by the Fundamental Research Funds for the Central Universities (No. 30918012204).

D. Wang et al. (Eds.): SPNCE 2020, LNICST 344, pp. 423–440, 2021.
https://doi.org/10.1007/978-3-030-66922-5_29

(SSE) [13] provides keyword search function while ensuring the security of cloud data. In an SSE scheme, the user encrypts files and (ind, keywords) pairs, then sends the encrypted data to the server; When the user wants to search the files containing keyword w, he generates a search token using the secret key and sends it to the server; After receiving the search token, it runs the match operation to obtain the files containing the keyword w, and then sends those files to the user. Finally, the user will decrypt the received data to get the files that he wants. Dynamic searchable symmetric encryption (DSSE) is more practical than SSE. It can support dynamic update operations. The user can add new files to the server or delete an old file in the server.

Some basic leakages are inevitable for SSE schemes as they support keyword search over encrypted data. Curtmola [4] proposed the definition of the leakage of access pattern and search pattern. DSSE scheme leaks more information than SSE scheme. In DSSE scheme, an adversary can inject files containing some special keywords into the server's database; Then the adversary can use old search queries to search those injected files and then obtain some useful information from the search result. This kind of attack is called file injection attack (FIA) [19]. The FSSE schemes can well resist file injection attack. Because in a FSSE scheme, old queries can not match the new file which not queried before. Bost [1] proposed a new idea for constructing FSSE schemes. Only trapdoor permutation is used to construct the FSSE scheme in Bost's scheme. In order to improve the practicability of FSSE scheme, some FSSE schemes [7,16] that support conjunctive-keyword search have been proposed. These schemes can search files containing multiple keywords. Most of the existing SSE schemes that support boolean queries leak the *result pattern* information [3]. This leakage allows the server to know which files contain part of the search keywords.

1.1 Our Contributions

The main contributions of this paper are as follows:

- We construct an efficient and secure FSSE scheme that supports conjunctive-keyword query and multi-user. Compared with the best current conjunctive-keyword FSSE scheme, our scheme has a better update and query efficiency. Our scheme can hide the number of keywords in a search query.
- Our scheme proves the feasibility of ASPE in constructing conjunctive-keyword FSSE scheme.
- We give the security analysis and detailed efficiency description of our scheme. We also implement our scheme using Java programming language to test its practicality.

1.2 Related Work

Song et al. [13] proposed the first SSE scheme, which searching over the ciphertext instead of using the index table. So its search complexity is linear with the number of documents. After that, in order to improve search efficiency, many schemes using an index table [2,3,5,8,9,15] have been proposed. Goh et al. [5] proposed the first index-based SSE scheme. Schemes using index table will greatly improve search efficiency. In addition, researchers designed some SSE schemes with more properties, and hence, is more practical. Kamara et al. [9] proposed a DSSE scheme that can achieve sub-linear search efficiency. To support conjunctive-keyword search, Golle et al. [6] first proposed conjunctive-keyword search scheme. But their scheme are not very efficient. It has linear complexity in the whole number of documents. After that, Cash et al. [3] proposed the first SSE scheme that can achieve sub-linear search efficiency and support boolean queries. Lai [11] proposed a scheme that solves the Keyword-Pair Result Pattern leakage of OXT and achieves almost the same efficiency as OXT. But Lai's scheme needs one more round communication in search protocol. Zhang et al. [19] gave a formalized definition of a very strong attack, named file injection attack. This attack can easily recover the keyword of a query. Stefanov et al. [15] first formalized the notion of forward security for SSE scheme. After Bost [1] proposed a very creative way to design a practical FSSE scheme, many practical FSSE schemes were successively proposed. Song et al. [14] improved efficiency based on Bost's scheme using pseudorandom permutation. After that, Zhang et al. [20] proposed a more efficient way to construct FSSE scheme than Song's scheme [14]. According to Zhang's idea, we can use only hash function to construct a FSSE scheme. Recently, a few FSSE schemes support conjunctive-keyword search [7,16] were proposed. But the cryptography tools they used are not efficient enough.

1.3 Organization

The rest of this paper is organized as follows. In Sect. 2, we describe the system model, threat model and design goals of our system. The notations and cryptographic primitives used in this paper are introduced in Sect. 3. In Sect. 4, we give the detailed structure of our scheme and describe how to deploy our scheme in a real scenario. We analyze the security of our scheme and compare the complexity with existing schemes and give the experimental result in Sect. 5 and 6, respectively. Finally, we give a brief conclusion in Sect. 7.

2   Problem Statement

In this section, we will introduce the system model, threat model and design goals of our system.

2.1   System Model

Figure 1 records an overview of our system model. There are three parties: server, private key generator and user. User can be both data owner and data user. These parties can be described as follows:

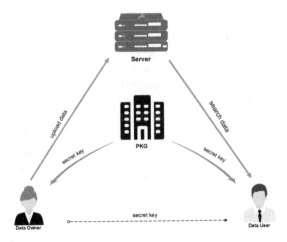

Fig. 1. System model

- **Server:** The server is a semi-honest provider of cloud storage services. It stores encrypted data sent by data owner and responses search operation honestly. But it will learn and try to get some plain text information
- **PKG:** PKG is a trusted center. It first generates system parameters, including the public parameter pp and the master key msk. PKG then uses the system parameters to generate private keys for each user. PKG will not leak his master key to anyone.
- **User:** All users are authenticated. They can both be data owner and data user. They can send their file to the server and search the files in the server. Users can communicate with each other for data sharing.

How the System Works. First, PKG generates the system parameters pp and msk and predefine the keyword space W. PKG then responds to the registration request of the authenticated user and generates a key for him. After receiving the key, the user generates a random number by himself and uses both the key from PKG and the random number as his secret key. After getting the secret key, the user can encrypt his data and then send the ciphertext to the server. When a user wants to search his own data containing keywords $W_{search} = \{w_1, \cdots, w_l\}$, he can use his secret key to generate a search token, and then sends the search token to the server. After receiving the search token, the server runs match operation using the token and return the result to the user. When a user U_1 wants to search the data of another user U_2, U_1 first sends a request to U_2, U_2 then share his secret key and a state st_c with U_1. U_1 can generate search token using the secret key and state, and then search files in the server.

2.2   Threat Model

The server in our system is considered semi-honest, that is, it will perform every operation honestly, but it will try to learn information about the ciphertext from each operation. The private key generator PKG is considered credible. It will not leak any information about system parameters and any user's secret key to other parties. Only an authenticated person can register as a user in the system. And each user will not leak his secret key to unauthenticated parties.

2.3 Design Goals

We design our system with the following goals:

- **Forward Privacy:** Newly added files would not cause previous queries to leak information.
- **Update and Search Efficiency:** Our scheme is more efficient than other existing conjunctive-keyword FSSE schemes both in update and search efficiency.
- **Scalability:** The method we use to construct our system can be easily applied to other single-keyword FSSE schemes.

3 Preliminaries

In this section, we present some notations and cryptographic primitives used in this paper. Table 1 is part of the notations we use in this paper.

Table 1. Notations and descriptions

Notation	Description
λ	Security parameter
N	The number of total keywords allowed by the system
$\mathcal{Z}_p$	A finite field, where p is a big prime integer
$\mathbb{G}_1, \mathbb{G}_t$	Two cyclic groups with prime order p
pp	The public parameter of PKG
msk	The master key of PKG
W	A predefined ordered keyword space
H_i	i-th hash function
F_i	i-th pseudorandom function
F_p	A pseudorandom function with range $\mathbb{Z}_p$
$\lvert \cdot \rvert$	Cardinality of set
Σ, T	Two maps
st_i	The ith state
ind	A document index
U_{id}	One user with identity id
$x \xleftarrow{\$} X$	Uniformly sample an element from a set X
$negl(\lambda)$	The negligible function in the security parameter λ
$PRPV(key, \boldsymbol{v})$	Randomly rearrange the vector's elements
op	The update operation, where $op \in \{add, del\}$
W_{search}	All keywords in a search query
W_{ind}	All keywords in a file ind

3.1 Bilinear Maps

Definition 1. *Let $\mathbb{G}_1$ and $\mathbb{G}_t$ be two cyclic groups with the same prime order p. Let g be a generator of group $\mathbb{G}_1$. $e : \mathbb{G}_1 \times \mathbb{G}_1 \to \mathbb{G}_t$ is a map from $\mathbb{G}_1$ to $\mathbb{G}_t$. e is a bilinear map if it has the following three properties:*

1. *Bilinearity: $\forall g_1, g_2 \in \mathbb{G}_1$ and $a, b \in \mathbb{Z}_p$, we have $e(g_1^a, g_2^b) = e(g_1, g_2)^{ab}$.*
2. *Non-degeneracy: If g is a generator of $\mathbb{G}_1$, then $e(g, g)/ \neq 1$, where 1 is the unity element of $\mathbb{G}_t$.*
3. *Computability: $\forall g_1, g_2 \in \mathbb{G}_1, e(g_1, g_2)$ can be computed in polynomial time.*

3.2 Dynamic Symmetric Searchable Encryption

We briefly introduce DSSE based on [1]. A database $DB = (ind_i, W_i)_{i=1}^D$ is a tuple of index/keywords pairs with $ind_i \in \{0,1\}^l$, $W_i \in \{0,1\}^*$, where D is the number of documents in DB. $W = \bigcup_{i=1}^D$ denote the total number of keywords in database DB. Let $N = \sum_{i=1}^D |W_i|$ be the number of document/keyword pairs. $DB(w)$ is the set of the document containing keyword w. Dynamic searchable encryption can be denoted by $\Pi = (Setup, Search, Update)$. Among them, $Setup$ is an algorithm, $Search$ and $Update$ are two protocols between server and client. Π can be described as follows:

- $Setup(DB)$ is used to initialize and start a system. It takes as input a plaintext database DB, and outputs (EDB, K, σ). EDB is the encrypted database, K is the secret key and σ is client's state.
- $Search(EDB, K, q, \sigma)$ is a protocol between the server and the user. The user first uses his secret key K, state σ and a search query $q = \{w_1, \cdots, w_l\}$ to generate a search token, then he sends the token to the server. Server runs the match algorithm which takes as input search token and its encrypted database EDB, then sends back the search result to the client.
- $Update(EDB, K, \sigma, op, in) = (Update_C(K, \sigma, op, in), Update_S(EDB))$ is a protocol between the user and the server. $Update_C$ is run by the user. The $Update_C$ takes as input the key K, state σ, operation op and input in parsed as the index ind and a set of keywords W_{ind} and output a update token. The $Update_S$ is run by the server and takes as input EDB. The operation op is taken from the set $\{add, del\}$, meaning the addition and deletion operations of a document/keyword pair.

3.3 Security Definition

It is difficult to obtain efficiency when design SSE schemes achieving high security. Therefore, when designing a SSE scheme, some information will be appropriately allowed to leak to get higher efficiency. Let $\mathcal{L} = \{\mathcal{L}_{Setup}, \mathcal{L}_{Update}, \mathcal{L}_{Search}\}$ be a set of predefined leakage functions. We can call a SSE scheme is secure if no more information is leaked than $\mathcal{L}$. Let $\Pi = (Setup, Update, Search)$ be a dynamic searchable symmetric encryption scheme, $\mathcal{A}$ be an adversary, λ be a security parameter. We can define the security of DSSE scheme by these two experiments [14]:

- **Real**$_{\mathcal{A}}^{\Pi}(\lambda)$: The adversary $\mathcal{A}$ chooses database DB, then the experiment runs $Setup(DB)$ and returns EDB to $\mathcal{A}$. $\mathcal{A}$ adaptively chooses and executes a list of query $\{q_i\}$. If q_i is a search query, then the experiment runs $Search(EDB, K, q_i, \sigma)$ and returns the result. If q_i is an update query, the experiment answers the query by running $Update(EDB, K, \sigma, op, in)$. Finally, the adversary $\mathcal{A}$ outputs a bit $b \in \{0, 1\}$.
- **Ideal**$_{\mathcal{A},\mathcal{S}}^{\Pi}(\lambda)$: The adversary $\mathcal{A}$ chooses a database DB, then the experiment runs $\mathcal{S}(\mathcal{L}_{Setup}(DB))$ with the leakage $\mathcal{L}_{Setup}(DB)$ and returns EDB to $\mathcal{A}$. Then the adversary adaptively chooses and executes a list of query $\{q_i\}$. If q_i is a search query, then the experiment runs $\mathcal{S}(\mathcal{L}_{Search}(q_i))$ with leakage $\mathcal{L}_{Search}(q_i)$ and returns the result. If q_i is an update query, the experiment runs $\mathcal{S}(\mathcal{L}_{Update}(q_i))$ with leakage $\mathcal{L}_{Update}(q_i)$. Finally, the adversary $\mathcal{A}$ outputs a bit $b \in \{0, 1\}$.

Definition 2. Π *is called* $\mathcal{L}$-*adaptively-secure SSE scheme, if for any probabilistic polynomial time (PPT) adversary* $\mathcal{A}$, *there exists a PPT simulator* $\mathcal{S}$ *such that:*

$$Pr(\boldsymbol{Real}_{\mathcal{A}}^{\Pi}(\lambda) = 1) - Pr(\boldsymbol{Ideal}_{\mathcal{A},\mathcal{S}}^{\Pi}(\lambda) = 1) \leq negl(\lambda)$$

where $negl(\lambda)$ *is a negligible function.*

Definition 3 *(Forward privacy [14]).* *A* $\mathcal{L}$-*adaptively-secure SSE scheme* Π *is forward secure if the update leakage function* $\mathcal{L}_{Update}$ *can be written as*

$$\mathcal{L}_{Update}(q_i) = (i, op_i, ind_i)$$

where $q_i = (ind_i, w_i, op_i)$ *is an update query,* op_i *is an update operation, and* ind_i *is a document index. This definition means that an update query leaks no information about the update keyword.*

3.4   Result Pattern Leakage

Result Pattern (RP) Leakage [11] leakage is a kind of leakage in SSE protocols, i.e. the files matching one search token. In conjunctive-keyword setting, the ideal RP leakage is the files matching all the keywords in one search operation, i.e. *the Whole Result Pattern (WRP)*. But most existing conjunctive-keyword search SSE scheme leak more information than WRP. For example, if user wants to search the files containing keywords $w_1, \cdots, w_t$. In OXT, there are three types of RP leakage:

- Single Keyword Result Pattern (SKRP), i.e. the server can know $DB(w_1)$
- Keyword-Pair Result Pattern (KPRP), i.e. the server can know $DB(w_1, w_i), 2 \leq i \leq t$
- Multiple Keyword Cross-Query Intersection Result Pattern(IP), i.e. the server can know $DB(w_1, \{w_i\})$ where $\{w_i\} \subseteq \{w_2, \cdots, w_t\}$

3.5 Asymmetric Scalar-Product-Preserving Encryption [17]

An ASPE scheme can be denoted by $\Pi = \{Setup, Encrypt, KeyGen, Decrypt\}$. Π can be described as follows:

- **Setup(λ).** This algorithm takes as input secure parameter λ and outputs the secret key K.
- **Encrypt(v, K).** This algorithm takes as input the message v, a vector, and the secret key K, and outputs the ciphertext c.
- **KeyGen(q, K).** This algorithm takes as input the query vector q and the secret key K, and outputs the search key d.
- **Decrypt(c, d).** This algorithm takes as input the ciphertext c and the search key d and outputs the inner product of v and q.

We call Π is a ASPE scheme if and only if Π satisfies the following conditions:

1. $v \cdot q = Decrypt(c, d)$ for any message vector v and any query vector q
2. $v_i \cdot v_j \neq Decrypt(c_i, c_j)$ for any message vector v_i and v_j.

4 Our Scheme

In this section, we give the construction of our scheme. We design our scheme based on [7] and [20] which are conjunctive-keyword FSSE scheme and verifiable FSSE scheme supporting single-keyword search, respectively. We use a more efficient cryptographic tool to integrate these two schemes and then construct our more efficient conjunctive-keyword FSSE scheme.

4.1 Construction

Let $\mathbb{G}_1$, $\mathbb{G}_t$ be two cyclic groups with the same prime order p. g is a generator of $\mathbb{G}_1$. Let $e : \mathbb{G}_1 \times \mathbb{G}_1 \to \mathbb{G}_t$ be a bilinear map. Let N be the cardinality of predefined ordered keyword space W, i.e. $N = |W|$. $PRPV(k, v)$ is a pseudo-random permutation. It can randomly reorder the elements of the vector v, and its reordering rules are determined by k. We use the ASPE scheme proposed in [17] to construct our scheme. We express op with a bit length, i.e. $op \in \{add, del\} = \{1, 0\}$. The details of our scheme are shown in the Algorithm 1. The scheme contains four parts: $Setup$, $Derive$, $Update$ and $Search$. They can be described as follows:

- $Setup(1^\lambda)$. This algorithm is run by PKG to initialize and start the system. First, PKG predefine an ordered keyword space W. With the cyclic groups $\mathbb{G}_1$, $\mathbb{G}_t$ and bilinear map e, PKG first randomly samples $(s, a) \xleftarrow{\$} \mathbb{Z}_p$ as its master key msk. Then PKG generates the public parameter $pp = \{N, p, g, g_1, H_1, H_2, H_3, F_p, F_1, F_2\}$ where $H_1, H_3 : \{0, 1\}^* \to \{0, 1\}^\lambda$, $H_2 : \{0, 1\}^* \to \{0, 1\}^{2\lambda+1}$, $F_i : \{0, 1\}^\lambda \times \{0, 1\}^* \to \{0, 1\}^\lambda$ and $F_p : \{0, 1\}^\lambda \times \{0, 1\}^* \to \mathbb{Z}_p$. Finally, PKG publishes the public parameter pp, and keeps the master key msk.

- $Derive(msk, U_{id})$. This is a protocol between the user and the PKG. First, an authenticated user U_{id} sends a registration request to PKG. After receiving the request, the PKG randomly generates two elements $(x, y) \xleftarrow{\$} \mathbb{Z}_p$ and computes $k_1 = g^x, k_2 = g^y$, and then sends (k_1, k_2) to U_{id}. After receiving the key (k_1, k_2), U_{id} randomly generates an integer $k_3 \xleftarrow{\$} \mathbb{Z}_p$. Finally, U_{id} keeps $sk_{id} = (k_1, k_2, k_3)$ as his secret key.

- $Update(ind, (w_{t_1}, w_{t_2}, \cdots, w_{t_l}), sk_{id}, \Sigma, T, op)$. This is a protocol between the user and the server where Σ is a map saved by the user to store user's latest local status, and T is a map saved by the server to store the encrypted keyword/file pairs. When a user U_{id} wants to update (add or delete) a file ind containing keywords $(w_{t_1}, w_{t_2}, \cdots, w_{t_l})$, he first generates the key k_s and then runs the $ASPE.Setup(seed)$ to get a secret key k_a for the ASPE scheme. Next, the user needs to initialize a vector $\boldsymbol{v}$ (line 4 to 8 in update protocol) where t_i represents the position of the keyword w_{t_i} in the keyword space W. Note that we set $\boldsymbol{v}[N+1] = -1$ and $\boldsymbol{v}[N+2] = 1$. Next, the user randomly shuffles the elements of the vector $\boldsymbol{v}$ by running $PRPV(k_s, \boldsymbol{v})$ and then encrypts the vector $\boldsymbol{v}'$ by running $ASPE.Encrypt(k_a, \boldsymbol{v}')$. Then he can generate a node for the chain corresponding to each keyword $w_i \in W_{ind}$. For each keyword $w_i \in W_{ind}$, the user needs to compute t_w, e and u first. Note that if $st_c = \Sigma[w_i]$ is $\perp$, we set $e = H_2(t_w||st_1) \oplus (\perp ||op||ind)$, otherwise we set $e = H_2(t_w||st_{c+1}) \oplus (st_c||op||ind)$. The symbol $\perp$ is used to mark the beginning of the current chain. After getting t_w, e, u, the user generates a key k_p to shuffle the elements of ciphertext vector $\boldsymbol{cv}$ and then hides $\boldsymbol{cv}'$ by doing $\boldsymbol{cv}' \oplus \boldsymbol{mc}$. Finally, the user sends the update token $u, e, \boldsymbol{mcv}$ to the server. After receiving the update token, the server stores it into the encrypted database T.

- $Search((w_{t_1} \wedge w_{t_2} \wedge \cdots \wedge w_{t_m}), sk_{id}, \Sigma, T)$. This is a protocol between the user and the server. When a user U_{id} wants to search the documents containing keywords $W_{search} = (w_{t_1}, w_{t_2}, \cdots, w_{t_m})$, he first searches for the st_c corresponding to w_{t_1} in map Σ. If there is no state st_c, i.e. $st_c = \perp$, the user can sure that there is no file containing the keyword w_{t_1} in the server. If $st_c \neq \perp$, the user generates k_s, k_p, k_a and a vector $\boldsymbol{q}$ like the update protocol. The difference is we set $\boldsymbol{q}[N+1] = m$ and $\boldsymbol{q}[N+2] = F_p(t_w, st_c)$ where $m = |W_{search}|$. Then the user shuffles the elements of the vector $\boldsymbol{q}$ and generates a search key by running $ASPE.KeyGen(k_a, \boldsymbol{q}')$. After shuffling the search key $\boldsymbol{cq}$ using k_p, the user sends the search token $t_w, st_c, \boldsymbol{cq}'$ to the server. After receiving the search token, the server uses t_w and st_c to traverse the encrypted keyword/file pair chain corresponding to w_{t_1}. For each node in the chain (representing an update operation), the server can get e and $\boldsymbol{mcv}$. Then the server can get (st_{c-1}, op, ind) from e where st_{c-1} is the previous state of keyword w_{t_1}. The server determines whether the update is performing an add operation or a delete operation through the op. If the current node is an add operation, the server calculates the ciphertext vector $\boldsymbol{cv}'$ and runs $ASPE.Decrypt(\boldsymbol{cv}', \boldsymbol{cq}')$ to get st_c' which is the inner product of $\boldsymbol{v}$ and $\boldsymbol{q}$. If $st_c' = F_p(t_w, st_c)$, then the server can determine that current file ind contains all the keywords in

W_{search} and then puts this file into the result set R. If the current node is a delete operation, the server checks whether the current file is included in the result R, and deletes if it contains. Finally, the server sends the result set R to the user.

Correctness. First, as long as k_p is the same, we have the following equation: $ASPE.Decrypt(cv', cq') = ASPE.Decrypt(cv, cq)$. Because if we use the same rules to shuffle the elements in two vectors, this does not change their inner product. Second, in the search protocol, the server can determine the current file ind contains all the keywords in W_{search} by $st_c' = F_p(t_w, st_c)$. Because if the file ind contains all the keywords in W_{search}, then $W_{search} \subset W_{ind}$. So there are m elements in the same position in v and q equal to 1. So the inner product of v and q is $m + (-m) + F_p(t_w, st_c) = F_p(t_w, st_c)$. If the file ind does not contain all the keywords in W_{search}, then there are less than m elements in the same position in v and q equal to 1. So the inner product of v and q is $m' + (-m) + F_p(t_w, st_c) \neq F_p(t_w, st_c)$, where $m' < m$.

4.2   How to Deploy

How to Support Multi-user. The basic idea we use to support multi-user is based on [18] which design a lightweight secret sharing method to share the secret key with another user. In our scheme, all keys can be generated by the key k_s. If a user U_1 wants to search his data stored in the cloud, he can generate the key k_s^1 using his secret key sk_1. With the k_s^1, he can generate the search token.

In the multi-user setting, let U_1 and U_2 be two different users with secret key $sk_1 = (g^{x_1}, g^{y_1}, k_3^1)$ and $sk_2 = (g^{x_2}, g^{y_2}, k_3^2)$. If user U_2 wants to search U_1's documents containing keywords $\{w_{t_1}, w_{t_2}, \cdots, w_{t_m}\}$, he needs U_1's key k_s^1 and the state $st_c = \Sigma[w_{t_1}]$. For the key k_s^1, U_2 can submit a request to U_1, then U_1 responds with $(g_1^{k_3^1}, g^{k_3^1})$. Then U_2 can calculate k_s^1, because of the following equation:

$$(e(g, k_2^1)/e(g_1, k_1^1))^{k_3^1} = e(g, g)^{k_3^1(y_1 - ax_1)} = e(g, g)^{k_3^1 s} = k_s^1$$

$$e(g^{k_3^1}, g^{y_2})/e(g_1^{k_3^1}, g^{x_2}) = e(g, g)^{k_3^1(y_2 - ax_2)} = e(g, g)^{k_3^1 s} = k_s^1$$

For st_c, U_1 can share it with U_2 through a public channel as follows: After getting the key k_s^1, U_2 generates a secret key of a symmetric encryption algorithm using k_s^1. U_1 also can generate the same secret key using his own key k_s^1. Then U_2 and U_1 can communicate securely.

How to Compress Vector Dimension. The disadvantage of our scheme 1 is that we need to predefined the keyword space W. This condition limits the practicality of our scheme. In addition, the efficiency of the ASPE scheme decreases as the keyword space expands, i.e. the dimension expands. In Algorithm 1, there are three points: the keyword space W can be sorted, the position of each keyword can be determined, and each document contains few keywords. So using a vector with dimension $|W|$ to record all the keywords' position in keywords

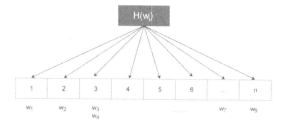

Fig. 2. Vector compression algorithm

space is extremely wasteful, because most elements of this vector is set to 0. We can save storage and computing resources by compressing this vector. The compression method is shown in Fig. 2. All we need to do is select a vector with a smaller dimension and then map the keywords to the elements of this vector. This compressing algorithm is a probabilistic algorithm. If you can accept errors (probably return a few extra files) with a small probability, you can use this method to compress the vector dimensions without losing too much accuracy.

5 Security Analysis

Informally, the security of our scheme depends on the security of the hash function, pseudorandom function and ASPE scheme. To support conjunctive-keyword search, we add an extra ciphertext into the update token, and then the server can do conjunctive-keyword search on the encrypted database. So as long as this extra ciphertext does not leak any information about the keywords of the newly added file, our scheme can guarantee the forward security. Let $W_{ind} = \{w_{t_1}, \cdots, w_{t_l}\}$ be the keywords in a file ind, and $W_{search} = \{w_{t_1}, \cdots, w_{t_m}\}$ be the keywords in a search query. For a search query q, the user always uses w_{t_1} to get t_w and st_c. We call this kind of keyword as $stag$. For a search query q, let $stag = Stag(q)$. Let $Hist = \{(DB_i, q_i)\}_{i=0}^Q$, where q_i is an update query or search query, DB_i is a snapshot of the database DB. For a search query W_{search}, let $W_s = \{w_i | w_i \in W_{search}, w_i \in W_{ind}\}$. If the ASPE scheme in [17] is secure, the leakage of our scheme can be described as follows:

1. Search Pattern:

$$sp(w_{t_1}) = \{i | Stag(q_i) = w_{t_1}, q_i \in Hist\}$$

2. Update History:

$$uh(w_{t_1}) = \{(i, op, ind_i) | q_i \text{ is a search query}, Stag(q_i) = w_{t_1}\}$$

3. Whole Result Pattern:

$$WRP(W_{search}) = \{ind_i | W_{search} \subseteq W_{ind_i}, q_i \text{ is a update query}\}$$

We can get the following theorem:

Theorem 1 *Let H_1, H_2, H_3 be three secure hash functions, and F, F_p be two secure pseudorandom function. We define the leakage function $\mathcal{L} = (\mathcal{L}_{setup}, \mathcal{L}_{search}, \mathcal{L}_{Update})$ as:*

$$\mathcal{L}_{setup} = \bot$$
$$\mathcal{L}_{search}(W_{search}) = (sp(w_{t_1}), uh(w_{t_1}), WRP(W_{search}))$$
$$\mathcal{L}_{Update}(i, op_i, W_{ind_i}, ind_i) = (i, op_i, ind_i)$$

Then the scheme 1 is is a $\mathcal{L}$-adaptively-secure dynamic SSE with forward privacy.

On conjunctive-keyword and single-user setting, our scheme has a similar structure to Hu's scheme [7], which has proven to be $\mathcal{L}$-adaptively-secure. So the security of our scheme on the single-user setting is based on the security of ASPE scheme. Detailed proof process proves that the scheme is indistinguishable from a simulated scheme, i.e. the security definition described in Sect. 2.

Unlike the cryptography tool used in Hu's scheme, the ASPE scheme is conceptually simple and efficient, but it doesn't have a high-security level. In fact, ASPE is proved to be not a CPA-secure scheme [12]. But this does not affect the forward security of our scheme. Because in update protocol, we masked cv' with mc, and mc is associated with st_{c+1}. Without the latest status st_{c+1}, the server can not get the correct ciphertext cv'. That guarantees the forward security of our scheme. If the adversary breaks the ASPE scheme, he can only get the vectors v' and q' in a old update and search query, where v' and q' are vectors reordered by $PRPV(\cdot, \cdot)$. So the adversary cannot obtain the plaintext of the keywords W_{ind} and W_{search}. But the RP leakage of this scheme will include the SKRP, KPRP and IP leakage.

6   Efficiency Analysis and Performance Evaluation

In this section, we are going to compare the complexity of our scheme with the enhanced scheme in [7] and the FOXT-B in [16], and code programs to test the efficiency of our scheme.

6.1   Efficiency Comparison

The notations used in this subsection are shown in Table 2.

The Table 3 shows the comparison result of communication complexity. And the Table 4 shows the comparison result of computational complexity.

6.2   Experiment Results

In this subsection, we use a set of data to test the efficiency of search and update protocol in our scheme. We run the program on a personal computer with Windows10 OS, Intel(R) Core(TM) i7-9750H CPU 2.60 GHz, and 4 GB of

Table 2. Notations and description

Notation	Description
l	The number of keywords in one file, i.e. $l = |W_{ind}|$
m	The number of keywords in one search query, i.e. $m = |W_{search}|$
N	The size of the predefined keyword space, i.e. $N = |W|$
c	The average size of $DB(w)$
$|\mathbb{Z}_p|$	The bit size of the element of $\mathbb{Z}_p$
$|\mathbb{G}_1|$	The bit size of the element of $\mathbb{G}_1$
P	The computation cost of a bilinear pairing operation
E	The computation cost of an exponentiation operation in cyclic group $\mathbb{G}_1$ and $\mathbb{G}_t$
H	The computation cost of hash function and pseudorandom function
$A.S$	The computation cost of $ASPE.Setup$
$A.E$	The computation cost of $ASPE.Encrypt$
$A.K$	The computation cost of $ASPE.KeyGen$
$A.D$	The computation cost of $ASPE.Decrypt$
π	The computation cost of trapdoor permutation function
$I.E$	The computation cost of $IPE.Encrypt$
$I.K$	The computation cost of $IPE.KeyGen$
$I.D$	The computation cost of $IPE.Decrypt$

Table 3. Comparison of communication complexity

Scheme	Update token	Search token
Hu's scheme	$(N+1) \cdot |\mathbb{G}_1| + 2\lambda$	$2(\lambda + |\mathbb{Z}_p|) + (N+1)|\mathbb{G}_1|$
FOXT-B	$2\lambda + |\mathbb{Z}_p| + |\mathbb{G}_1|$	$2\lambda + \mathbb{Z}_p + c(m-1)|\mathbb{G}_1|$
Our scheme	$3\lambda + 1 + 2(N+2) \cdot |\mathbb{Z}_p|$	$2\lambda + 2(N+2) \cdot |\mathbb{Z}_p|$

As we know, $|\mathbb{G}_1|$ is much larger than $\mathbb{Z}_p$. According to the result of this table, if N is not small, the size of update token in our scheme is small than Hu's schemes and large than FOXT-B, and the size of search token in our scheme is small than Hu's schemes and FOXT-B.

DDR4 RAM. We use the JPBC library for Java and choose the type A pairing to implement the pairing function. Let $\lambda = 256$ be the security parameter. We choose SM3 as hash functions H_i and let $F_i(key, data) = Hash(key||data)$. According to the complexity analysis in Subsect. 6.1, N, l and c are the main factors affecting the efficiency of update and search protocol. So we test how the efficiency change with these variables. The results are described as follows.

Table 4. Comparison of computational complexity

Scheme	Update.client	Search.client	Search.server
Hu's scheme	$l(4H + \pi) + I.E$	$H + I.K$	$c(3H + \pi + I.D)$
FOXT-B	$l(5H + \pi + E)$	$H + c(m-1)E$	$c(H + m \cdot E + \pi)$
Our scheme	$2P + E + 5l \cdot H + A.S + A.E$	$2P + E + 3H + A.K$	$c(4H + A.D)$

According to the result of this table, our scheme is more efficient than FOXT-B and Hu's enhanced scheme. Because ASPE scheme is much more efficient than IPE scheme, and we do not need to compute a lot of exponentiation operation in the cyclic group. The efficiency comparison between ASPE and IPE are shown in the Table 5.

Table 5. The efficiency of IPE and ASPE

N	Client side in search		Server side in search		Client side in update	
	I.K	A.K	I.D	A.D	I.E	A.E
5	0.8 ms	0.0003 ms	9.9 ms	0.00008 ms	2.6 ms	0.00037 ms
10	1.2 ms	0.0007 ms	24.1 ms	0.00016 ms	4.5 ms	0.00081 ms
30	2.4 ms	0.0045 ms	67.1 ms	0.00023 ms	12.5 ms	0.00514 ms
50	4.0 ms	0.0117 ms	110.2 ms	0.00030 ms	20.7 ms	0.01343 ms
100	9.8 ms	0.0446 ms	217.8 ms	0.00058 ms	43.2 ms	0.05144 ms
250	40.9 ms	0.2624 ms	540.9 ms	0.00128 ms	124.4 ms	0.03192 ms
500	140.6 ms	1.0726 ms	1100 ms	0.00229 ms	310.5 ms	1.86808 ms
750	303.7 ms	2.3842 ms	1600 ms	0.00334 ms	555.9 ms	4.46054 ms

The testing result of IPE is come from paper [10]. This comparison is reasonable because the operating environment in [10] is better than ours.

Figure 3 shows how the efficiency of the update protocol changes with the variables l and N. In these two subgraph, we let $N = 100$ for the left one and $l = 10$ for the right one, respectively. Note that the second curve is close to a straight line, but it is not, because of the computation complexity of ASPE.Encrypt is $\mathcal{O}(n^2)$. This result satisfies the complexity analysis in the Table 4. Figure 4 shows how the efficiency of the search protocol changes with the variables c and N. In these two subgraph, we let $N = 100$ for the left one and $c = 100$ for the right one, respectively. The second curve also is close to a straight line, but it is not, because of the computation complexity of ASPE.KeyGen is $\mathcal{O}(n^2)$. This result also satisfies the complexity analysis in the Table 4.

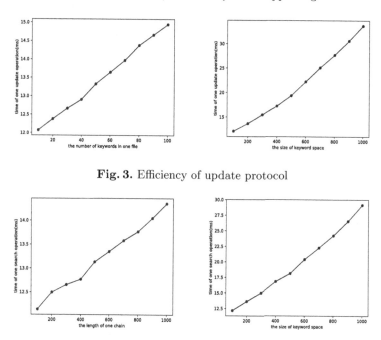

Fig. 3. Efficiency of update protocol

Fig. 4. Efficiency of search protocol

These test results prove that the efficiency of update protocol is linear with l, and search protocol is linear with c. Both the efficiency of update and search protocol is linear with N^2. However, due to the fast running speed of ASPE, if N grows not very large, the growth of N has little effect on the efficiency of update and search protocol.

Algorithm 1. Conjunctive-keyword FSSE using ASPE Scheme

$Setup(1^\lambda)$

1: $(s,a) \xleftarrow{\$} \mathbb{Z}_p$
2: $g_1 = g^a$
3: $msk = \{s,a\}$
4: $pp = \{N,p,g,g_1,H_1,H_2,H_3,F_p,F_1,F_2\}$
5: PKG keeps msk as master key

$Derive(msk,U_{id})$

PKG:
1: $(x,y) \xleftarrow{\$} \mathbb{Z}_p$, where x,y satisfy $y = ax + s(mod\ p)$
2: $k_1 = g^x, k_2 = g^y$
3: send k_1,k_2 to the user
   Client:
4: $k_3 \xleftarrow{\$} \mathbb{Z}_p$
5: the user keeps $sk_{id} = (k_1,k_2,k_3)$

$Update(ind,(w_{t_1},w_{t_2},\cdots,w_{t_l}),sk_{id},\Sigma,T)$

Client:
1: $k_s = (e(g,k_2)/e(g_1,k_1))^{k_3}$
2: $seed = H_1(k_s)$
3: $k_a = ASPE.Setup(seed)$
4: $v = 0$ where $v \in \mathbb{Z}_p^{N+2}$
5: for $i \in \{t_1,\cdots,t_l,N+2\}$ do
6:     $v[i] = 1$
7: end for
8: $v[N+1] = -1$
9: $v' = PRPV(k_s,v)$
10: $cv = ASPE.Encrypt(k_a,v')$
11: for $i \in \{t_1,\cdots,t_l\}$ do
12:     $t_w = F_1(k_s,w_i)$
13:     $st_c = \Sigma[w_i]$
14:     if $st_c = \perp$ then
15:         $st_1 \xleftarrow{\$} \{0,1\}^\lambda$
16:         $e = H_2(t_w||st_1) \oplus (\perp ||op||ind)$
17:     else
18:         $st_{c+1} \xleftarrow{\$} \{0,1\}^\lambda$
19:         $e = H_2(t_w||st_{c+1}) \oplus (st_c||op||ind)$
20:     end if
21:     $\Sigma[w_i] = st_{c+1}$
22:     $u = H_3(t_w||st_{c+1})$
23:     $k_p = F_2(k_s,w_i)$
24:     $cv' = PRPV(k_p,cv)$
25:     $mc = F_p(st_{c+1},t_w)$
26:     $mcv = cv' \oplus mc$
27:     send (u,e,mcv) to the server
28: end for
   Server:
29: $T[u] = (e,mcv)$

$Search((w_{t_1} \wedge w_{t_2} \wedge \cdots \wedge w_{t_m}),sk_{id},\Sigma,T)$

Client:
1: $st_c = \Sigma[w_{t_1}]$
2: if $st_c = \perp$ then
3:     return ϕ
4: end if
5: $k_s = (e(g,k_2)/e(g_1,k_1))^{k_3}$
6: $k_p = F_2(k_s,w_{t_1})$
7: $seed = H_1(k_s)$
8: $k_a = ASPE.Setup(seed)$
9: $t_w = F_1(k_s,w_{t_1})$
10: $q = 0$ where $q \in \mathbb{Z}_p^{N+2}$
11: for $i \in \{t_1,\cdots,t_m\}$ do
12:     $q[i] = 1$
13: end for
14: $q[N+1] = m$, $q[N+2] = F_p(t_w,st_c)$
15: $q' = PRPV(k_s,q)$
16: $cq = ASPE.KeyGen(k_a,q')$
17: $cq' = PRPV(k_p,cq)$
18: send (t_w,st_c,cq') to the server
   Server:
19: initialize two empty set R,Δ
20: while $st_c \neq \perp$ do
21:     $u = H_3(t_w||st_c)$
22:     $(e,mcv) = T[u]$
23:     $(st_{c-1},op,ind) = H_2(t_w||st_c) \oplus e$
24:     if $op = del$ then
25:         $\Delta = \Delta \cup \{ind\}$
26:     else
27:         if $ind \in \Delta$ then
28:             $\Delta = \Delta \setminus \{ind\}$
29:         else
30:             $mc = F_p(st_c,t_w)$
31:             $cv' = mcv \oplus mc$
32:             $st_c' = ASPE.Decrypt(cv',cq')$
33:             if $st_c' = F_p(t_w,st_c)$ then
34:                 $R = R \cup \{ind\}$
35:             end if
36:         end if
37:     end if
38: end while
39: send R to the client

7 Conclusion

In this paper, we analyze the feasibility of using the ASPE scheme to support multi-keyword search in FSSE scheme and construct a conjunctive-keyword FSSE scheme supporting multi-user. We analyze the security and efficiency of our scheme and how to deploy our scheme in the real scenario. Compared with existing conjunctive-keyword FSSE scheme, our scheme is more efficient and practical. For future work, we will try to design a more efficient, lower communication and storage overhead conjunctive-keyword FSSE scheme using Shamir threshold scheme.

References

1. Bost, R.: Σ oφoς forward secure searchable encryption. In: Proceedings of the 2016 ACM SIGSAC Conference on Computer and Communications Security, pp. 1143–1154 (2016)
2. Bost, R., Minaud, B., Ohrimenko, O.: Forward and backward private searchable encryption from constrained cryptographic primitives. In: Proceedings of the 2017 ACM SIGSAC Conference on Computer and Communications Security, pp. 1465–1482 (2017)
3. Cash, D., Jarecki, S., Jutla, C., Krawczyk, H., Roşu, M.-C., Steiner, M.: Highly-scalable searchable symmetric encryption with support for Boolean queries. In: Canetti, R., Garay, J.A. (eds.) CRYPTO 2013. LNCS, vol. 8042, pp. 353–373. Springer, Heidelberg (2013). https://doi.org/10.1007/978-3-642-40041-4_20
4. Curtmola, R., Garay, J., Kamara, S., Ostrovsky, R.: Searchable symmetric encryption: improved definitions and efficient constructions. J. Comput. Secur. 19(5), 895–934 (2011)
5. Goh, E.J., et al.: Secure indexes. IACR Cryptol. ePrint Arch. 2003, 216 (2003)
6. Golle, P., Staddon, J., Waters, B.: Secure conjunctive keyword search over encrypted data. In: Jakobsson, M., Yung, M., Zhou, J. (eds.) ACNS 2004. LNCS, vol. 3089, pp. 31–45. Springer, Heidelberg (2004). https://doi.org/10.1007/978-3-540-24852-1_3
7. Hu, C., et al.: Forward secure conjunctive-keyword searchable encryption. IEEE Access 7, 35035–35048 (2019)
8. Kamara, S., Papamanthou, C.: Parallel and dynamic searchable symmetric encryption. In: Sadeghi, A.-R. (ed.) FC 2013. LNCS, vol. 7859, pp. 258–274. Springer, Heidelberg (2013). https://doi.org/10.1007/978-3-642-39884-1_22
9. Kamara, S., Papamanthou, C., Roeder, T.: Dynamic searchable symmetric encryption. In: Proceedings of the 2012 ACM Conference on Computer and Communications Security, pp. 965–976 (2012)
10. Kim, S., Lewi, K., Mandal, A., Montgomery, H., Roy, A., Wu, D.J.: Function-hiding inner product encryption is practical. In: Catalano, D., De Prisco, R. (eds.) SCN 2018. LNCS, vol. 11035, pp. 544–562. Springer, Cham (2018). https://doi.org/10.1007/978-3-319-98113-0_29
11. Lai, S., et al.: Result pattern hiding searchable encryption for conjunctive queries. In: Proceedings of the 2018 ACM SIGSAC Conference on Computer and Communications Security, pp. 745–762 (2018)

12. Lin, W., Wang, K., Zhang, Z., Chen, H.: Revisiting security risks of asymmetric scalar product preserving encryption and its variants. In: 2017 IEEE 37th International Conference on Distributed Computing Systems (ICDCS), pp. 1116–1125. IEEE (2017)

13. Song, D.X., Wagner, D., Perrig, A.: Practical techniques for searches on encrypted data. In: Proceeding 2000 IEEE Symposium on Security and Privacy, S&P 2000, pp. 44–55. IEEE (2000)

14. Song, X., Dong, C., Yuan, D., Xu, Q., Zhao, M.: Forward private searchable symmetric encryption with optimized I/O efficiency. IEEE Trans. Depend. Secure Comput. **17**, 912–927 (2018)

15. Stefanov, E., Papamanthou, C., Shi, E.: Practical dynamic searchable encryption with small leakage. NDSS **71**, 72–75 (2014)

16. Wang, Y., Wang, J., Sun, S., Miao, M., Chen, X.: Toward forward secure SSE supporting conjunctive keyword search. IEEE Access **7**, 142762–142772 (2019)

17. Wong, W.K., Cheung, D.W.L., Kao, B., Mamoulis, N.: Secure KNN computation on encrypted databases. In: Proceedings of the 2009 ACM SIGMOD International Conference on Management of Data, pp. 139–152 (2009)

18. Xu, L., Xu, C., Liu, Z., Wang, Y., Wang, J.: Enabling comparable search over encrypted data for IoT with privacy-preserving. CMC-Comput. Mater. Continua **109**(2), 537–554 (2019)

19. Zhang, Y., Katz, J., Papamanthou, C.: All your queries are belong to us: the power of file-injection attacks on searchable encryption. In: 25th {USENIX} Security Symposium, {USENIX} Security 2016, pp. 707–720 (2016)

20. Zhang, Z., Wang, J., Wang, Y., Su, Y., Chen, X.: Towards efficient verifiable forward secure searchable symmetric encryption. In: Sako, K., Schneider, S., Ryan, P.Y.A. (eds.) ESORICS 2019. LNCS, vol. 11736, pp. 304–321. Springer, Cham (2019). https://doi.org/10.1007/978-3-030-29962-0_15

A³BAC: Attribute-Based Access Control Model with Anonymous Access

Runnan Zhang[1], Gang Liu[1(✉)], Shancang Li[2], Yongheng Wei[1], and Quan Wang[1]

[1] School of Computer Science and Technology, Xidian University, Xi'an 710071, China
gliu_xd@163.com
[2] Department of Computer Science and Creative Technologies,
University of the West of England, Bristol BS16 1QY, UK

Abstract. Researchers believe that anonymous access can protect private information even if it does not store in authorization organization. The current solution supports anonymous access by using a certificate instead of a subject identity or Attribute-Based Encryption. In a solution using a certificate, access may be linked to the certificate, which poses a risk of re-identification. The encryption of objects based on attributes limits the types of objects. The ABAC with anonymous access proposed in this paper called A³BAC inherits the features of the ABAC model, such as fine-grained authorization, policy flexibility, and unlimited object types. By combining HABS, it strengthens the identity-less of ABAC, so that the access does not involve a unique identification, reducing the risk of subject identity re-identification. It is a secure anonymous access framework.

Keywords: Access control · ABAC · Anonymous access · HABS

1 Introduction

In Internet of Things (IoT) and Mobile Internet of Things (MIoT), there are many problems like forensics [1] and consensus algorithm [12]. However, anonymous access control is rarely studied in the MIoT environment [7]. Some scholars have studied the access control technology supporting anonymous authorization [9–11]. The methods mainly use attributes to encrypt objects to support anonymous authorization in the access control model.

Ahuja R. and Mohanty S. K. [3] proposed a scalable Attribute-Based Encryption (ABE) scheme in the cloud environment. The scheme generates a hierarchical attribute private key for users through hierarchical authority and hierarchical Ciphertext Policy Attribute-Based Encryption (CP-ABE) algorithm. Yuen T. H., Liu J. K., Man. H. A., et al. [2] proposed an anonymous ABAC model based on an anonymous certificate to support k-times anonymous authorization for cloud services. Given the serious harm of privacy disclosure in the Personal Health Records (PHR) system, Pussewalage H. S. G. and Oleshchuk V. A. [4] proposed an anonymous ABAC scheme based on ABE and proxy re-encryption to protect privacy.

© ICST Institute for Computer Sciences, Social Informatics and Telecommunications Engineering 2021
Published by Springer Nature Switzerland AG 2021. All Rights Reserved
D. Wang et al. (Eds.): SPNCE 2020, LNICST 344, pp. 441–447, 2021.
https://doi.org/10.1007/978-3-030-66922-5_30

The above method supports anonymous access by using a certificate instead of a subject identity or ABE. There are some defects in these methods. In some schemes, the certificate of the subject is unique, so the subject's access may be linked to its certificate. The attacker can re-identify the identity of the subject through the access linked to the subject. Once the re-identification is successful, it will cause unexpected privacy disclosure [5]. The encryption of objects based on attributes limits the types of objects. It is friendly to access control of objects that can be moved from server to client, such as video files. It is unfriendly to access control of objects that cannot be moved to the client, such as web services, which needs additional mechanisms to provide access control and increases the complexity of its implementation. And using an ABE-based algorithm to access objects often needs to download the object or generate tokens, which increases the load of the network.

In this paper, we propose an ABAC model that supports anonymous access called A^3BAC. By combining HABS (homomorphic attribute-based signatures) and transferring some functions of ABAC to the attribute authority and audit institution, we strengthen the identity-less of ABAC, so that its authorization is no longer dependent on identity. A^3BAC does not use unique certificates and is friendly for all types of objects. It inherits the features of fine-grained access control, flexible policy, and unlimited object type of ABAC. It overcomes the defect that the ABAC framework does not support anonymous access, and provides an audit that improves security. Extending the ABAC framework, giving it anonymous access and audit support, can expand ABAC's scope of application and improve its practicality [14].

Our Contribution:

1. This paper proposes an ABAC framework that supports anonymous access and fine-grained control of attributes by subjects.
2. The workflow of A^3BAC and the HABS algorithm that each entity needs to execute is elaborated.

2   Preliminary

2.1   ABAC Model

While there is currently no single agreed-upon model or standardization of ABAC, there are commonly accepted high-level definitions and descriptions of its function. One such high-level description is presented in [6]:

Attribute-Based Access Control: An access control method where subject requests to perform operations on objects are granted or denied based on assigned attributes of the subject, assigned attributes of the object, environmental conditions, and a set of policies that are specified in terms of those attributes and conditions.

Most researchers agree with the ABAC framework shown in Fig. 1 [15]. policy Enforce Point (PEP) performs preliminary verification of access requests and executes access control decisions; Policy Decision Point (PDP) evaluates access requests based on acquired attributes and policies; Policy Information Point (PIP) manages attributes of all subjects, objects, and environment; Policy Attribute Point (PAP) manages and maintains a policy library. In ABAC, when a subject accesses an object, it sends an access request to an authorized organization. The process of evaluating access requests by the ABAC framework used by most researchers is shown in Fig. 1.

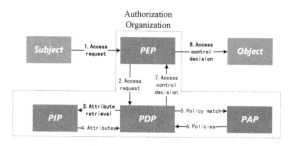

Fig. 1. ABAC framework

The access request used by the ABAC framework shown in Fig. 1 must contain the unique identifier of the subject and object. The basis for retrieving attributes is the unique identifier of the subject and object contained in the access request.

2.2 HABS Algorithm

HABS is an anonymous certification scheme based on the Attribute-Based Signatures (ABS). The ABS is designed for the user to sign a message with fine-grained control over identifying information, and it does not support the properties required for anonymous certification [8]. HABS has a clear identification of missing properties to serve anonymous certification objectives. HABS supports a flexible selective disclosure mechanism at no extra computation cost [13], which is inherited from the expressiveness of ABS for defining access policies.

HABS relies on four procedures based on the inspector, subject, issuer and verifier.

The *system initialization procedure* derives a global parameter containing the inspector's public key and pairs of public and private keys for the subject, inspector, and issuer. HABS.Setup and HABS.KeyGen is executed by the trusted third party.

The *credential issuing procedure* issues a certified credential for the subject based on its attributes. The HABS.Issue algorithm is executed by the issuer. The HABS.Obtain algorithm is executed by the subject.

The *verifying procedure* enables the verifier to check that a subject is authorized to access an object with respect to some access policy. As such, the verifier has first to send a random message to the subject. Second, the user signs the received message based on his credential. In a nutshell, the subject signs the received message based on the subset of his attributes that satisfy the signature predicate. The user finally sends his signature to the verifier who checks the resulting signature. The HABS.Show algorithm is executed by the subject. The HABS.Verify algorithm is executed by the verifier.

The HABS supports the *inspection procedure* performed by a separate and trusted entity referred to as the inspector. It relies on two algorithms namely HABS.trace and HABS.judge needed to identify the subject and give proof of judgment. They are executed by the inspector.

HABS supports a flexible selective disclosure mechanism that allows the subject to sign the message with a subset of its attributes. The verifier only obtains the subset of attributes after successful verification and does not reveal the identity of the subject. In addition, this scheme ensures the unlinkability between sessions while maintaining the anonymity of the subject.

3 A^3BAC Framework

A^3BAC is an access control framework with anonymous access and audit capabilities. A^3BAC does not use unique certificates and is friendly for all types of objects. It inherits the characteristics of fine-grained access control, flexible policy and unlimited object type of ABAC.

3.1 A^3BAC Framework

A^3BAC shown in Fig. 2 is a direct extension of the ABAC framework, which inherits many advantages of ABAC, such as fine-grained access control, flexible policy and unlimited object. The functions of PDP, PEP, and PAP in the A^3BAC framework are consistent with the corresponding modules in the framework shown in Fig. 1. The main functions of other modules are as Fig. 2.

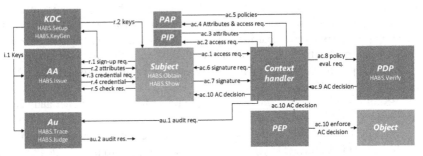

Fig. 2. A^3BAC framework; the green module is provided by a trusted third party, the blue module is provided by the authorization organization; the yellow module is the subject and object; req. means request; res. means result; eval. means evaluation. (Color figure online)

- PIP: is only responsible for managing object attributes and environmental attributes.
- Context handler: repackage and redirect requests based on context.
- Key Distribution Center (KDC): A trusted third party responsible for generating and distributing keys.
- Attribute Authority (AA): AA is responsible for managing all subject attributes. Subject obtains a certificate generated by AA.
- Audit Authority (Au): In A^3BAC, Au is responsible for auditing subject access.

3.2 Workflow of A³BAC

The ABAC model has four workflows, which are initialization, registration, anonymous access and audit. Before the system starts running, the initialization workflow is executed:

(i.1) is HABS *system initialization procedure.*
After the system is initialized, the system is started. Registration, anonymous access, and audit are all performed at runtime. When the subject enters the system for the first time, it executes the registration workflow:
(r.1-5) is HABS *credential issuing procedure.*
When the system is running, the subject anonymously accesses the object:
(ac.1-5) is the same as the ABAC except that they do not contain subject attributes.
(ac.6) The context handler generates the signature predicate γ and the random message for the policies in the policy set, and packages them into signature requests and send signature requests to the subject one by one.
(ac.7) After receiving the signature request, the subject selects the appropriate subset of attributes and runs HABS.Show to sign the random message in the signature request and sends the signature to the Context handler.
(ac.8) After receiving the signature, the Context handler packages the signature with the corresponding policy, object attributes, and environment attributes into a policy evaluation request and sends it to the PDP.
(ac.9) The PDP evaluates the policies based on the attributes in the requests. Then, PDP combines the evaluation results of these policies to form an access control (AC) decision and return it to the Context handler.
(ac.10) The Context handler sends the access control decision to the subject. If the decision is deny, the workflow ends. If the decision is permit, it is forwarded to PEP and PEP enforces the access control decision.
When abnormal access of the subject is discovered, the audit workflow can be started:
(au.1) The context handler sends an audit request contains the signature and access request to Au.
(au.2) Au evaluates the audit request. If the request is denied, the audit decision will be output directly. If the audit decision is permit, the HABS.Trace and HABS.Judge algorithm is executed based on the information in the audit request to expose the corresponding subject identity. Then Au packages the subject identity and the evidence as to the audit decision and outputs it.

Obviously, the subject identity is not involved in A³BAC's access workflow. The subject-related information obtained by the context handler or PDP in the access workflow is the signature. The signature only involves the predicate, the random message and a subset of subject attributes. Thus, A³BAC's access workflow is identity independent, and A³BAC's access is anonymous.

4   Analysis

Kaaniche N. and Laurent M. [7] proved the correctness, unforgeability, anonymity and the anonymity removal of the HABS algorithm. The security feature of A³BAC is the

Table 1. Solution comparison

Literature	Anonymous access	Fine-grained AC	Policy flexibility	Audit	Restricted object type
[2]	Y	Y	N	N	N
[3]	Y	Y	N	N	Y
[4]	Y	Y	N	N	Y
[9]	Y	Y	N	Y	N
[10]	Y	N	N	Y	N
[11]	Y	Y	N	N	Y
A^3BAC	Y	Y	Y	Y	N

same as the HABS algorithm. And A^3BAC also has these properties. The comparison between A^3BAC and the existing solution is shown in Table 1.

The main problem of the existing solution is that it cannot balance anonymous access, auditing, restricted object type, fine-grained access control, and policy flexibility. Anonymous access, auditing involves security issues; restricted object type, fine-grained access control and policy flexibility involve availability. In other words, it is difficult for the access control framework to balance security and availability. A^3BAC has both these characteristics by combining HABS and ABAC.

5 Conclusion

Existing anonymous access solutions have the problems of subject re-identification and constraints on the types of objects. The ABAC with anonymous access proposed in this paper called A^3BAC inherits the features of the ABAC model, such as fine-grained authorization, policy flexibility, and unlimited object types. By combining HABS, it strengthens the identity-less of ABAC, so that the access does not involve a unique identification, reducing the risk of subject identity re-identification. The A^3BAC inherits HABS anonymity, unforgeability, unlinkability, anonymous removal, auditing and other features, ensuring the security of the model.

In future work, we will implement and evaluate A^3BAC. A^3BAC will be deployed and its performance will be tested. We will also evaluate the anonymity guarantee provided by A^3BAC to the subject.

References

1. Li, S., Choo, K.R., Sun, Q., et al.: IoT forensics: Amazon echo as a use case. IEEE Internet Things 6(4), 6487–6497 (2019)
2. Yuen, T.H., Liu, J.K., Man, H.A., et al.: k-times attribute-based anonymous access control for cloud computing. IEEE Trans. Comput. 64(9), 2595–2608 (2015)

3. Ahuja, R., Mohanty, S.K.: A scalable attribute-based access control scheme with flexible delegation cum sharing of access privileges for cloud storage. IEEE Trans. Cloud Comput. **PP**(99), 1 (2017)

4. Pussewalage, H.S.G., Oleshchuk, V.: A patient-centric attribute based access control scheme for secure sharing of personal health records using cloud computing. In: IEEE, International Conference on Collaboration and Internet Computing, pp. 46–53 (2017)

5. Sweeney, L.: k-anonymity. Int. J. Uncertainty Fuzziness Knowl.-Based Syst. **10**(05), 557–570 (2008)

6. Hu, V., Ferraiolo, D., Kuhn, R., et al.: Guide to attribute based access control (ABAC) definition and considerations. ITLB (2014)

7. Li, S., Da Xu, L., Zhao, S.: 5G internet of things: a survey. J. Ind. Inf. Integr. **10**, 1–9 (2018)

8. Maji, H.K., Prabhakaran, M., Rosulek, M.: Attribute-based signatures. In: International Conference on Topics in Cryptology: CT-RSA, pp. 376–392. Springer-Verlag (2011)

9. Backes, M., Camenisch, J., Sommer, D.: Anonymous yet accountable access control. In: [ACM WPES] Proceedings of ACM Workshop on Privacy in the Electronic Society, p. 40 (2005)

10. Yao, X., Liu, H., Ning, H., et al.: Anonymous credential-based access control scheme for clouds. IEEE Cloud Comput. **2**(4), 34–43 (2015)

11. Zhang, Y., Li, J., Chen, X., et al.: Anonymous attribute-based proxy re-encryption for access control in cloud computing. Secur. Commun. Netw. **9**(14), 2397–2411 (2016)

12. Li, S., Zhao, S., Yang, P., et al.: Distributed consensus algorithm for events detection in cyber-physical systems. IEEE Internet Things J. **6**(2), 2299–2308 (2019)

13. Kaaniche, N., Laurent, M.: Attribute-based signatures for supporting anonymous certification. In: Computer Security – ESORICS 2016 (2016). Li, S., Xu, L.D., Zhao, S., et al.

14. Servos, D., Osborn, S.L.: Current research and open problems in attribute-based access control. ACM Comput. Surv. **49**(4), 1–45 (2017)

15. Wang, X.M., Fu, H., Zhang, L.C.: Research progress on attribute-based access control. Acta Electronica Sinica **38**(7), 1660–1667 (2010)

Blockchain-Based Decentralized Privacy-Preserving Data Aggregation (BDPDA) Scheme for SmartGrid

Hongbin Fan[1], Yining Liu[2(✉)], and Zhixin Zeng[2]

[1] College of Software and Communication Engineering, XiangNan University, Chenzhou 423000, China
[2] School of Information and Communication, Guilin University of Electronic Technology, Guilin 541004, China
ynliu@guet.edu.cn

Abstract. Smart grid is the next-generation grid that combines advanced power technology and modern communication technology. Smart meters face serious security challenges such as the leakage of user privacy and the absence of trusted third parties. Blockchain provides a viable solution that can use its key technologies to solve these problems. In blockchain technology, there is no necessary need of the third party in the energy supply sector. We introduce decentralization into smart grid, and a blockchain-based data aggregation scheme is designed. Due to the transparency of data in blockchain, the privacy of users may be disclosed. Therefore, our scheme adopts Paillier cryptosystem algorithm to encrypt the user's electricity consumption data, realizing the confidentiality of electricity consumption data, which is convenient for billing and power regulation. Through performance analysis of the scheme, it shows that the scheme has better security and better functions.

Keywords: Blockchain · Decentralized · Data aggregation · Privacy-preserving · Smart grid

1 Introduction

The traditional power grid adopts centralized power generation to meet the power demand. Therefore, the security of the centralized way is too dependent on the control center or the trusted third party. The smart grid based on the physical grid system using sensor measurement technology, communication technology, control technology and computer technology [1]. The information flow between suppliers and users in smart grid is bidirectional, while the traditional power grid adopts the unidirectional centralized system. Users can control the intelligent use of household appliances and equipment at any time according to the floating situation of electricity price in different time periods. Suppliers can automatically monitor the grid, prevent power outages, optimize grid performance, etc. However, the process of smart grid power consumption data collection may lead to the leakage of user privacy information [2, 3].

© ICST Institute for Computer Sciences, Social Informatics and Telecommunications Engineering 2021
Published by Springer Nature Switzerland AG 2021. All Rights Reserved
D. Wang et al. (Eds.): SPNCE 2020, LNICST 344, pp. 448–459, 2021.
https://doi.org/10.1007/978-3-030-66922-5_31

Blockchain technology has attained significant attention recently and provides a number of ways for reliable processing and storage of data in a decentralized manner. It is now recognized that trustless decentralized energy production and transfer using blockchain technology is a promising approach. This work proposes a novel blockchain-based decentralized green energy distribution system for trustless reliable energy exchanges in a smart grid. The proof of distribution problem in a decentralized environment is first formalized. Finally, a decentralized green energy distribution smart-grid case study is presented to demonstrate the utility of the system in real-life situations.

The main contributions of this paper are as follows:

1) Decentralization: the proposed scheme is reliable as it does not rely on a trusted third party or central authority and all processes are conducted in a decentralized manner through blockchain nodes.

2) Data integrity: since the previous smart meter data aggregation methods are centralized, unlike blockchain, which allows request verification through hash mechanism. BLS signature and Paillier encryption are based on bilinear pairing, which guarantees the security and integrity of message transmission.

3) Mining node selection: The smart meters select a mining node through leader election algorithm, the mining node records smart meters' data into the blockchain.

Note that the original idea has been presented in the original conference paper. Compared with the original conference paper, this paper adds an author who participated in the revision of the paper. In the current version, the system model diagram and Blockchain structure diagram have been modified, and the graph description of leader selection algorithm and the Paillier cryptosystem introduction are added to make it easier to understand. The algorithm in this paper is compared with the existing algorithm in detail, which better reflects the advantages of the algorithm in this paper.

The rest of this paper is organized as follows. Section 2, we introduce the previous work in privacy-preserving data aggregation. Section 3, Blockchain, bilinear pairing, Boneh-Lynn-Shacham Short Signature, and the Paillier cryptosystem are given. System model and design goals are introduced in Sect. 4. In Sect. 5, our scheme is described in detail. The security analysis is in Sect. 6. In Sect. 7, the performance of our scheme is evaluated. We conclude our research in Sect. 8.

2   Related Work

In order to protect the privacy of users in smart meter data aggregation, many scholars provide various schemes.

Li et al. [4] proposed a privacy-preserving multi-subset data aggregation scheme (PPMA), their scheme based on Paillier cryptosystem, which enables the aggregation of electricity consumption data of different ranges. Liu et al. [5] proposed a privacy-preserving data aggregation without any TTP. In this scheme construct a virtual aggregation area for users with a certain degree of trust to shield the data of a single user. Guan et al. [6] proposed adjust the aggregation threshold according to the energy consumption information and time period of each specific residential area to ensure the privacy of personal data during the aggregation process, while supporting fault tolerance. Karampour et al. [7] proposed use Paillier encryption system and AV net mask to realize the aggregation of privacy protection data in smart grid can effectively protect the privacy of user data without any security channel. However, the above research methods do not consider the trusted environment.

To achieve a trusted environment, several studies used blockchain as privacy-preserving method for data aggregation. Guan et al. [8] proposed a privacy-preserving data aggregation scheme for power grid communications. The study divided users into different groups and each group has a private blockchain. The study uses multiple pseudonyms to hide users' identity. In this scheme, key management center (KMC) is used to generate multiple public and private keys for users, which does not realize decentralization.

3 Preliminaries

In this section, we briefly introduce the necessary background.

A. Blockchain

Blockchain technology was first proposed in 2008 by Satoshi Nakamoto for Bitcoin [9]. Blockchain technology has been widely used in payment, Internet of things, healthcare, finance and so on [10]. Blockchain is a decentralized distributed ledger database maintained by network-wide nodes [11], which comprising a chain of different data blocks in a chronological order. All hash data added to the block is immutable. Blockchain is a new application mode of consensus mechanism, distributed data storage, encryption algorithm and so on. Its key technologies include block structure, Merkle tree, P2P network, hash function, timestamp, asymmetric encryption mechanism, etc. [12].

B. Boneh-Lynn-Shacham Short Signature

Boneh-Lynn-Shacham (BLS) Short Signature [13] scheme is a typical bilinear pairing scheme. The scheme uses a SHA-256 hash function $H_1 : \{0, 1\}^* \rightarrow G_1$ and g is a random generator of G_1, and a bilinear map $e : G_1 \times G_1 \rightarrow G_2$. The BLS signature scheme is divided into three phases:

1) Key generation: The secret key $x \in Z_q^*$, and compute the public key $PK = x \cdot g$.
2) Signature: The plaintext $m \in G_1$, and compute the signature $\sigma = x \cdot H(m)$
3) Verification: If $e(\sigma, g) = e(H(m), PK)$, then the signature is verified. Otherwise fails.

C. Paillier cryptosystem

Paillier cryptosystem [14] is an additive homomorphic encryption system that allows computation of encrypted data. The additive homomorphic encryption property can directly calculate the encryption of their sum from the multiplication of the encrypted values of some data, thus effectively protecting the privacy of the data. It includes the following three algorithms:

1) Key generation: Randomly select two large primes p and q, where $|p| = |q| = |\kappa|$. Then calculate $\lambda = lcm(p - 1, q - 1)$. Defined a function $L(v) = \frac{v-1}{N}$, where N = pq. Choose a generator $g \in Z_{N^2}^*$, and calculate $\mu = (L(g^\lambda \bmod N^2))^{-1} \bmod N$. The public key is (N, g), and the corresponding private key is (λ, μ).

2) Encryption: Given a message $m \in Z_N$, choose a random number $r \in Z_N^*$. $gcd(r, N) = 1$, The ciphertext is calculated as $C = Enc(m) = g^m \cdot r^N \bmod N^2$.

3) Decryption: Given the ciphertext $C \in Z_N$, The corresponding message is decrypted as $m = Dec(C) = L(C^\lambda \bmod N^2) \cdot \mu \bmod N$.

4 System Model and Design Goals

A. System model

The system model of BDPDA demonstrated in Fig. 1 consists of operation center (OC) and smart meter (SM) in the residential area (RA). In our scheme, we mainly focuses on remove the control center and the trusted third party while protecting the data privacy of the user's smart meter.

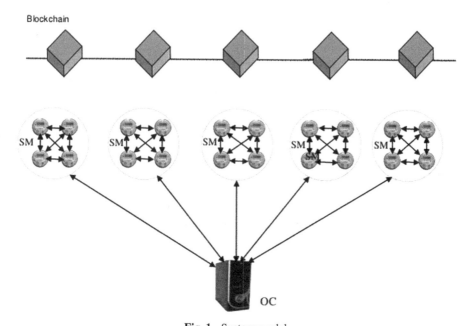

Fig. 1. System model

1) Operation center (OC)

The operation center reads the real-time power consumption data aggregated by the mining nodes of each block from the private blockchain for billing, power consumption trend analysis, adjustment of power generation plans, and dynamic pricing. To increase efficiency, each administrative district will establish its own OC.

2) Smart meter (SM)

A SM is an electricity meter for each user's site in the residential area. The smart meter regularly and simultaneously (e.g. once 15 min) collects the power consumption data of each user's household electrical equipment. Peer-to-peer (P2P) communication is used between SMs. Each residential area uses leader election algorithm to select a smart meter from the smart meters as the mining node (MN), Then each residential area constructs a block through MN.MN is responsible for generating system parameters, authenticates the legitimacy of the data transmitted by the smart meter and aggregates the encrypted data. Then, SM encrypts all kinds of collected data and uploads it to the MN after a short period of time.

B. Design goals

To solve the issues mentioned above, ensure the integrity and privacy of users' power consumption data while decentralizing and de-trusted third parties, the design goals include four aspects.

1) Privacy-preservation. Neither OC nor any other user has access to the user's data in the residential area. An external adversary cannot obtain the user's power consumption data, even if he knows the cipher text and public key. When the adversary and OC collude with each other, they can't get the power consumption data of a single user's smart meter.
2) Decentralizing. The BDPDA scheme does not need a trusted third party or central authority.

5   The Proposed Scheme

In this section, the data aggregation of decentralized smart grid based on blockchain is proposed. The notations are listed in Table 1.

A. System Initialization

OC collects electricity consumption data of L residential areas. There are n smart meters in RAj. Through leader election algorithm, selects a SM as a mining node from the n SMs in RAj, then constructs the jth block, where MNj is the root of the Merkle tree in the jth block. The state change of leader election algorithm is shown in Fig. 2. The consumption data of SMs in RAj is aggregated to MN through Merkle tree. The structure of Blockchain is shown in Fig. 3.

Table 1. Notations

Symbol	Definition
g1, g2	A generator of G
RA_j	The Jth residential area
m_i	Power consumption data of the ith smart meter in RA_j
n	Number of smart meters in the jth residential area
H1	Hash functions: H1 : $\{0,1\}^* \rightarrow$ G
L	Number of residential areas
SM_i	Smart meter in jth residential area
MN_j	Mining node of the jth residential area
M_j	The aggregated electricity consumption data of the j-th residential areas
$\parallel$	Concatenation operation

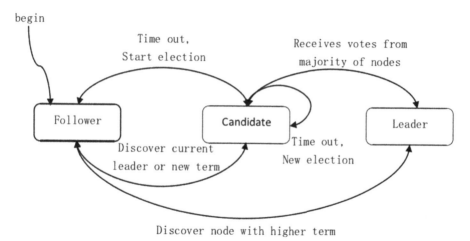

Fig. 2. State transition model of leader election algorithm

MN_j runs Bilinear parameter generator $Gen(\kappa)$ to generate (q, g_1, G_1, G_2, e), and g_1 is a generator of G_1. MN calculates Paillier cryptosystem public key (N, g_2), corresponding private key (λ, μ), $g_2 \in Z_{N^2}^*$. MN choose a SHA-256 hash function H_1 and a secure cryptographic hash function H_2, where $H_1 : \{0, 1\}^* \rightarrow G_1, H_2 : \{0, 1\}^* \rightarrow \{0, 1\}^\kappa$. MN_j publishes the system public parameter $\{q, g_1, g_2, G_1, G_2, e, N, H_1\}$.

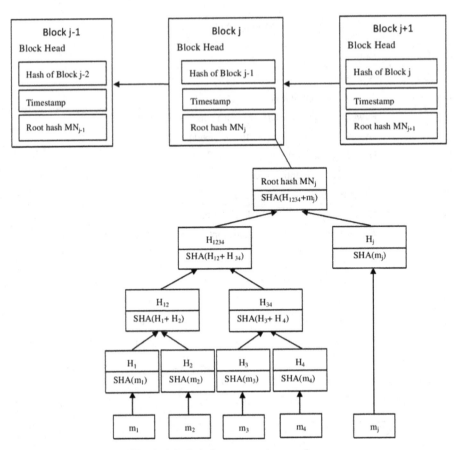

Fig. 3. Blockchain structure in our scheme

B. Ciphertext generation

1) Step 1: SM_i selects a random number $x_i \in Z_q^*$ as the private key and computes the corresponding public key $PK_i = x_i \cdot g_1$.

2) Step 2: SM_i collects electricity consumption data at timestamp T, and computes the Hash value H_2 (T), then selects a random number $r_i \in Z_N^*$ to generate ciphertext: $C_i = g_2^{m_i} \times (r_i \times H_2(T))^N \mod N^2$.

3) Step 3: SM_i generates the BLS short signature $\sigma_i = x_i \cdot H_1(C_i \| PK_i \| Ts_i)$, Ts_i is the current timestamp to prevent replay attack.

4) Step 4: SM_i sends $C_i \| PK_i \| Ts_i \| \sigma_i$ to MN through the Merkle tree.

C. Ciphertext aggregation

1) Step 1: After receives $C_i\|PK_i\|Ts_i\|\sigma_i$, MN_j verifies whether $e(\sigma_i, g_1) \overset{?}{=} e(H_1(C_i\|PK_i\|Ts_i), PK_i)$

hold, the signature is valid and MN_j will accept SM_i's ciphertext. In order to make the verification more efficient, MN_j adopts batch verification.

2) Step 2: MN_j aggregates the ciphertext.

$$C = \prod_{i=1}^{n} C_i = \prod_{i=1}^{n} g_2 \cdot (r_i \cdot H_2(T))^N \bmod N^2 = g_2^{\sum_{i=1}^{n} m_i} \cdot \prod_{i=1}^{n} (r_i \cdot H_2(T))^N \bmod N^2$$

D. Ciphertext Decryption

MN_j uses the private key (λ, μ) to decrypt the aggregated ciphertext to obtain the aggregated electricity consumption data M_j of the j-th residential district.

$$M_j = L(C^\lambda \bmod N^2) \cdot \mu \bmod N = \sum_{i=1}^{n} m_i$$

E. Data reading

MN_j generates the j+1th block, and adds the j-th block to the blockchain after the j-1 block. OC obtains the power consumption data through the public key read blockchain.

6 Security Analysis

A. Privacy-preserving

When an external attacker invades a smart meter, only the ciphertext Ci sent by a smart meter can be obtained. Even if the malicious user intercepts the ciphertext Ci, because he/she does not know the decryption key λ of the Paillier encryption algorithm, he/she cannot decrypt the ciphertext Ci to obtain the power consumption data of a single user. The power consumption data of a single smart meter is not disclosed, so as to protect the privacy of users.

B. Decentralized

In our scheme, the blockchain can be implemented without a trusted third party or central authority, the availability and reliability of data is guaranteed by MN election. No single organization can control or run SM. P2P network is adopted among smart meters to realize decentralization. The whole process does not rely on a trusted third party to make our solution more reliable and convenient.

7 Performance Evaluation

The performance of BDPDA is evaluated in this section, including the computation complexity of SM and OC, and the communication overhead.

A. Computation complexity

Compared with multiplication operation and exponentiation operation, Leader election and Hash operation is negligible. In the BDPDA scheme, the computations in the data aggregation process mainly include three phases, data encryption, batch verification and aggregation, decryption. We denote the computational cost of an exponentiation operation and a multiplication operation, by T_{exp}, T_{mul}, respectively. The computation complexities of the major entities in the system are as show in Table 2.

Table 2. Comparing computation complexity between the proposed scheme and other schemes

Scheme Ref.	Overhead SM	Overhead GW	Overhead CC	Overhead MN
[4]	$3T_{exp} + 4T_{mul}$	nT_{mul}	$T_{exp} + (4n + 3)T_{mul}$	–
[6]	$4T_{exp} + 3T_{mul}$	$3T_{exp} + (2n + 1)T_{mul}$	$3T_{exp} + 2T_{mul}$	–
[7]	$2T_{exp} + nT_{mul}$	nT_{mul}	$T_{exp} + 3 - T_{mul}$	–
BDPDA	$2T_{exp} + 4T_{mul}$	–		$T_{exp} + (n + 1)T_{mul}$

We conduct the experiments with the cpabe0.10 [15] library on a 3.0 GHz-processor and a 2 GB memory PC. As shown in Fig. 4, our scheme has advantage in computational overhead compared with PPMA, EFFECT and Karampour's schemes.

B. Communication overhead

The communication of the proposed BDPDA scheme is only SMi to MNj. Suppose that SMi generates a 2048-bit ciphertext Ci and chooses 160-bit Z_N^*. Table 3 shows the communication overhead of our scheme compared with the other three schemes with n users. It is obvious that our BDPDA scheme has a lower total communication cost than other schemes.

C. Comparison with Existing Schemes

This section describes the comparison of the proposed scheme with the existing schemes. The comparison results show that schemes [4, 6] and [7] are not based on blockchain and cannot achieve decentralization. Although scheme [8] is based on blockchain, it uses Key management center to generate public and private keys, relying on trusted

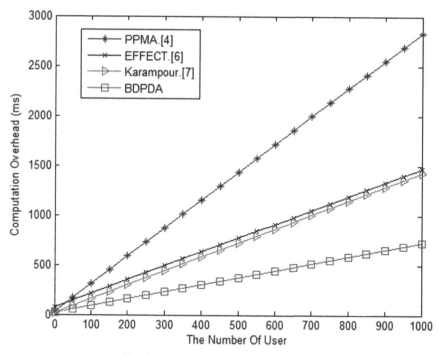

Fig. 4. Comparison of computational cost

Table 3. Comparing communication cost between the proposed scheme and other schemes

Scheme Ref.	SM-to-SM (bit)	SM-to-GW (bit)	GW-to-CC (bit)	SM-to-MN(bit)
[4]	–	2048n	2048	–
[6]	–	2048n	2048	–
[7]	$n(2048(n-1))$	2048n	2048	–
BDPDA	–	–	–	2048n

third party, so decentralization cannot be achieved. Therefore, as shown in Table 4, we can see that the scheme proposed in this paper can protect user privacy while achieving decentralization.

Table 4. Comparison between proposed scheme and other related schemes

Security requirements	[4]	[6]	[7]	[8]	Our scheme
Blockchain-Based	No	No	No	Yes	Yes
Decentralization	No	No	No	No	Yes
Privacy	Yes	Yes	Yes	Yes	Yes
Non-repudiation	No	Yes	No	Yes	Yes
Confidentiality	Yes	Yes	Yes	Yes	Yes
Data Integrity	Yes	Yes	Yes	Yes	Yes
Replay attack resistance	No	Yes	Yes	Yes	Yes
Data unforgeability	No	Yes	Yes	Yes	Yes

8 Conclusion

In this paper, BDPDA scheme for smart grid is proposed, the blockchain is used to implement the decentralization of data privacy protection for the smart grid. The smart meters select a mining node through election algorithm, the mining node records smart meters' data into the blockchain. BLS signature and Paillier encryption are based on bilinear pairing, which guarantees the security and integrity of message transmission. The security analysis has been proven that our mechanism meets the requirements of privacy protection and security of smart grids. The performance evaluation shows that our scheme has advantage compared with some popular data aggregation schemes in computational efficiency. As future work, we will study the combination of blockchain and other algorithms to aggregate multidimensional data and fault tolerance.

References

1. Fang, X., Misra, S., Xue, G., Yang, D.: Smart grid-the new and improved power grid: a survey. IEEE Commun. Surv. Tutor. **14**(4), 944–980 (2011)
2. Xue, K.P., Li, S.H., Hong, J.N., et al.: Two-cloud secure database for numeric-related SQL range queries with privacy preserving. IEEE Trans. Inf. Forensics Secur. **12**, 1596–1608 (2017)
3. Wu, J., Dong, M., Ota, K., Liang, L., Zhou, Z.: Securing distributed storage for social internet of things using regenerating code and Blom key agreement. Peer-to-Peer Netw. Appl. **8**, 1133–1142 (2015). https://doi.org/10.1007/s12083-014-0286-y
4. Li, S., Xue, K., Yang, Q., Hong, P.: PPMA: privacy-preserving multisubset data aggregation in smart grid. IEEE Trans. Ind. Inf. **14**, 462–471 (2018)
5. Liu, Y., Guo, W., Fan, C., Chang, L., Cheng, C.: A practical privacy-preserving data aggregation (3PDA) scheme for smart grid. IEEE Trans. Ind. Inform. **15**, 1767–1774 (2018)
6. Guan, Z., Zhang, Y., Zhu, L., Wu, L., Yu, S.: EFFECT: an efficient flexible privacy-preserving data aggregation scheme with authentication in smart grid. Sci. China Inf. Sci. **62**(3), 32103 (2019). https://doi.org/10.1007/s11432-018-9451-y

7. Karampour, A., Ashouri-Talouki, M., Ladani, B.T.: An efficient privacy-preserving data aggregation scheme in smart grid. In: 2019 27th Iranian Conference on Electrical Engineering (ICEE), pp. 1967–1971. IEEE (2019)

8. Guan, Z.T., et al.: Privacy-preserving and efficient aggregation based on blockchain for power grid communications in smart communities. IEEE Commun. Mag. **56**(7), 82–88 (2018)

9. Nakamoto, S. Bitcoin: a peer-to-peer electronic cash system (2008, consulted)

10. Crosby, M., Pattanayak, P., Verma, S., et al.: Blockchain technology: beyond bitcoin. Appl. Innov. **2**(6–10), 71 (2016)

11. Yuan, Y., Wang, F.-Y.: Parallel blockchain: concept, methods and issues. Acta Autom. Sinica **43**(10), 1703–1712 (2017)

12. Xie, Q.H.: Research on blockchain technology and financial business innovation. Financ. Dev. Res. **5**, 77–82 (2017)

13. Boneh, D., Lynn, B., Shacham, H.: Short signatures from the Weil pairing. In: Boyd, C. (ed.) ASIACRYPT 2001. LNCS, vol. 2248, pp. 514–532. Springer, Heidelberg (2001). https://doi.org/10.1007/3-540-45682-1_30

14. Paillier, P.: Public-key cryptosystems based on composite degree residuosity classes. In: Stern, J. (ed.) EUROCRYPT 1999. LNCS, vol. 1592, pp. 223–238. Springer, Heidelberg (1999). https://doi.org/10.1007/3-540-48910-X_16

15. Bethencourt, J.: Advanced Crypto Software Collection: The CPABE Toolkit (2018). http://acsc.cs.utexas.edu/cpabe/

Author Index

Printed in the United States
By Bookmasters